Wiley Series in Statistics in Practice

Advisory Editor, Marian Scott, University of Glasgow, Scotland, UK

Founding Editor, Vic Barnett, Nottingham Trent University, UK

Statistics in Practice is an important international series of texts that provide detailed coverage of statistical concepts, methods and worked case studies in specific fields of investigation and study.

With sound motivation and many worked practical examples, the books show in down-to-earth terms how to select and use an appropriate range of statistical techniques in a particular practical field within each title's special topic area.

The books provide statistical support for professionals and research workers across a range of employment fields and research environments. Subject areas covered include medicine and pharmaceutics; industry, finance and commerce; public services; the earth and environmental sciences; and so on.

The books also provide support to students studying statistical courses applied to the above areas. The demand for graduates to be equipped for the work environment has led to such courses becoming increasingly prevalent at universities and colleges.

It is our aim to present judiciously chosen and well-written workbooks to meet everyday practical needs. Feedback of views from readers will be most valuable to monitor the success of this aim.

A complete list of titles in this series appears at the end of the volume.

Wiley Series in Statistics in Practice

Advisory Editor, Marian Scott, University of Glasgow, Scotland, UK

Founding Editor, Vic Barnett, Nottingham Trent University, UK

Human and Biological Sciences

Brown and Prescott · Applied Mixed Models in Medicine

Ellenberg, Fleming and DeMets · Data Monitoring Committees in Clinical Trials: A Practical Perspective

Lawson, Browne and Vidal Rodeiro · Disease Mapping With WinBUGS and MLwiN

Lui · Statistical Estimation of Epidemiological Risk

*Marubini and Valsecchi · Analysing Survival Data from Clinical Trials and Observation Studies

Parmigiani · Modeling in Medical Decision Making: A Bayesian Approach

Senn · Cross-over Trials in Clinical Research, Second Edition

Senn · Statistical Issues in Drug Development

Spiegelhalter, Abrams and Myles · Bayesian Approaches to Clinical Trials and Health-Care Evaluation

Turner · New Drug Development: Design, Methodology, and Analysis

Whitehead · Design and Analysis of Sequential Clinical Trials, Revised Second Edition

Whitehead · Meta-Analysis of Controlled Clinical Trials

Zhou, Zhou, Liu and Ding · Applied Missing Data Analysis in the Health Sciences

Earth and Environmental Sciences

Buck, Cavanagh and Litton · Bayesian Approach to Interpreting Archaeological Data

Cooke · Uncertainty Modeling in Dose Response: Bench Testing Environmental Toxicity

Gibbons, Bhaumik, and Aryal · Statistical Methods for Groundwater Monitoring, Second Edition

Glasbey and Horgan · Image Analysis in the Biological Sciences

Helsel · Nondetects and Data Analysis: Statistics for Censored Environmental Data

Helsel · Statistics for Censored Environmental Data Using Minitab* and R, Second Edition

McBride · Using Statistical Methods for Water Quality Management: Issues, Problems and Solutions

Ofungwu · Statistical Applications for Environmental Analysis and Risk Assessment

Webster and Oliver · Geostatistics for Environmental Scientists

Industry, Commerce and Finance

Aitken and Taroni · Statistics and the Evaluation of Evidence for Forensic Scientists, Second Edition

Brandimarte · Numerical Methods in Finance and Economics: A MATLAB-Based Introduction, Second Edition

Brandimarte and Zotteri · Introduction to Distribution Logistics

Chan and Wong · Simulation Techniques in Financial Risk Management, Second Edition

Jank · Statistical Methods in eCommerce Research

Jank and Shmueli · Modeling Online Auctions

Lehtonen and Pahkinen · Practical Methods for Design and Analysis of Complex Surveys, Second Edition

Lloyd · Data Driven Business Decisions

Ohser and Mücklich · Statistical Analysis of Microstructures in Materials Science

Rausand · Risk Assessment: Theory, Methods, and Applications

Network Meta-Analysis for Decision-Making

Sofia Dias
University of Bristol
Bristol, UK

A. E. Ades
University of Bristol
Bristol, UK

Nicky J. Welton
University of Bristol
Bristol, UK

Jeroen P. Jansen
Precision Health Economics
Oakland, CA

Alexander J. Sutton
University of Leicester
Leicester, UK

This edition first published 2018
© 2018 John Wiley & Sons Ltd

Registered Offices
John Wiley & Sons, Inc., 111 River Street, Hoboken, NJ 07030, USA
John Wiley & Sons Ltd, The Atrium, Southern Gate, Chichester, West Sussex, PO19 8SQ, UK

Editorial Office
9600 Garsington Road, Oxford, OX4 2DQ, UK

For details of our global editorial offices, customer services, and more information about Wiley products visit us at www.wiley.com.

Wiley also publishes its books in a variety of electronic formats and by print-on-demand. Some content that appears in standard print versions of this book may not be available in other formats.

Library of Congress Cataloging-in-Publication Data

Names: Dias, Sofia, 1977- author.
Title: Network meta-analysis for decision-making / by Sofia Dias, University of Bristol, Bristol, UK [and four others].
Description: Hoboken, NJ : Wiley, 2018. | Includes bibliographical references and index. |
Identifiers: LCCN 2017036441 (print) | LCCN 2017048849 (ebook) | ISBN 9781118951712 (pdf) | ISBN 9781118951729 (epub) | ISBN 9781118647509 (cloth)
Subjects: LCSH: Meta-analysis. | Mathematical analysis.
Classification: LCC R853.M48 (ebook) | LCC R853.M48 N484 2018 (print) | DDC 610.72/7–dc23
LC record available at https://lccn.loc.gov/2017036441

Cover Design: Wiley
Cover Image: © SergeyNivens/Gettyimages

Set in 10/12pt Warnock by SPi Global, Pondicherry, India

10   9   8   7   6   5   4   3   2   1

Contents

Preface

1. Who Is This Book for?

This book is intended for anyone who has an interest in the synthesis, or 'pooling', of evidence from randomised controlled trials (RCTs) and particularly in the statistical methods for network meta analysis. A standard meta-analysis is used to pool information from trials that compare two interventions, while network meta-analysis extends this to the comparison of any number of interventions.

Network meta-analysis is one of the core methodologies of what has been called comparative effectiveness research (Iglehart, 2009), and, in view of the prime role accorded to trial evidence over other forms of evidence on comparative efficacy, it might be considered to be the most important.

The core material in this book is largely based on a 3-day course that we have been running for several years. Based on the spectrum of participants we see on our course, we believe the book will engage a broad range of professionals and academics. Firstly, it should appeal to all statisticians who have an interest in evidence synthesis, whether from a methodological viewpoint or because they are involved in applied work arising from systematic reviews, including the work of the Cochrane Collaboration.

Secondly, the methods are an essential component of health technology assessment (HTA) and are routinely used in submissions not only to re-imbursement agencies such as the National Institute for Health and Care Excellence (NICE) in England but also, increasingly, to similar organisations worldwide, including the Canadian Agency for Drugs and Technologies in Health, the US Agency for Healthcare Research and Quality and the Institute for Quality and Efficiency in Health Care in Germany. Health economists involved in HTA in academia and those working in pharmaceutical companies, or for the consultancy firms who assist them in making submissions to these bodies, comprise the largest single professional group for whom this book is intended.

Clinical guidelines are also making increasing use of network meta-analysis, and statisticians and health economists working with medical colleges on guideline development represent a third group who will find this book highly relevant.

Finally, the book will also be of interest, we believe, to those whose role is to manage systematic reviews, clinical guideline development or HTA exercises and those responsible at a strategic level for determining the methodological approaches that should underpin these activities. For these readers, who may not be interested in the technical details, the book sets out the assumptions of network meta-analysis, its properties, when it is appropriate and when it is not.

The book can be used in a variety of ways to suit different backgrounds and interests, and we suggest some different routes through the book at the end of the preface.

2. The Decision-Making Context

The contents of this book have their origins in the methodology guidance that was produced for submissions to NICE. This is the body in England and Wales responsible for deciding which new pharmaceuticals are to be used in the National Health Service. This context has shaped the methods from the beginning.

First and foremost, the book is about an approach to evidence synthesis that is specifically intended for decision-making. It assumes that the purpose of every synthesis is to answer the question 'for this pre-identified population of patients, which treatment is "best"?' Such decisions can be made on any one of a range of grounds: efficacy alone, some balance of efficacy and side effects, perhaps through multi-criteria decision analysis (MCDA) or cost-effectiveness. At NICE, decisions are based on efficacy and cost-effectiveness, but whatever criteria are used, the decision-making context impacts evidence synthesis methodology in several ways.

Firstly, the decision maker must have in mind a quite specific target population, not simply patients with a particular medical condition but also patients who have reached a certain point in their natural history or in their referral pathway. These factors influence a clinician's choice of treatment for an individual patient, and we should therefore expect them to impact how the evidence base and the decision options are identified. Similarly, the candidate interventions must also be characterised specifically, bearing in mind the dose, mode of delivery and concomitant treatments. Each variant has a different effect and also a different cost, both of which might be taken into account in any formal decision-making process. It has long been recognised that trial inclusion criteria for the decision-making context will tend to be more narrowly drawn than those for the broader kinds of synthesis that aim for what may be best described as a 'summary' of the literature (Eccles et al., 2001). In a similar vein Rubin (1990) has distinguished between evidence synthesis as 'science' and evidence synthesis as 'summary'. The common use of random effects models to average over important heterogeneity has attracted particular criticism (Greenland, 1994a, 1994b).

Recognising the centrality of this issue, the Cochrane Handbook (Higgins and Green, 2008) states: 'meta-analysis should only be considered when a group of studies is sufficiently homogeneous in terms of participants, interventions and outcomes to provide a meaningful summary'. However, perhaps because of the overriding focus on scouring the literature to secure 'complete' ascertainment of trial evidence, this advice is not always followed in practice. For example, an overview of treatments for enuresis (Russell and Kiddoo, 2006) put together studies on treatments for younger, treatment-naïve children, with studies on older children who had failed on standard interventions. Not surprisingly, extreme levels of statistical heterogeneity were observed, reflecting the clinical heterogeneity of the populations included (Caldwell et al., 2010). This throws doubt on any attempt to achieve a clinically meaningful answer to the question 'which treatment is best?' based on such a heterogeneous body of evidence. Similarly, one cannot meaningfully assess the efficacy of biologics in rheumatoid arthritis by combining trials on first-time use of biologics with trials on patients who have failed on biologic therapy (Singh et al., 2009). These two groups of patients require different decisions based on analyses of different sets of trials. A virtually endless list of examples could be cited. The key point is that the immediate effect of the decision-making perspective, in contrast to the systematic review perspective, is to greatly reduce the clinical heterogeneity of the trial populations under consideration.

The decision-making context has also made the adoption of Bayesian Markov chain Monte Carlo (MCMC) methods almost inevitable. The preferred form of cost-effectiveness analysis at NICE is based on probabilistic decision analysis (Doubilet et al., 1985; Critchfield and Willard, 1986; Claxton et al., 2005b). Uncertainty in parameters arising from statistical sampling error and other sources of uncertainty can be propagated through the decision model to be reflected in uncertainty in the decision. The decision itself is made on a 'balance of evidence' basis: it is an 'optimal' decision, given the available evidence, but not necessarily the 'correct' decision, because it is made under uncertainty.

Simulation from Bayesian posterior distributions therefore gives a 'one-step' solution, allowing proper statistical estimation and inference to be embedded within a probabilistic decision analysis, an approach sometimes called 'comprehensive decision analysis' (Parmigiani and Kamlet, 1993; Samsa et al., 1999; Parmigiani, 2002; Cooper et al., 2003; Spiegelhalter, 2003). This fits perfectly not only with cost-effectiveness analyses where the decision maker seeks to maximise the expected net benefit, seen as monetarised health gain minus costs (Claxton and Posnett, 1996; Stinnett and Mullahy, 1998), but also with decision analyses based on maximising any objective function. Throughout the book we have used the flexible and freely available WinBUGS software (Lunn et al., 2009) to carry out the MCMC computations required.

3. Transparency and Consistency of Method

Decisions on which intervention is 'best' are increasingly decisions that are made in public. They are scrutinised by manufacturers, bodies representing the health professions, ministries of health and patient organisations, often under the full view of the media. As a result, these decisions, and by extension the technical methods on which they are based, must be transparent, open to debate and capable of being applied in a consistent and fair way across a very wide range of circumstances. In the specific context of NICE, there is not only a scrupulous attention to process (National Institute for Health and Clinical Excellence, 2009b, 2009c) and method (National Institute for Health and Care Excellence, 2013a) but also an explicit rationale for the many societal judgements that are implicit in any health guidance (National Institute for Health and Clinical Excellence, 2008c).

This places quite profound constraints on the properties that methods for evidence synthesis need to have. It encourages us to adopt the same underlying models, the same way of evaluating model fit and the same model diagnostics, regardless of the form of the outcome. It also encourages us to develop methods that will give the same answers, whether trials report the outcomes on each arm, or just the difference between arms, or whether they report the number of events and person-years exposure or the number of patients reaching the endpoint in a given period of time. Similarly, we should aim to cope with situations where results from different trials are reported in more than one format. Meeting these objectives is greatly facilitated by the generalised linear modelling framework introduced in Chapter 4 and by the use of shared parameter models. The extraordinary flexibility of MCMC software pays ample dividends here, as shared parameter models cannot be readily estimated by frequentist methods.

An even stronger requirement is that the same methods of analysis are used whether there are just two interventions to be compared or whether there are three, four or more. Similarly, the same methods should be used for two-arm trials or for multi-arm trials, that is, trials with three or more arms. For example, manufacturers of treatments B and C, which have both been compared with placebo A in RCTs, would accept that recommendations might have to change, following the addition of new evidence from B versus C trials, but not if this was because a different kind of model had been used to synthesise the data. The methods used throughout this book can be applied consistently: a single software code is capable of analysing any connected network of trials, with two or more treatments, consisting of any combination of indirect comparisons, pairwise comparisons and multi-arm trials, without distinction. This is not a property shared by several other approaches (Lumley, 2002; Jackson et al., 2014), in which fundamentally different models are proposed for networks with different structures. This is not to deny, of course, that these models could be useful in other circumstances, such as when checking assumptions.

4. Some Historical Background

The term 'network meta-analysis', from Lumley (2002), is relatively recent. Other terms used include mixed treatment comparisons (MTCs) or multiple-treatments meta-analysis (MTM). The idea of network meta-analysis as an extension of simple pairwise meta-analysis goes back at least to David Eddy's Confidence Profile Method (CPM) (Eddy et al., 1992), particularly to a study of tissue plasminogen activator (t-Pa) and streptokinase involving both indirect comparisons and what was termed a 'chain of evidence' (Eddy, 1989) (see Chapter 11). Another early example, from 1998, was a four-treatment network on smoking cessation from Vic Hasselblad (1998), another originator of the CPM. This much studied dataset has been used by many investigators and lecturers to illustrate their models, and it continues to do service in this book.

A second strand of work can be found in the influential *Handbook of Research Synthesis* edited by Cooper and Hedges (1994), which came from an educational and social psychology tradition, rather than medicine. A chapter by Gleser and Olkin (1994) describes a method for combining data from trials of several interventions, including some multi-arm studies.

Two other groups, working independently of both the CPM and of the educational psychology statisticians, also published seminal papers. Heiner Bucher, working with Gordon Guyatt and others, discussed 'indirect comparisons' (Bucher et al., 1997a, 1997b). A year earlier, Julian Higgins and Anne Whitehead (1996) proposed two separate ideas. One was to use external data to 'borrow strength' to inform the heterogeneity parameter in a random effects models, and the other to pool 'direct' and 'indirect' evidence on a particular comparison, subject to an assumption that they both estimated the same parameter. The statistical model that they proposed is essentially identical to the core model used throughout this book.

In the course of time, we have re-parameterised the Higgins and Whitehead model in several ways (Lu and Ades, 2004, 2006; Ades et al., 2006), adapting it to changes in WinBUGS and recoding it to make it as general as possible. The form in which it is used throughout this book is the same as in the NICE

Decision Support Unit Technical Support Documents (Dias et al., 2011a, 2011b, 2011c, 2011d, 2011e), which were written to support the NICE guide to the methods of technology appraisal in its 2008 (National Institute for Health and Clinical Excellence, 2008a) and 2013 editions (National Institute for Health and Care Excellence, 2013a). These documents were reproduced in shorter form as a series of papers in the journal Medical Decision Making (Dias et al., 2013a, 2013b, 2013c, 2013d). Finally, it is worth mentioning that the same coding with different variable names was adopted by the International Society for Pharmacoeconomics and Outcomes Research (ISPOR) Task Force (Hoaglin et al., 2011; Jansen et al., 2011).

5. How to Use This Book

This book is designed to be used in a variety of different ways to suit readers with different interests and backgrounds. The first chapter 'Introduction to Evidence Synthesis' may serve as a miniature version of the core chapters in the book, touching on the assumptions of network meta-analysis, its key properties, showing what it does that other approaches cannot do and mentioning some of the policy implications. These themes are all picked up in greater detail in later chapters.

The second chapter, 'The core model' introduces the key network meta-analysis concepts for binomial data. It not only sets out the core model in some detail for dichotomous data but also shows readers how to set up data and obtain outputs from WinBUGS. Chapter 3 introduces the basic tools for model comparison, model critique, leverage analysis and outlier detection. In Chapter 4 a generic network meta-analysis model is presented based on generalised linear models (McCullagh and Nelder, 1989). This brings out the highly modular nature of the WinBUGS coding: once readers are familiar with the basic structure of the core model and its coding, they will be able to adapt the model to carry out the same operations on other outcomes, or an extended set of operations, such as meta-regression and inconsistency analysis. This modularity is exploited throughout this book to expand the core model from Chapters 2 and 4 in subsequent chapters. A large number of example WinBUGS codes are provided as online material. To complete the core material, Chapter 5 discusses the conceptual and computational issues that arise when a network meta-analysis is used to inform a cost-effectiveness analysis or other model for decision making. Chapter 6 briefly suggests some workarounds that can be useful in sparse data or zero cell count situations.

From our perspective (see Chapter 12), all forms of heterogeneity are undesirable. Chapters 7, 8 and 9 look at the issues arising from specific types of heterogeneity: inconsistency, covariate adjustment through meta-regression and bias adjustment. Although not listed as 'core material', readers should not regard these chapters as optional extras: a thorough understanding of these issues is fundamental to securing valid and interpretable results.

Chapters 10 and 11 take synthesis methodology into a further level of complexity, looking at the possibilities for network meta-analysis of survival outcomes based on data in life-table form and synthesis within- and between-trials reporting multiple (correlated) outcomes.

A final chapter on the validity of network meta-analysis attempts to address the many concerns about this method that have appeared in the literature. In this final chapter we try to clarify the assumptions of network meta-analysis and the terminology that has been used; we review empirical studies of the consistency assumption and present some 'thought experiments' to throw light on both the origins of 'bias' in evidence synthesis and the differences – if any – between direct and indirect evidence.

For the more experienced WinBUGS user, the book can be used as a primer on Bayesian synthesis of trial data using WinBUGS, or as a source of WinBUGS code that will suit the vast majority of situations usually encountered, with little or no amendment. The chapters on bias models, survival analysis and multiple outcomes may be read simply out of interest, or as a source of ideas the reader might wish to develop further.

For readers who wish to learn how to use network meta-analysis in practice, but who are not familiar with WinBUGS, a grounding in Bayesian methods (Spiegelhalter et al., 2004) and WinBUGS (Spiegelhalter et al., 2007) is essential. The WinBUGS manual, its tutorial, the BLOCKER and other examples in the manual are key resources, as is The Bugs Book (Lunn et al., 2013). Such readers should start with pairwise meta-analysis of binomial data (Chapter 2). They should assure themselves that they understand the WinBUGS outputs and that they are fully aware of the technical details that have to be attended to, such as convergence and the use of a suitably large post-convergence sample. Once all this has been mastered, more complex structures and other outcomes can be attempted (Chapter 4).

Throughout the book more difficult sections are asterisked (*). These sections are not essential, but may be of interest to the more advanced reader.

There are several routes through the book. For a health economist engaged in decision modelling, working with a network meta-analysis generated by, for example, a statistician, Chapter 5 will be particularly relevant. A project manager supervising a synthesis or an HTA exercise, or someone involved in methodological guidance, will want to read the relevant parts of the core material and the heterogeneity sections, ignoring the WinBUGS coding and algebra, glance through the extensions to complex outcomes and give careful attention to Chapter 12 on validity.

We have tried, above all, not to be prescriptive. We have also avoided the modern trend of issuing 'guidance'. Instead, our priority is to be absolutely clear about the properties that alternative models have. We have already set out some of the properties that we believe that evidence synthesis models need to have in the context of transparent, rational, decision making. In the course of the book, we will add further to this list of desiderata. Readers can decide for themselves how important they feel these properties are and can then make an informed choice.

Finally, we would like to acknowledge the many colleagues and collaborators who have contributed to, supported, disagreed with and inspired our work. It is not possible to mention them all, but we would like to record our special thanks to statisticians Keith Abrams, Debbi Caldwell, Julian Higgins, Ian White, Chris Schmid and Tom Trikalinos; to health economists Nicola Cooper, Mark Sculpher, Karl Claxton and Neil Hawkins; in the commercial and government sectors, to Jeanni van Loon at Mapi Group, Rachael Fleurence at PCORI, Meindert Boyson and Carole Longson at NICE and Tammy Clifford at CADTH and above all to our dear colleague Guobing Lu.

List of Abbreviations

1-YS	1-year survival
Acc t-PA	accelerated tissue plasminogen activator
ACE	angiotensin-converting enzyme
ACR	American College of Rheumatology
AFT	accelerated failure time
AIC	Akaike information criterion
AMI	acute myocardial infarction
ANOVA	analysis of variance
APSAC	Anistreplase
ARB	angiotensin-receptor blockers
BASDAI	Bath Ankylosing Spondylitis Disease Activity Index
CADTH	Canadian Agency for Drugs and Technologies in Health
CC	conventional care
CCB	calcium-channel blockers
CEA	cost-effectiveness analysis
CEAC	cost-effectiveness acceptability curve
CI	confidence interval
Cloglog	complementary log–log
CP	coronary patency
CPM	Confidence Profile Method
CrI	credible interval
CTDa	cyclophosphamide–thalidomide–dexamethasone attenuated
CZP	certolizumab pegol
DAG	directed acyclic graph
DIC	deviance information criterion
DMARD	disease-modifying anti-rheumatic drug
EOGBS	early-onset (neonatal) group B streptococcus
EQ-5D	EuroQol 5D
FP	fractional polynomial
GBS	group B streptococcus
GLM	generalised linear model

GRADE	Grading of Recommendations Assessment, Development and Evaluation
HDT-SCT	high-dose chemotherapy and stem cell transplantation
HIV	human immunodeficiency virus
HR	hazard ratio
HRQoL	health-related quality of life
HTA	health technology assessment
ICDF	inconsistency degrees of freedom
IMOR	informative missingness odds ratio
IPD	individual participant data
ISPOR	International Society for Pharmacoeconomics and Outcomes Research
IVAP	Intravenous antibacterial prophylaxis
IVSK	intravenous streptokinase
LDL	low-density lipoprotein
LHR	log hazard ratio
MAR	missing at random
MatC	maternal carriage
MC	Monte Carlo
MCDA	multi-criteria decision analysis
MCMC	Markov chain Monte Carlo
MLE	maximum likelihood estimate
MP	melphalan–prednisone
MPT	melphalan–prednisone–thalidomide
MPV	melphalan–prednisone–bortezomib
MTC	mixed treatment comparisons
MTM	multiple-treatments meta-analysis
MTX	methotrexate
MVN	multivariate normal
NC	neonatal carriage
NICE	National Institute for Health and Care Excellence
NNT	numbers needed to treat
OLS	ordinary least squares
OR	odds ratio
OS	overall survival
PASI	psoriasis area severity index
PFS	progression-free survival
PH	proportional hazards
PICO	population, interventions, comparators, outcome
PRISMA	preferred reporting items for systematic reviews and meta-analyses
PTCA	percutaneous transluminal coronary angioplasty
QALY	quality-adjusted life years
RA	rheumatoid arthritis

RCT	randomised controlled trial
RD	risk difference
RE	random effects
r-PA	reteplase
RR	relative risk
SK	streptokinase
SMD	standardised mean difference
STW	successfully treated week
TB	tuberculosis
TF	treatment failure
TIVA	total intravenous anaesthesia
TNK	tenecteplase
t-PA	tissue plasminogen activator
UK	urokinase
UME	unrelated mean (relative) effects
UTW	unsuccessfully treated week
VCV	variance covariance

About the Companion Website

Don't forget to visit the companion website for this book:

www.wiley.com/go/dias/networkmeta-analysis

There you will find valuable material designed to enhance your learning, including:

1) Main text and exercises, which contains all code used in the book and files required for the exercises
2) Solutions, with solution code.

1

Introduction to Evidence Synthesis

1.1 Introduction

This chapter gives a broad overview of indirect comparisons and network meta-analysis. Using a worked example, we look at what these methods are for, what they can do, the core assumptions, the kinds of output they can produce and some of the implications for decision-making. These topics are picked up and covered in more detail in later chapters.

We assume that the evidence that has been assembled for meta-analysis has been identified by a protocol-driven systematic review. Although the methods of systematic review are not covered in this book, it will become clear that the underlying assumptions of meta-analysis put certain constraints on the conduct of the systematic review. This will be discussed further, particularly in Chapter 12 on the validity of network meta-analysis. We also assume that only randomised controlled trials (RCTs) will be included in the synthesis, although we include some comments on how observational studies might be included in Chapter 9 on bias models.

We begin by discussing the purpose of indirect comparisons and network meta-analysis, which is to form the basis for coherent, evidence-based treatment decisions. We then look at some simple methods for networks involving three treatments to illustrate the concepts. Next a worked example of a larger network meta-analysis is presented to show the main properties of the method. After mentioning the assumptions made by network meta-analysis, we briefly look at the question of which trials should be included in a network meta-analysis. This turns into a discussion of the participants, intervention, comparators, outcomes (PICO), the standard 'script' that defines inclusion criteria in systematic reviews (Sackett et al., 2000; Higgins and Green, 2008). We briefly consider whether PICO definitions should be reconsidered for the purposes of network meta-analysis.

Network Meta-Analysis for Decision-Making, First Edition. Sofia Dias, A. E. Ades, Nicky J. Welton, Jeroen P. Jansen and Alexander J. Sutton.
© 2018 John Wiley & Sons Ltd. Published 2018 by John Wiley & Sons Ltd.
Companion website: www.wiley.com/go/dias/networkmeta-analysis

1.2 Why Indirect Comparisons and Network Meta-Analysis?

While pairwise meta-analysis seeks to combine evidence from trials comparing two treatments, A and B, indirect comparisons involve more than two treatments. A variety of structures can be found, some of which are shown in Figure 1.1. More complex structures are also common and will be presented in later chapters. The first use of indirect comparisons was by Bucher et al. (1997a, 1997b). Their objective was to draw inferences about the relative treatment effects of treatments B and C, using data from B versus A (AB) and C versus A (AC) trials. An indirect estimate of d_{BC}, the effect of C relative to B, is formed by subtracting the direct estimate of d_{AB} from the direct estimate of d_{AC}:

$$\hat{d}_{BC}^{Ind} = \hat{d}_{AC}^{Dir} - \hat{d}_{AB}^{Dir} \tag{1.1}$$

While it is true that the indirect comparison provides an estimate of d_{BC}, in practice this may not be the main reason for looking at the AC and AB trials. More often than not, the objective is to consider which is the best of the three treatments. To make such a comparison, based on one AB trial and one AC trial, we have to make the assumption that, although the event rates in the A arm may have been different in the AB and AC trials, the treatment effects, that is, the differences between arms, that are seen in one trial would – subject to sampling error – also be seen in the other. It is for this reason that some

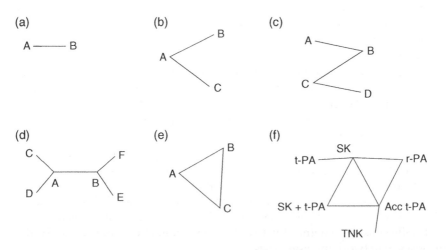

Figure 1.1 Some examples of connected networks: (a) a pairwise comparison, (b) an indirect comparison via reference treatment A, (c) a 'snake' indirect comparison structure, (d) another indirect comparison structure, (e) a simple triangle network and (f) a network of evidence on thrombolytics drugs for acute myocardial infarction (Boland et al., 2003).

authors have referred to this as an 'adjusted' indirect comparison (Glenny et al., 2005), because it takes account of the fact that the AB and AC trials will not necessarily have the same absolute event rate on the A arm. But there is probably no need for this terminology, as the idea of an 'unadjusted' indirect comparison, based on the absolute event rates, has never been seriously considered in the (pairwise) meta-analysis literature as it breaks the fundamental property of randomisation. (A form of unadjusted indirect comparison has, however, been proposed for use in network meta-analysis under the heading 'arm-based models' (Hong et al., 2013, 2015, 2016). These models are reviewed in Chapter 5.)

The assumption that treatment differences on a suitable scale are relatively stable across trials, while absolute event rates can vary, is a fundamental assumption in clinical trials and meta-analysis.

The introduction of 'loops' of evidence into a network represents an additional level of complexity. Examples in Figure 1.1e and f include at least one closed loop of evidence. In the triangle structure of Figure 1.1c, for example, we have two sources of evidence on d_{BC}, direct evidence from the BC trials and indirect evidence from the AB and AC trials. Loops have a special significance as they mean that we are in a position to combine the direct and indirect evidence on d_{BC} and form a single pooled estimate. We are also in a position to assess the degree of consistency (or agreement) between the direct and indirect estimates.

Although we continue to use the terminology 'direct' and 'indirect' throughout the book, these words only have meaning when referring to a specific treatment contrast (we use the term 'contrast' to refer to a pairwise comparison between two treatments). As the network grows in complexity, networks involving over 20 treatments are not uncommon (Corbett et al., 2013; Kriston et al., 2014; Mayo-Wilson et al., 2014), and particularly in a decision-making context, the terms 'direct' and 'indirect' lose all meaning, as evidence that is indirect for one contrast is direct for another, and – depending on network structure – evidence on a single contrast may impact on estimates throughout the whole network. The challenge facing the investigator is first to assemble *all* the trials comparing *any* of the treatments of interest in the target population and second to use the totality of this trial evidence to determine an internally consistent set of estimated relative treatment effects between all treatments. Thus, every trial in the network can be considered as contributing 'direct' evidence towards the decision problem. Furthermore, if we set aside for the moment the results of the trial, its size, and its position in the network structure, we can say that every trial is equally relevant and has an equal chance of influencing the decision.

The primary objective of network meta-analysis is, therefore, simply to combine all the available data in a coherent and internally consistent way. To see what this means, consider the triangle structure of Figure 1.1e. One way to analyse these data would be to carry out three separate meta-analyses on the AB, AC and BC sets of trials to generate three estimates: $\hat{d}_{AB}^{Dir}, \hat{d}_{AC}^{Dir}, \hat{d}_{BC}^{Dir}$. However, these estimates cannot in themselves form a coherent basis for deciding which treatment is the best. Suppose we had three boys: John, Paul and George. John is 2 inches taller than Paul, and Paul is 3 inches taller than George. From this we know that John is 5 inches taller than George. If we were not able to make this inference, if John could be 6 or 2 inches taller than George, then we would not be able to decide who was tallest.

What is needed, therefore, is not the original, unrelated estimates $\hat{d}_{AB}^{Dir}, \hat{d}_{AC}^{Dir}, \hat{d}_{BC}^{Dir}$, but three coherent estimates $\hat{d}_{AB}^{Coh}, \hat{d}_{AC}^{Coh}, \hat{d}_{BC}^{Coh}$ that have the property: $\hat{d}_{AC}^{Coh} = \hat{d}_{AB}^{Coh} + \hat{d}_{BC}^{Coh}$. With John = A, Paul = B and George = C, we can deduce that the AC difference is 5, the sum of the AB and BC differences. Network meta-analysis is no more than a technical way of processing the trial-specific treatment effects to arrive at the coherent estimates that are required for coherent decisions about which treatment is best while correctly reflecting parameter uncertainty. However, it has to be appreciated that the coherent estimates represent a view of the data that makes some assumptions. These are set out in Section 1.5.

A network meta-analysis can be carried out whenever the included trials and treatments form a connected network, that is, a network where there is a path linking each treatment to every other (Figure 1.1). In that sense, pairwise meta-analysis and indirect comparisons are special cases of simple network meta-analysis. See Chapter 2 for more details on this.

1.3 Some Simple Methods

The first explicit use of indirect comparisons was due to Bucher et al. (1997a, 1997b). An indirect estimate for the treatment effect of C relative to B is formed by subtracting the estimate of B relative to A from the estimate of C relative to A, as in equation (1.1). Before going further, note that we keep to a convention that d_{XY} is the additional effect of Y relative to X and that we always keep treatment indices in numerical or alphabetical order, d_{XY} or d_{35}; $-d_{YX}$ is equivalent to d_{XY}, but we never refer to the former to avoid confusion. A helpful way of remembering how the notation works is to draw the line:

Now consider the relative effect of B compared to A as the 'distance' in relative effect terms between points A and B on the line (hence the notation d_{AB}). This immediately clarifies why the indirect estimate of d_{BC} is calculated as $d_{AC} - d_{AB}$ in equation (1.1), and not $d_{AB} - d_{AC}$. The variance of the indirect estimate is the sum of the variances of the two direct estimates, as each is derived from different trials including different patients and therefore independent:

$$V_{BC}^{Ind} = V_{AC}^{Dir} + V_{AB}^{Dir} \qquad (1.2)$$

See Exercise 1.1. Notice that equation (1.2) tells us that indirect evidence will always have a higher variance than any of its component parts. As the number of links increases, as in Figure 1.1c, the variance of the indirect estimate will become very large (see Exercise 1.3).

We can extend this by pooling the indirect and direct estimates into a single combined estimate. In the triangle structure of Figure 1.1c, we have three sets of trials on AB, AC and BC, and we now conduct three separate pairwise meta-analyses to generate the three pairwise pooled estimates $\hat{d}_{AB}^{Dir}, \hat{d}_{AC}^{Dir}, \hat{d}_{BC}^{Dir}$ and their variances $V_{AB}^{Dir}, V_{AC}^{Dir}, V_{BC}^{Dir}$. We therefore have two independent sources of evidence on d_{BC}, one direct and the other indirect. This invites us to pool the two estimates into a single combined estimate, for example, using inverse-variance weighting:

$$\hat{d}_{BC}^{Pooled} = \frac{w_{BC}^{Dir} \hat{d}_{BC}^{Dir} + w_{BC}^{Ind} \hat{d}_{BC}^{Ind}}{w_{BC}^{Dir} + w_{BC}^{Ind}}, \quad \text{where} \quad w_{BC}^{Dir} = \frac{1}{V_{BC}^{Dir}}$$

$$\text{and} \quad w_{BC}^{Ind} = \frac{1}{V_{BC}^{Ind}} = \frac{1}{V_{AB}^{Dir} + V_{AC}^{Dir}} \qquad (1.3)$$

See Exercise 1.2. Before accepting this combined estimate, it will be important to check that the direct and indirect estimates of d_{BC} are consistent with each other. This is explored in Chapter 7.

The aforementioned procedure gives us a pooled estimate of d_{BC}, but it does not immediately yield the set of coherent estimates of all three parameters. We can generate these by simply repeating the process for the other edges. The pooled estimates $\hat{d}_{AB}^{Pooled}, \hat{d}_{AC}^{Pooled}$ and $\hat{d}_{BC}^{Pooled}$ will have the required property of coherence, but we can arrive at these estimates more efficiently in a single step using suitable computer programs. Indeed, the simple approach to indirect and mixed treatment comparisons in equations (1.2) and (1.3) is extremely limited: as we go from 3 to 4, 5, 10, 20 or 30 treatments, the number of pairwise comparisons that can be made increases from 3 to 6 and then 15, 45, 190 and finally 445. (For S treatments, there are $S(S-1)/2$ possible contrasts.) It is evident that distinguishing between direct and indirect evidence in the context of a decision on which of several treatments is best is not only a pointless exercise but also a very tedious one. What is needed is a method that can put all this evidence together to produce the coherent estimates required and some methods for assessing the consistency of evidence from different sources.

1.4 An Example of a Network Meta-Analysis

We now develop a worked example of a network meta-analysis. This is taken from a health technology assessment report (Boland et al., 2003) that was published in 2003, based on an analysis of thrombolytic treatments carried out for NICE, but before the use of network meta-analysis had become a routine feature of NICE technology appraisals. The evidence structure (Figure 1.1f) comprises six active treatments: streptokinase (SK), tissue-plasminogen activator (t-PA), accelerated tissue-plasminogen activator (Acc t-PA), SK + t-PA, tenecteplase (TNK) and reteplase (r-PA). Table 1.1 is a logically ordered layout of the evidence structure that makes it easy to see that the majority of the evidence is on trials that compare SK with t-PA. In the original report the authors presented their findings as shown in Table 1.2, which shows the results of four separate pairwise meta-analyses, each undertaken independently, and therefore not having the required property of coherence.

In the absence of a single coherent analysis, the authors were obliged to put forward their conclusions in terms of the pairwise analyses:

> ... streptokinase is as effective as non-accelerated alteplase ... tenecteplase is as effective as accelerated alteplase ... reteplase is at least as effective as streptokinase ... [is] streptokinase as effective as, or inferior to accelerated alteplase ... [is] reteplase as effective as accelerated alteplase or not ... (Boland et al., 2003).

It is difficult to conclude from statements of this sort which treatment performs best. Indeed, given that each estimate is subject to sampling error, under some circumstances, it could prove *impossible* to identify the best treatment using this approach.

Table 1.1 The thrombolytics dataset, 14 trials, six treatments (data from Boland et al., 2003): streptokinase (SK), tissue-plasminogen activator (t-PA), accelerated tissue-plasminogen activator (Acc t-PA), tenecteplase (TNK), and reteplase (r-PA) 14 trials.

RCTs	SK	t-PA	Acc t-PA	SK + t-PA	r-PA	TNK
8	✓	✓				
1	✓		✓	✓		
1	✓			✓		
1	✓				✓	
2			✓		✓	
1			✓			✓

Fifteen possible pairwise comparisons (a tick represents comparisons made in RCTs). The network is shown in Figure 1.1f.

Table 1.2 Findings from the HTA report on thrombolytics drugs (Boland et al., 2003).

Treatment comparison	Trials	Odds ratio	95% CIs
SK vs t-PA	8	1.00	0.94, 1.06
Acc t-PA vs TNK	1	0.99	0.88, 1.13
Acc t-PA vs r-PA	2	1.24	0.61, 2.53
r-PA vs SK	1	0.94	0.79, 1.12

Table 1.3 Thrombolytics example, fixed effect analysis: odds ratios (posterior medians and 95% CrI).

	SK	t-PA	Acc t-PA	t-PA + SK	r-PA	TNK
SK	–	1.00	0.86	0.96	0.95	
		(0.94, 1.06)	(0.78, 0.94)	(0.87, 1.05)	(0.79, 1.12)	
t-PA	1.00	–				
	(0.94, 1.06)					
Acc t-PA	0.86	0.87	–	1.12	1.02	1.01
	(0.79, 0.94)	(0.78, 0.96)		(1.00, 1.25)	(0.90, 1.16)	(0.88, 1.14)
t-PA + SK	0.96	0.96	1.12	–		
	(0.88, 1.05)	(0.86, 1.08)	(1.00, 1.24)			
r-PA	0.90	0.91	1.04	0.94	–	
	(0.80, 1.02)	(0.79, 1.03)	(0.94, 1.16)	(0.82, 1.08)		
TNK	0.87	0.87	1.01	0.90	0.96	–
	(0.75, 1.01)	(0.74, 1.03)	(0.89, 1.14)	(0.77, 1.14)	(0.82, 1.14)	

Upper right triangle gives the ORs from pairwise comparisons (column treatment relative to row), lower left triangle ORs from the network meta-analysis (row treatment relative to column).

Table 1.3 sets out the full set of estimated median odds ratio (OR) estimates from a fixed effects network meta-analysis in the lower left triangle and the median ORs from the unrelated pairwise analyses in the upper right triangle, along with their 95% credible intervals. Details on how the network estimates were obtained are given in Chapter 2. The network meta-analysis fills in each of the 15 possible pairwise contrasts. See Exercise 1.4.

It is always informative to look at the relationship between the pairwise estimates and the network estimates. One critical question is whether the direct evidence is consistent with the indirect evidence on each contrast; this cannot be answered from this summary table, but formal methods to detect inconsistency are introduced in Chapter 7. However, there are a number of useful 'reality' checks that can be carried out. In some parts of the network, for example, the network meta-analysis does not appear to have had much effect: the OR of t-PA relative to SK and its credible interval are entirely unchanged. Examination of the network structure (Figure 1.1f) immediately explains why: the (SK, t-PA) edge is an isolated spur on the network and the SK versus t-PA trials are the sole source of data on this contrast. On the other hand, the network estimate of the (SK, r-PA) comparison has a tighter credible interval than the pairwise estimate, and this is due to the additional indirect evidence contributed by the (SK, Acc t-PA) trials. However, we do not see the same effect with the (SK, SK + t-PA) comparison because this contrast is informed only by a single three-arm trial so that the 'loop' formed by the (SK + t-PA, Acc t-PA) and (SK, Acc t-PA) edges does not constitute additional information. The way that data in one part of the network affects estimates in other parts is determined by the network structure and the quantity of evidence on different edges. Investigators need to be able to 'read' the table in this way to gain an understanding of the main 'drivers' of the results.

The best treatment is Acc t-PA (OR relative to SK 0.86), with TNK a close second (OR 0.87). Following the accepted principles of pairwise meta-analysis, we distinguish between the *relative* treatment effects on mortality, which are pooled in evidence synthesis and are expressed here as ORs, and the absolute mortality on specific treatments. Our view is that absolute effects for given treatments must be considered quite separately, because although trials are essential to inform relative effects, they are not necessarily the best source of information on absolute effects of treatments. This is dealt with in detail in Chapter 5. However, if we can assume a distribution for the absolute 35-day mortality on reference treatment SK, which is appropriate for our target populations, we can use the relative effect estimates from the network meta-analysis to infer the absolute mortality on all the treatments. This is shown in Table 1.4. Given a mortality on SK of 8.33%, we expect 7.29% mortality on Acc t-PA and 7.34% on TNK.

Also shown in Table 1.4 are the posterior median ranks of each treatment (rank '1' indicates lowest mortality) and their credible intervals. The posterior distribution of the rankings and the probability that each treatment is 'best', which is derived from the rankings, are evidently quite unstable. Indeed, the reason for looking at these statistics is only to make oneself aware of the statistical uncertainty in the treatment differences. Decisions should never be based on these probability statements. For example, if a decision was to be made strictly on the grounds of the effect of treatment on a 35-day survival, regardless of cost or side effects, the treatment chosen should be the one that was associated with the lowest expected mortality, in this case Acc t-PA, as shown by the ranking of the posterior *expected* effects in Table 1.4. This is not necessarily the same as the treatment that is most likely to be associated with the lowest mortality, which in this case is TNK (see Exercise 1.5). Similarly,

Table 1.4 Thrombolytics example, fixed effect analysis: posterior summaries.

	Odds ratios		Rank			35 day mortality %		Probability best (%)
	Median	95% CrI	Mean	Median	95% CrI	Mean	95% CrI	
SK	Reference		5.3	5	(4, 6)	8.33	(4.9, 13.1)	0
t-PA	1.00	(0.94, 1.06)	5.1	5	(3, 6)	8.30	(4.9, 13.1)	0.2
Acc t-PA	0.87	(0.79, 0.94)	1.7	2	(1, 3)	7.29	(4.3, 11.6)	41
SK+t-PA	0.96	(0.88, 1.05)	4.1	4	(2, 6)	8.04	(4.7, 12.7)	1.3
r-PA	0.90	(0.80, 1.01)	2.7	3	(1, 5)	7.59	(4.4, 12.1)	16
TNK	0.87	(0.75, 1.01)	2.1	2	(1, 5)	7.34	(4.2, 11.8)	43

Odds ratios, treatment rankings with credible intervals, probability 'best', and absolute mortality. The odds ratios are taken from Table 1.3.

decisions based on net benefit would be based on the treatment with the highest expected net benefit, not on the treatment that is most likely to have the highest net benefit (see Chapter 5). These probability statements can be seen as a way of avoiding the complexity of 15 pairwise significance tests, but should be interpreted with caution. In this case, under the fixed effects model, the data appear to rule out SK, t-PA and SK + t-PA, but do not distinguish between Acc t-PA, TNK, and '3rd place' r-PA.

1.5 Assumptions Made by Indirect Comparisons and Network Meta-Analysis

The assumptions made by network meta-analysis can be stated in several ways. First, in a fixed effects model, one is assuming that the d_{AB} effect that is estimated in the AB trials is the same as the d_{AB} effect that would be estimated in the AC, AD, BC and CD trials *if* these trials had included treatments A and B. With a random effects model, one is assuming that the trial-specific effect $\delta_{i,AB}$, which is estimated in trial i comparing AB, is a sample from a random effects distribution with mean d_{AB} and variance σ_{AB}^2, that all the other AB trials are estimating effects from this same distribution and that if A and B were included in the AC, AD, BC and CD trials, they too would be estimating parameters that are random samples from the same random effects distribution (see Chapter 2 for further details).

Another way to express this is to imagine that all the trials were in fact multi-arm trials of treatments A, B, C and D, but only a subset of the treatments are reported and the remaining treatments are missing at random (Little and

Rubin, 2002). This does not mean that every treatment is equally likely to be entered into trials, nor even that missingness is unrelated to efficacy. What is required is that missingness mechanism operates with no regard to the true *relative* efficacy of the treatments (see Chapter 12).

These assumptions are often called the 'consistency assumptions'. Contrary to what is often stated (Song et al., 2009), the consistency assumptions are *not* additional assumptions made by network meta-analysis, but they follow from more basic assumptions underlying all meta-analyses (Lu and Ades, 2009), as explained in Chapter 2 and explored further in Chapter 12. Even so, all the doubts that have been expressed about indirect comparisons and network meta-analysis can be seen as doubts about this assumption. For example:

> Between-trial comparisons [indirect comparisons] are unreliable. Patient populations may differ in their responsiveness to treatment.... Thus an apparently more effective treatment may ... have been tested in a more responsive population (Cranney et al., 2002).
>
> Placebo controlled trials lacking an active control give little useful information about comparative effectiveness.... Such information cannot reliably be obtained from cross-study [indirect] comparisons, as the conditions of the studies may have been quite different (ICH Expert Working Group, 2000).

The warnings 'patient populations may differ' or 'the conditions of the study may have been quite different' constitute a general warning that there may be unrecognised effect modifiers, present in the AB trials but not in the AC trials, so that a BC comparison based on this information may be confounded. This is unquestionably correct. But it is not clear that indirect comparisons and network meta-analysis are really different in this aspect from pairwise meta-analysis, where unrecognised effect modifiers are also present, as proven by the frequent need to fit random effects models (Engels et al., 2000), often with extremely high levels of between-study variation (Turner et al., 2012).

Putting this in another way, while there is no doubt that indirect comparisons may be subject to biases arising from failures in internal or external validity, equation (1.1) tells us that indirect evidence (on the left hand side) can only be biased if direct evidence (on the right hand side) is biased. We can unpick this paradox only by reminding ourselves that the concept of 'bias' is relative to a target parameter. For example, suppose our objective is to estimate the effect of C relative to B in patients who are naïve on both B and C, say, $d_{BC}^{(N)}$. We need to be sure that the included AB and AC trials are based on patients who were indeed naïve to both these treatments. If we are satisfied that $\hat{d}_{AB}^{Dir}$ and $\hat{d}_{AC}^{Dir}$ are unbiased estimates of $d_{AB}^{(N)}$ and $d_{AC}^{(N)}$ relating to patients who are naïve on both B and C, we can be satisfied that $\hat{d}_{BC}^{(N)Ind} = \hat{d}_{AC}^{(N)Dir} - \hat{d}_{AB}^{(N)Dir}$ is also unbiased for this population. If, however, the AC trials are performed on patients who have failed on B, we may think that the direct estimate, $\hat{d}_{AC}^{Dir}$, is 'biased' with respect to the target population, in this case the indirect estimate will inherit this bias.

This, of course, highlights the importance of paying careful attention to which evidence is included in a network meta-analysis (see Chapter 12), or, alternatively, it invites us to consider some form of covariate adjustment (Chapter 8) or bias modelling (Chapter 9). Although we return to this theme in the later chapters, readers may find it useful to see the direction of travel in advance.

The trial data that are available for evidence synthesis are likely, in many instances, to suffer from a range of imperfections. Trials may be only partially *relevant* (external bias) because the patient population is mixed, or the clinical settings are different. Further, the conduct of the trial may have characteristics, such as lack of blinding or allocation concealment, that have been shown to be associated with compromised internal validity (internal bias) (Schulz et al., 1995). This has given rise to a range of methods centred on trial 'quality'. One approach has been the Cochrane Risk of Bias tool (Lundh and Gotzsche, 2008). Another is the Grading of Recommendations Assessment, Development and Evaluation (GRADE) system of assessing the quality of pairwise meta-analyses on a set of pre-determined criteria (Guyatt et al., 2008), extended to deliver quality ratings on network meta-analyses (Puhan et al., 2014; Salanti et al., 2014). These approaches are discussed in more detail in Chapter 12.

Our approach will be somewhat different. Rather than carrying out a network meta-analysis on trial results that we suspect are biased and rating our results from such analysis as unreliable, we would prefer to address any problems with the evidence in advance so as to produce the best possible estimates, whose reliability is appropriately reflected in their credible intervals.

Our strategy to achieve this has several planks. First, knowing that network meta-analysis makes certain assumptions, the most important step is to do everything possible to ensure that the data that are assembled to answer the decision question actually meet those assumptions (see Chapter 12). Second, if there is variation in effect modifiers across the trials in the network, we can use meta-regression (see Chapter 8) to carry out covariate adjustment. Third, if there are potential biases associated with explicit markers of trial 'quality', bias adjustment methods can be applied (see Chapter 9). These can take the form of a regression adjustment, or use can be made of a growing literature that quantifies the extent of potential bias in trials (Wood et al., 2008; Savovic et al., 2012a, 2012b).

1.6 Which Trials to Include in a Network

The existence of methods that can compare multiple treatments, based on all the RCT evidence that compares any of them with any of the others, raises a series of further, more strategic issues. When evaluating whether a new product should be adopted by a health service or whether it should be reimbursed by insurance, one approach is to compare the new therapy to one or more standard therapies and issue a recommendation based on whether it is clinically more effective, or more cost-effective, than standard treatments. Another option is to evaluate the new product against all the standard comparators *and* against all the new competitor products, and then choose the best.

These two procedures appear to be very different, and they clearly have different implications for manufacturers, as the 'choose the best' strategy has the effect of removing all but one of a new set of treatments from the field, and yet both decisions could be based on the same set of trials, and indeed on the same network meta-analysis. This is because the 'evidence space' does not have to coincide with the 'decision space'. The procedure of ranking and choosing the best treatment does not have to be applied to *all* the treatments; it can apply only to the ones that are relevant for the decision, and it is simpler and probably fairer to all concerned, for both decisions to have the same evidential base.

1.6.1 The Need for a Unique Set of Trials

A similar principle lay behind the decision in the methods guide by the National Institute for Health and Clinical Excellence (NICE, 2008b) to permit for the first time the inclusion in evidence networks of treatments that were not of intrinsic interest themselves, but which formed a link between otherwise disconnected networks (Figure 1.2a). If there was no treatment G that had been trialled with at least one of the treatments A, B and C *and* one of the treatments X and Y, then there would be no way of using network meta-analysis to compare X or Y with A, B or C. However, once a treatment G that links the two networks has been identified (Figure 1.2b), are there other treatments that could have made the required connection? If there are, these should also be included, to prevent any suggestions of 'gaming' with the network structure (choosing comparisons that flattered one particular product over the others). The addition of treatment H (Figure 1.2c) illustrates this situation. Finally, if the comparator set is A, B, C, D, G, H, X and Y, we must ask: have all the trials comparing any two or more of these treatments been included? These are no less 'relevant' to the decision problem than the others and should therefore also be included (Figure 1.2d).

To ensure a transparent process that removes any temptation to manipulate trial inclusion/exclusion criteria, a procedure is required, which identifies a

Figure 1.2 (a) An unconnected network; (b) use of an intermediate treatment to link a network (dashed lines); (c) other intermediate treatments that could also be used as links; and (d) incorporation of all trials comparing the enlarged set of treatments.

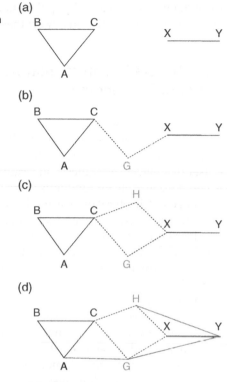

unique set of trials. Following the illustration in Figure 1.2, we suggest the following algorithm:

1) Identify all the comparators of interest.
2) Identify all the trials that compare two or more of the comparators in the population of interest.
3) Remove trial arms that are not comparators of interest from trials with more than two arms. If this forms a connected network, stop.
4) If this is not a connected network,
   a) Identify further treatments that have been compared with *any* treatments in both sub-networks (including any additional arms that may have been removed in step 3).
   b) Add these to the comparators of interest.
   c) Carry out steps (2) and (3) with the larger set of comparators.

This procedure identifies a minimum set of trials that form a connected network while still being unique. Of course, this does not necessarily guarantee that a connected network can be obtained (see Section 1.7). In addition, a degree of subjective judgement remains as there can always be debate about the relation between a trial population, often poorly described, and the target 'population of interest', as in any synthesis.

1.7   The Definition of Treatments and Outcomes: Network Connectivity

This raises the question: supposing no connected network of trial evidence can be formed, can one use one-arm studies? Can one use observational studies? Rather than raise false hopes in the reader's mind, we make it clear from the outset that evidence synthesis on treatment effects, as conceived in this book, is based exclusively on randomised trial evidence. Nevertheless, the methods discussed in Chapter 9 on bias models can be applied to observational data on treatment efficacy. These methods force the user to be absolutely explicit about the distribution of potential bias. Similarly, the entire book follows the tradition of meta-analytic methods that concerns the statistical combination of *relative* treatment effects, which rules one-arm studies out of bounds – except when they are used to inform a natural history model. These points are taken up in Chapter 5.

With these provisos, the issue of network connectedness cannot be separated from how treatments and outcomes are defined. In traditional pairwise meta-analysis, this has been determined by the PICO script (Sackett et al., 2000). In the era of network meta-analysis, we suggest in Sections 1.7.1–1.7.3 that there is a need to liberalise the I, C and O constraints on study inclusion while tightening up quite considerably on the P.

1.7.1   Lumping and Splitting

With pairwise meta-analyses it remains common to 'lump' treatments together. Sometimes this is done so there will be enough trials to justify a quantitative synthesis (Gotzsche, 2000). One example of this is a comparison of thrombolytics treatments for acute myocardial infarction against the surgical procedure percutaneous transluminal coronary angioplasty (PTCA) (Keeley et al., 2003b). This analysis, which lumps the thrombolytics treatments (Figure 1.1f), found that PTCA was (statistically) significantly superior to thrombolytics taken as a class. However, in subsequent published correspondence, a 'splitter' objected that this was not the relevant comparison; instead one should compare PTCA with the *best* of the thrombolytics treatments, which was Acc t-PA (Fresco et al., 2003). On this comparison – based on very few trials – PTCA was superior, but this was no longer statistically significant. Revealingly, the original author, evidently a committed 'lumper', declared herself 'mystified' by the criticism of 'lumping trials together' because 'meta-analyses, by definition, pool data' (Keeley et al., 2003a).

In a network meta-analysis one can accept that different thrombolytics may not all be equally effective while at the same time retaining statistical power by combining direct and indirect evidence. A network meta-analysis of the same dataset demonstrated a statistically significant advantage of PTCA over Acc t-PA (Caldwell et al., 2005). See Chapters 2, 3 and 7 for more details on this example.

The default network meta-analysis, therefore, treats every intervention, every dose and every treatment combination a separate 'treatment'. On the other hand, the finer the distinction between treatments, the greater the risk of disconnected networks. There are three ways in which this can be ameliorated: dose response models, treatment combination models or class effects models. The effect of these techniques is to reduce – sometimes dramatically – the number of effective parameters that need to be estimated, and this impacts powerfully on the connectivity of the network. These models are described further in Chapter 8.

1.7.2 Relationships Between Multiple Outcomes

Another way to create more connections is to include more than one 'outcome' in the same synthesis. Trials very frequently report different outcomes at different follow-up times, and some report more than one outcome at more than one time. However, if we restrict ourselves to one particular outcome, or one particular time, we will produce evidence networks that are sparse, poorly connected, or even disconnected. If, on the other hand, the different outcomes can be included in the same analysis, a more connected network is available. What is required is an explicit model for how the treatment effects at different times or on different outcomes are related. However, this is not a trivial undertaking. Chapter 11 gives a number of examples showing how clinical and logical relationships between outcomes can be exploited to provide a single, coherent analysis of all the treatments on all the outcomes and at the same time creating connected networks from unconnected components.

1.7.3 How Large Should a Network Be?

There is no theoretical limit to how large a network can be, and a question that is often asked is 'how large *should* a network be?' In theory, the larger the network, the more robust the conclusions will be, in the sense that the conclusions will become increasingly insensitive to the inclusion or exclusion of any particular trial. However, there is little doubt that the larger the network becomes, the greater the risk of clinical heterogeneity in the trial populations, particularly as one reaches further back to include older trials. This is because new treatments tend to be introduced for more severely ill patients. As time goes on more data on side effects accumulates, and eventually clinicians start to feel it might benefit less ill patients who are then included in later trials.

For this reason it may often turn out that larger networks will fail to produce more precise estimates: greater clinical heterogeneity may lead to greater statistical heterogeneity in treatment effects, larger between-trial variances in random effects models and therefore lower precision in estimates of mean treatment effects (Cooper et al., 2011). It may also increase the risk of inconsistency, as older treatments will have been compared on more severely ill populations than newer treatments. Thus different sources of evidence (direct and indirect) may conflict.

1.8 Summary

In this introductory chapter we have looked at why indirect comparisons and network meta-analysis are needed. We have emphasised that their key property is that they produce an internally coherent set of estimates, which is essential for rational decision-making, whether based on efficacy of cost-effectiveness. We have observed that the simpler methods, which can be applied to triangular networks, have limited scope, and they become completely unfeasible as the networks grow in complexity. We have set out results from a worked example and shown how to interpret them and how comparison with pairwise meta-analysis results can help us understand the main 'drivers' of the results. Finally we have set out the key assumptions in an informal way and looked at some of the strategic policy issues that the existence of network meta-analysis immediately raises: what treatments should be compared, which trials should be included, and how large should a network of comparisons be.

As we work through the book, many of these themes will be picked up again and covered in more detail.

1.9 Exercises

1.1 This table gives results, in the form of log hazard ratios from a study of virologic suppression following highly active antiretroviral therapy for HIV (Chou et al., 2006). Obtain an indirect estimate for $\hat{d}_{BC}$.

	Log hazard ratio	95% CI	Standard error
$\hat{d}_{BC}$	0.47	0.27, 0.67	0.10
$\hat{d}_{AB}$	2.79	1.69, 3.89	0.56
$\hat{d}_{AC}$	1.42	−0.76, 2.08	0.34

1.2 Pool the direct and indirect estimates on BC from the previous exercise. Comment on the validity of the pooled estimate.

1.3 Find an indirect estimate for $\hat{d}_{AD}$ and its standard error using the data below. (Hint: draw the treatment network first.)

	Difference in median survival	Standard error
$\hat{d}_{AB}$	−2.8	1.42
$\hat{d}_{BC}$	2.7	1.24
$\hat{d}_{CD}$	3.0	1.20

1.4 The table below shows posterior mean and median odds ratio from two sets of analyses on a triangle network of three treatments A, B and C. One of these analyses represents pairwise summaries from unrelated meta-analyses, while the other comes from a network meta analysis. Which is which?

	Analysis 1		Analysis 2	
	Mean	Median	Mean	Median
OR_{AB}	1.773	1.624	1.669	1.405
OR_{AC}	2.378	2.289	2.444	2.341
OR_{BC}	1.540	1.411	1.249	0.9538

1.5 *In Section 1.4 it was noted that the treatment with the highest expected treatment effect may not always be the same as the treatment that is most likely to have the highest treatment effect. Under what circumstances can this happen?

2

The Core Model

2.1   Bayesian Meta-Analysis

There is a substantial literature on statistical methods for meta-analysis, going back to methods for combination of results from two-by-two tables and the introduction of random effects meta-analysis (DerSimonian and Laird, 1986), an important benchmark in the development of the field. Over the years methodological and software advances have contributed to the widespread use of meta-analytic techniques. A series of instructional texts and reviews have appeared (Cooper and Hedges, 1994; Smith et al., 1995; Egger et al., 2001; Sutton and Abrams, 2001; Higgins and Green, 2008; Sutton and Higgins, 2008), but there have been only a few attempts to produce a comprehensive guide to the statistical theory behind meta-analysis (Whitehead and Whitehead, 1991; Whitehead, 2002).

We present a single unified framework for evidence synthesis of aggregate data from RCTs that delivers an internally consistent set of estimates while respecting the randomisation in the evidence (Glenny et al., 2005). The core models presented in this chapter can synthesise data from pairwise meta-analysis, indirect and mixed treatment comparisons, that is, network meta-analysis (NMA) with or without multi-arm trials (i.e. trials with more than two arms), without distinction. Indeed, pairwise meta-analysis and indirect comparisons are special cases of NMA, and the general NMA model and WinBUGS code presented here instantiate that.

We take a Bayesian approach to synthesis using Markov chain Monte Carlo (MCMC) implemented in WinBUGS (Lunn et al., 2013) mainly due to the flexibility of the Bayesian approach, ease of implementation of models and parameter estimation in WinBUGS and its natural extension to decision modelling (see Preface). While we do not describe Bayesian methods, MCMC or the basics of running WinBUGS in this book, basic knowledge of all these concepts is required to fully understand the code presented. However, the underlying principles and concepts described can still be understood without detailed

Network Meta-Analysis for Decision-Making, First Edition. Sofia Dias, A. E. Ades, Nicky J. Welton, Jeroen P. Jansen and Alexander J. Sutton.
© 2018 John Wiley & Sons Ltd. Published 2018 by John Wiley & Sons Ltd.
Companion website: www.wiley.com/go/dias/networkmeta-analysis

knowledge of Bayesian methods, MCMC or WinBUGS. Further references and a guide on how to read the book are given in the Preface.

We begin by presenting the core Bayesian models for meta-analysis of aggregate data presented as the number of events observed in a fixed and known number of individuals – commonly referred to as binary, dichotomous or binomial data. We describe fixed and random treatment effects models and show how the core models, based on the model proposed by Smith et al. (1995) for pairwise meta-analysis, are immediately applicable to indirect comparisons and NMA with multi-arm trials without the need for any further extension. This general framework is then applied to other types of data in Chapter 4.

2.2 Development of the Core Models

Consider a set of M trials comparing two treatments, 1 and 2, in a pre-specified target patient population, which are to be synthesised in a meta-analysis. A fixed effects analysis would assume that each study i generates an estimate of the same parameter d_{12}, the relative effect of treatment 2 compared with 1 on some scale, subject to sampling error. In a random effects model, each study i provides an estimate of the study-specific treatment effects $\delta_{i,12}$, the relative effect of treatment 2 compared with 1 on some scale in trial i, which are assumed not to be equal but rather *exchangeable*. This means that all $\delta_{i,12}$ are 'similar' in a way that assumes that the trial labels, i, attached to the treatment effects $\delta_{i,12}$ are irrelevant. In other words, the information that the trials provide is independent of the order in which they were carried out, over the population of interest (Bernardo and Smith, 1994). The exchangeability assumption is equivalent to saying that the trial-specific treatment effects come from a common distribution, and this is often termed the 'random effects distribution'. In a Bayesian framework, the study-specific treatment effects, δ_i, are estimated for each study and are often termed the 'shrunken' estimates. They estimate the 'true' treatment effect for each trial and therefore 'shrink' the observed trial effects towards the random effects mean. See Chapters 3 and 8 for further details. The normal distribution is often chosen, so that

$$\delta_{i,12} \sim \mathrm{Normal}\left(d_{12}, \sigma_{12}^2\right) \tag{2.1}$$

with d_{12} and σ_{12}^2 representing the mean and variance, respectively. However, as we will see, any other suitable distribution could be chosen instead, and this could be implemented in WinBUGS. Note that the mean of the random effects distribution is the pooled effect of treatment 2 compared with 1, which is the main parameter of interest in a meta-analysis. It follows that the fixed effects model is a special case, obtained by setting the variance to zero, implying that $\delta_{i,12} = d_{12}$ for all i.

In a Bayesian framework the parameters to be estimated are given prior distributions. In general we will want the observed data from the RCTs to be the main, overwhelming, influence on the estimated treatment effects and will therefore consider non-informative or minimally informative prior distributions, wherever possible. The degree of information in a prior distribution is intimately related to the range of plausible values for a parameter. For the pooled treatment effect d_{12}, we assume that it has been specified on a continuous scale and that it can take any value between plus and minus infinities (see Section 2.2.1). An appropriate (non-informative) prior distribution is then

$$d_{12} \sim \text{Normal}\left(0,100^2\right) \tag{2.2}$$

where the variance is very large, meaning that the distribution is essentially flat over the plausible range of values for the treatment effect.

For the between-study heterogeneity variance σ_{12}^2, the chosen prior distribution must be constrained to give only positive values, and a reasonable upper bound will depend on the expected range of observed treatment effects. We will choose uniform prior distributions with lower bound at zero and a suitable upper bound for the outcome measure being considered. In Section 2.3.2 alternative prior distributions are discussed.

2.2.1 Worked Example: Meta-Analysis of Binomial Data

Caldwell et al. (2005) extended the thrombolytics network presented in Chapter 1 by adding studies from another systematic review on the same patient population, which also included an extra treatment (Keeley et al., 2003b). This extended network is presented in Figure 2.1.

The data available are the number of mortalities by day 35, out of the total number of patients in each arm of the 36 included trials (Table 2.1). We will start by specifying the fixed and random effects two-treatment (pairwise) meta-analysis models and WinBUGS code for PTCA compared with Acc t-PA and will then extend the models and code to incorporate all available treatments and trials.

2.2.1.1 Model Specification: Two Treatments

Eleven trials compared PTCA with Acc t-PA (data shown in the last 11 rows of Table 2.1). We will define Acc t-PA as our reference or 'control' treatment, treatment 1, and PTCA will be treatment 2. Defining r_{ik} as the number of events (deaths), out of the total number of patients in each arm, n_{ik}, for arm k of trial i, we assume that the data generation process follows a binomial likelihood, that is

$$r_{ik} \sim \text{Binomial}\left(p_{ik},n_{ik}\right) \tag{2.3}$$

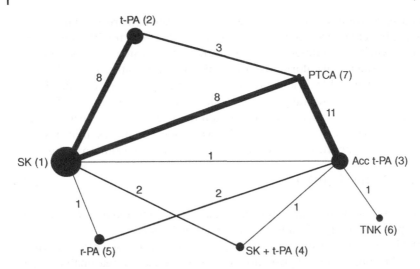

Figure 2.1 Extended thrombolytics example: network plot. Seven treatments are compared (data from Caldwell et al., 2005): Streptokinase (SK), tissue-plasminogen activator (t-PA), accelerated tissue-plasminogen activator (Acc t-PA), reteplase (r-PA), tenecteplase (TNK), percutaneous transluminal coronary angioplasty (PTCA). The numbers on the lines and line thickness represent the number of studies making those comparisons, the widths of the circles are proportional to the number of patients randomised to each treatment, and the numbers in brackets are the treatment codes used in WinBUGS.

where p_{ik} represents the probability of an event in arm k of trial i ($i = 1, \ldots,$ 11; $k = 1, 2$).

Since the parameters of interest, p_{ik}, are probabilities and therefore can only take values between 0 and 1, a transformation (link function) is used that maps these probabilities into a continuous measure that can take any value between plus and minus infinities. The most commonly used link function for the probability parameter of a binomial likelihood is the logit link function (see also Chapter 4). The probabilities of success p_{ik} are modelled on the logit scale as

$$\text{logit}\left(p_{ik}\right) = \mu_i + \delta_{i,1k} \tag{2.4}$$

In this setup, μ_i are trial-specific baselines, representing the **log odds** of the outcome in the 'control' treatment (i.e. the treatment in arm 1), and $\delta_{i,1k}$ are the trial-specific **log odds ratios** (LORs) of an event on the treatment in arm k compared with the treatment in arm 1 (which in this case is always treatment 1), which can take any value between plus and minus infinities. We set $\delta_{i,11}$, the relative effect of treatment 1 compared with itself, to zero, $\delta_{i,11} = 0$, which implies that the treatment effects $\delta_{i,1k}$ are only estimated when $k > 1$. Equation (2.4) can therefore also be written as

$$\text{logit}\left(p_{i1}\right) = \mu_i$$
$$\text{logit}\left(p_{i2}\right) = \mu_i + \delta_{i,12}$$

Table 2.1 Extended thrombolytics example.

Study ID	Year	Number of arms	Arm 1	Arm 2	Arm 3	Arm 1		Arm 2		Arm 3	
						Events	Patients	Events	Patients	Events	Patients
GUSTO-1	1993	3	SK	Acc t-PA	SK+t-PA	1,472	20,251	652	10,396	723	10,374
ECSG	1985	2	SK	t-PA		3	65	3	64		
TIMI-1	1987	2	SK	t-PA		12	159	7	157		
PAIMS	1989	2	SK	t-PA		7	85	4	86		
White	1989	2	SK	t-PA		10	135	5	135		
GISSI-2	1990	2	SK	t-PA		887	10,396	929	10,372		
Cherng	1992	2	SK	t-PA		5	63	2	59		
ISIS-3	1992	2	SK	t-PA		1,455	13,780	1,418	13,746		
CI	1993	2	SK	t-PA		9	130	6	123		
KAMIT	1991	2	SK	SK+t-PA		4	107	6	109		
INJECT	1995	2	SK	r-PA		285	3,004	270	3,006		
Zijlstra	1993	2	SK	PTCA		11	149	2	152		
Riberio	1993	2	SK	PTCA		1	50	3	50		
Grinfeld	1996	2	SK	PTCA		8	58	5	54		
Zijlstra	1997	2	SK	PTCA		1	53	1	47		
Akhras	1997	2	SK	PTCA		4	45	0	42		
Widimsky	2000	2	SK	PTCA		14	99	7	101		
DeBoer	2002	2	SK	PTCA		9	41	3	46		

(Continued)

Table 2.1 (Continued)

Study ID	Year	Number of arms	Arm 1	Arm 2	Arm 3	Arm 1		Arm 2		Arm 3	
						Events	Patients	Events	Patients	Events	Patients
Widimsky	2002	2	SK	PTCA		42	421	29	429		
DeWood	1990	2	t-PA	PTCA		2	44	3	46		
Grines	1993	2	t-PA	PTCA		13	200	5	195		
Gibbons	1993	2	t-PA	PTCA		2	56	2	47		
RAPID-2	1996	2	Acc t-PA	r-PA		13	155	7	169		
GUSTO-3	1997	2	Acc t-PA	r-PA		356	4,921	757	10,138		
ASSENT-2	1999	2	Acc t-PA	TNK		522	8,488	523	8,461		
Ribichini	1996	2	Acc t-PA	PTCA		3	55	1	55		
Garcia	1997	2	Acc t-PA	PTCA		10	94	3	95		
GUSTO-2	1997	2	Acc t-PA	PTCA		40	573	32	565		
Vermeer	1999	2	Acc t-PA	PTCA		5	75	5	75		
Schomig	2000	2	Acc t-PA	PTCA		5	69	3	71		
LeMay	2001	2	Acc t-PA	PTCA		2	61	3	62		
Bonnefoy	2002	2	Acc t-PA	PTCA		19	419	20	421		
Andersen	2002	2	Acc t-PA	PTCA		59	782	52	790		
Kastrati	2002	2	Acc t-PA	PTCA		5	81	2	81		
Aversano	2002	2	Acc t-PA	PTCA		16	226	12	225		
Grines	2002	2	Acc t-PA	PTCA		8	66	6	71		

Data from Caldwell et al. (2005).
Events are the number of deaths by day 35. Treatment definitions are given in Figure 2.1.

For a random effects model, the LORs for trial i, $\delta_{i,12}$, obtained as logit(p_{i2}) − logit(p_{i1}), come from the random effects distribution in equation (2.1). For a fixed effects model, equation (2.4) is replaced by

$$\text{logit}\left(p_{ik}\right) = \mu_i + d_{1k} \tag{2.5}$$

where $d_{11} = 0$. The pooled LOR d_{12} is assigned the prior distribution in equation (2.2), and the prior distribution for the between-trial heterogeneity standard deviation is

$$\sigma_{12} \sim \text{Uniform}\left(0,\, 2\right) \tag{2.6}$$

where the upper limit of 2 is quite large on the log odds scale (see Section 2.3.2). As noted previously, care is required when specifying prior distributions for this parameter as they can be unintentionally very informative if the upper bound is not sufficiently large. Checks should be made to ensure that the posterior distribution is not unduly influenced by the upper limit chosen (see Section 2.3.2).

Note that although data are available for each arm of each trial, we are only interested in pooling the **relative** treatment effects measured in each trial as this is what RCTs are designed to inform. Thus pooling occurs at the relative effect level (i.e. we pool the LOR). An important feature of *all* the meta-analytic models presented here is that the trial-specific baselines μ_i are regarded as nuisance parameters that are estimated in the model but are of no further interest (see also Chapters 5 and 6). Therefore they will need to be given (non-informative) unrelated prior distributions:

$$\mu_i \sim \text{Normal}(0,\, 100^2) \tag{2.7}$$

An alternative is to place a second hierarchical model on the trial-specific baselines, or to put a bivariate normal model on both the baselines and treatment effects (van Houwelingen et al., 1993, 2002; Warn et al., 2002). However, unless this model is correct, the estimated relative treatment effects will be biased (Senn, 2010; Dias et al., 2013c; Senn et al., 2013). Our approach is therefore more conservative, in keeping with the widely used frequentist meta-analysis methods in which relative effect estimates are treated as data and study-specific baselines eliminated entirely (see also Chapter 5).

2.2.1.2 WinBUGS Implementation: Two Treatments
The implementation in WinBUGS of the fixed effects model described in equations (2.2), (2.3), (2.5) and (2.7) is as follows:

WinBUGS code for pairwise meta-analysis: Binomial likelihood, logit link, fixed effect model.

The # symbol is used for comments – text after this symbol is ignored by WinBUGS. Note that WinBUGS specifies the normal distribution in terms of its mean and precision.

The full code with data and initial values is presented in the online file *Ch2_FE_Bi_logit_pair.odc.*

```
# Binomial likelihood, logit link
# pairwise meta-analysis (2 treatments)
# Fixed effect model
model{                          # *** PROGRAM STARTS
for(i in 1:ns){                 # LOOP THROUGH STUDIES
    mu[i] ~ dnorm(0,.0001)      # vague priors for all trial baselines
    for (k in 1:2) {            # LOOP THROUGH ARMS
        r[i,k] ~ dbin(p[i,k],n[i,k])  # binomial likelihood
        logit(p[i,k]) <- mu[i] + d[k] # model for linear predictor
    }
}
d[1]<- 0          # treatment effect is zero for reference treatment
d[2] ~ dnorm(0,.0001) # vague prior for treatment effect
}                               # *** PROGRAM ENDS
```

As with all WinBUGS models, the likelihood describing the data generating process needs to be specified along with a 'model' and prior distributions. For each study i and for each study arm k, the data are in the form of events r[i,k] out of total patients n[i,k], which are described as coming from a binomial distribution with probability p[i,k]: r[i,k] ~ dbin(p[i,k],n[i,k]). Since multiple trials are available, each with two arms, the likelihood needs to be specified multiple times, and this is done by enclosing it within a loop where i goes from 1 to ns, the number of included studies (given as data), and a loop covering all treatment arms, k = 1 and 2. This specifies the appropriate likelihood for each arm of each study. Also within the i and k loops is the main part of the meta-analysis model, termed the linear predictor (see Chapter 4), which translates equation (2.5) into code. In the model, d[k] represents the fixed treatment effect of treatment k compared with treatment 1, that is, d[k] = d_{1k}. Prior distributions for the trial-specific baselines (nuisance parameters) mu[i] are specified for each trial within the i loop. Finally, d[1] is set to zero and the prior distribution for d[2] is specified (equation (2.2)).

In WinBUGS, the model must be checked for correct syntax by clicking on 'check model' in the **Specification Tool** obtained from the menu **Model→Specification....** With the model checked, data must be loaded for the programme to compile. The data to load has two components: a list specifying the number of studies ns (in the example, ns = 11) and the main body of data in a column or vector format where r[,1] and n[,1] are the numerators and denominators for treatment 1 and r[,2] and n[,2] the numerators and denominators for treatment 2, respectively. For this example these values are taken from the last 11 rows of Table 2.1. Text is included after the hash symbol (#) for ease of reference to the original data source and to facilitate model diagnostics but is ignored by WinBUGS. Both sets of data need to be loaded for the model to run. This can be done in WinBUGS by clicking 'load data' in the Specification Tool for each type of data in turn:

```
# Data (Extended Thrombolytics example - 2 treatments)
list(ns=11)
r[,1]   n[,1]   r[,2]   n[,2]   #   Study ID
3       55      1       55      #   Ribichini 1996
10      94      3       95      #   Garcia 1997
40      573     32      565     #   GUSTO-2 1997
5       75      5       75      #   Vermeer 1999
5       69      3       71      #   Schomig 2000
2       61      3       62      #   LeMay 2001
19      419     20      421     #   Bonnefoy 2002
59      782     52      790     #   Andersen 2002
5       81      2       81      #   Kastrati 2002
16      226     12      225     #   Aversano 2002
8       66      6       71      #   Grines 2002
END
```

Once the data are loaded, the number of chains to run can be set in the appropriate box and the model can be complied by clicking 'compile'. Appropriate initial values need to be given to all parameters with a distribution (except the data) for the model to run. We will run three chains with the following initial values:

```
# Initial values
#chain 1
list(d=c( NA, 0),   mu=c(0,0,0,0,0,     0,0,0,0,0,      0))
#chain 2
list(d=c( NA, -1), mu=c(-3,-3,-3,-3,-3, -3,-3,-3,-3,-3, -3))
#chain 3
list(d=c( NA, 2),   mu=c(-3,5,-1,-3,7,   -3,-4,-3,-3,0,  -7))
```

Note that parameter d is a vector with two components: d[1], which is fixed (set to zero) and therefore does not require initial values to be specified, and d[2], which is assigned a prior distribution and therefore requires an initial value. The initial value for the first component of vector d is therefore set as 'NA', which denotes a missing value in WinBUGS. The trial-specific baselines mu are also defined as vectors with length equal to the number of studies (in this case 11); therefore eleven initial values need to be specified. Initial values are loaded by clicking '**load inits**' for each chain in turn. Any set of plausible initial values can be given and any number of chains can be run simultaneously. Plausible initial values are those that are allowed by each parameter's prior distribution, but very extreme values should be avoided as they can produce numerical errors. Note that once the model has converged, the initial values will have no influence on the posterior distributions so they are not meant to reflect any prior knowledge on the likely values of the parameters.

We recommend running at least two chains with very different starting values in order to properly assess convergence and robustness of results to the starting values (Welton et al., 2012; Lunn et al., 2013). However, we often choose to run three chains to allow a greater variety of starting values to be tried. The model can then be run from the **Model→Update...** menu. Parameters can be monitored, convergence checked and outputs obtained using the **Inference→Samples...** menu. For further details see the WinBUGS help menu and Lunn et al. (2013).

In this example, convergence was achieved by 10,000 iterations. A further 20,000 iterations were run on three chains, giving a posterior sample of 60,000 values, on which all results are based. Posterior summaries for d[2], the pooled LOR of mortality on treatment 2 (PTCA) compared with treatment 1 (Acc t-PA), are given as follows:

node	mean	sd	MC error	2.5%	median	97.5%	start	sample
d[2]	−0.2336	0.1178	7.668E-4	−0.4654	−0.2333	−0.002876	10001	60000

The columns labelled 'mean' and 'sd' give the mean and standard deviation of the posterior distribution of d[2]. The column labelled 'MC error' shows the Monte Carlo standard error of the samples, which will decrease as the number of iterations increases and should typically be small. A common rule of thumb is to ensure that the MC error is less than 5% of the posterior standard deviation (Lunn et al., 2013). In the example, 5% of the standard deviation is $0.1178 \times 0.05 = 0.00589$, while the MC error $= 0.0007668$, which is considerably smaller. We can therefore be satisfied that sufficient posterior samples have been used for inference.

Bounds for the 95% credible interval (CrI) are given by the 2.5 and 97.5% quantiles, presented in the columns labelled '2.5%' and '97.5%', respectively. The median of the posterior distribution is given in the column labelled 'median', and finally the columns labelled 'start' and 'sample' give the starting iteration and the total number of values the results are based on. The full posterior distribution of d[2] is shown in Figure 2.2.

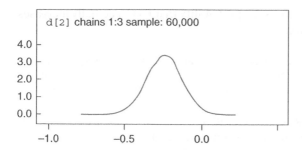

Figure 2.2 Extended thrombolitics example (pairwise meta-analysis): posterior distribution of d[2], the log odds ratio of PTCA compared with Acc t-PA, from the fixed effects model – from WinBUGS.

In general we summarise the posterior distribution by its median and 95% CrI, but note that for (approximately) symmetric distributions such as in Figure 2.2, the median and the mean will be very similar. In the example, using two decimal places, mean = median = −0.23. Hence we will say that the posterior median (and mean) of the LOR of mortality is −0.23 with 95% CrI (−0.47, −0.003), which is wholly negative, meaning that the odds of mortality on PTCA are lower than on Acc t-PA. In addition, Figure 2.2 shows that the probability that this LOR is positive is small.

However, usually we want to report the odds ratio (OR) of mortality with its credible interval. This can be easily achieved in WinBUGS by adding the following code before the last closing brace, which states that the OR is obtained by exponentiating the LOR:

```
or <- exp(d[2])
```

In addition, we may want to quantify the probability that the LOR is positive, that is, that PTCA does not reduce mortality compared with Acc t-PA. This can be done by adding the following code:

```
prob.harm <- step(d[2])
```

where prob.harm will hold the required probability using the step function, which returns the value 1 when its argument is greater than or equal to zero. Averaging the number of times d[2] ≥ 0 over all iterations (post-convergence) will give the probability that d[2] ≥ 0, that is, the probability that the OR for mortality is greater than 1.

Monitoring or will give posterior summaries for the OR of mortality. Monitoring prob.harm provides several summaries, but only the posterior mean is of interest as it represents the probability that PTCA actually increases mortality compared with Acc t-PA:

node	mean	sd	MC error	2.5%	median	97.5%	start	sample
or	0.7971	0.09416	6.144E-4	0.6279	0.7919	0.9971	10001	60000
prob.harm	0.02362	0.1519	7.52E-4	0.0	0.0	0.0	10001	60000

The posterior median OR is 0.79 with 95% CrI (0.63, 0.997), indicating that PTCA reduces mortality by about 20% compared with Acc t-PA, although the upper bound of the CrI is nearly 1, suggesting the possibility of no effect. The probability that PTCA actually increases mortality is small at 0.024 (note that only the column headed 'mean' is interpretable for this parameter).

The implementation of the random effects model described in equations (2.1), (2.2), (2.3), (2.4), (2.6) and (2.7) is as follows:

WinBUGS code for Binomial likelihood, logit link, random effects model, two treatments.

The # symbol is used for comments – text after this symbol is ignored by WinBUGS. Note that WinBUGS specifies the normal distribution in terms of its mean and precision.

The full code with data and initial values is presented in the online file *Ch2_RE_Bi_logit_pair.odc*.

```
# Binomial likelihood, logit link
# pairwise meta-analysis (2 treatments)
# Random effects model
model{                          # *** PROGRAM STARTS
for(i in 1:ns){                 # LOOP THROUGH STUDIES
    delta[i,1] <- 0             # treatment effect is zero for control arm
    mu[i] ~ dnorm(0,.0001)      # vague priors for all trial baselines
    for (k in 1:2) {            # LOOP THROUGH ARMS
        r[i,k] ~ dbin(p[i,k],n[i,k])  # binomial likelihood
        logit(p[i,k]) <- mu[i] + delta[i,k]  # model for linear
                                             predictor
    }
    delta[i,2] ~ dnorm(d[2],tau)  # trial-specific LOR distributions
}
d[1]<- 0                        # treatment effect is zero for reference
                                  treatment
d[2] ~ dnorm(0,.0001)           # vague prior for treatment effect
sd ~ dunif(0,2)                 # vague prior for between-trial SD
tau <- pow(sd,-2)               # between-trial precision = (1/between-
                                  trial variance)
}                               # *** PROGRAM ENDS
```

Most of the lines in the random effects code are the same as for the fixed effects model. In particular the likelihood and prior distributions for d and mu are the same, as are all the loops. Differences between the two sets of code are highlighted in bold. Since we are now using the model in equation (2.4), we need to set delta[i,1] to zero for each study and specify the random effects distribution in equation (2.1), as well as the prior distribution for the between-study standard deviation, sd. Note that WinBUGS describes the normal distribution in terms of its mean and precision; therefore we need to add the extra variable tau, which is defined as the inverse of the between-study variance, that is, the precision. Redundant subscripts have been dropped in the code.

The model needs to be checked, data loaded, compiled and initial values given, as before. The data structure is exactly the same as for the fixed effects model, but we now need to add an initial value for the between-study heterogeneity parameter sd, which is a single number (scalar):

```
# Initial values
#chain 1
list(d=c( NA,  0), sd=1, mu=c(0,0,0,0,0,        0,0,0,0,0,        0))
#chain 2
list(d=c( NA, -1), sd=0.1, mu=c(-3,-3,-3,-3,-3,   -3,-3,-3,-3,-3,  -3))
#chain 3
list(d=c( NA,  2), sd=0.5, mu=c(-3,5,-1,-3,7,      -3,-4,-3,-3,0,   -7))
```

Any value can be chosen for sd as long as it is within the bounds of the prior distribution, in this case between zero and two (but not actually zero or two). Note that we have not specified initial values for the study-specific treatment effects delta. Although as a general rule initial values should be specified for all nodes assigned a distribution (except the data), in this case we can omit the initial values for delta and allow WinBUGS to generate them by clicking the '**gen inits**' button *after* loading the initial values for all the chains. This is because the initial values that will be generated for delta will be chosen from the random effects distribution that is specified in terms of d and tau, to which we have given sensible (i.e. not very extreme) initial values, ensuring that values for delta will also not be very extreme. The same code can be added to calculate the OR and probability of harm.

Results from running both the fixed and random effects models are shown in Table 2.2. All results are based on 20,000 iterations on three chains, after a burn-in of 10,000.

Results for the fixed and random effects model are similar, although the latter produces wider credible intervals, since it allows for between-trial heterogeneity. However, the posterior median of the between-study heterogeneity is relatively small, and the lower bound for its 95% CrI is very close to zero. Its full posterior distribution is shown in Figure 2.3, and we can see that it does not resemble a uniform distribution, suggesting it has been suitably updated from the prior distribution, which was not too restrictive in this case (see Section 2.3.2 for more details).

Table 2.2 Extended thrombolytics example (pairwise meta-analysis): results from fixed and random effects meta-analyses of mortality on PTCA compared with Acc t-PA.

	Odds ratio			Heterogeneity	
	Median	95% CrI	Pr(harm)	Median	95% CrI
Fixed effect	0.79	(0.63, 1.00)	0.02	–	–
Random effects	0.77	(0.55, 1.02)	0.03	0.15	(0.01, 0.64)

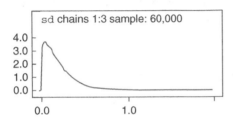

Figure 2.3 Extended thrombolytics example (pairwise meta-analysis): posterior distribution of the between-study standard deviation (sd) for the meta-analysis of PTCA and Acc t-PA – from WinBUGS.

In Chapter 3, we will discuss how to assess global model fit and how to choose between random and fixed effects models, but for now we note that the posterior distribution for the heterogeneity is mainly concentrated around small values.

2.2.2 Extension to Indirect Comparisons and Network Meta-Analysis

We started by defining a set of M trials over which the study-specific treatment effects of treatment 2 compared with treatment 1, $\delta_{i,12}$, were exchangeable with mean d_{12} and variance σ_{12}^2. We now suppose that, within the same set of trials (i.e. trials that are relevant to the same research question), comparisons of treatments 1 and 3 are also made. To carry out a random effects meta-analysis of treatment 1 versus 3, we would now assume that the study-specific treatment effects of treatment 3 compared with treatment 1, $\delta_{i,13}$, are also exchangeable such that $\delta_{i,13} \sim \text{Normal}(d_{13}, \sigma_{13}^2)$. From the transitivity relation $\delta_{i,23} = \delta_{i,13} - \delta_{i,12}$, it can be shown that the study-specific treatment effects of treatment 3 compared with 2, $\delta_{i,23}$, are also exchangeable:

$$\delta_{i,23} \sim \text{Normal}\left(d_{23}, \sigma_{23}^2\right)$$

It can further be shown (Lu and Ades, 2009) that this implies

$$d_{23} = d_{13} - d_{12}$$

and

$$\sigma_{23}^2 = \sigma_{12}^2 + \sigma_{13}^2 - 2\rho_{23}^{(1)}\sigma_{12}\sigma_{13}$$

where $\rho_{23}^{(1)}$ represents the correlation between the relative effects of treatment 3 compared with treatment 1 and the relative effect of treatment 2 compared with treatment 1 (Lu and Ades, 2009).

Note the relationship between the standard assumptions of pairwise meta-analysis and those required for indirect and mixed treatment comparisons. For separate random effects pairwise meta-analyses, we need to assume exchangeability of the effects $\delta_{i,12}$ over the 1 versus 2 trials and also of the effects $\delta_{i,13}$ over the 1 versus 3 trials. For NMA, we must assume the exchangeability of both treatment effects over both 1 versus 2 *and* 1 versus 3 trials. The theory extends readily to additional treatments $k = 4, 5, \dots, S$. In each case we must assume the exchangeability of the δ's across the **entire** set of trials. Then the within-trial transitivity relation is enough to imply the exchangeability of all the treatment effects $\delta_{i,xy}$. The *consistency equations* (Lu and Ades, 2006, 2009)

$$d_{23} = d_{13} - d_{12}$$
$$d_{24} = d_{14} - d_{12}$$
$$\vdots$$
$$d_{(S-1),S} = d_{1S} - d_{1,(S-1)}$$

are also therefore implied (Section 1.5). These assumptions are required by indirect comparisons and NMA, but given that we are assuming that all trials are relevant to the same research question, they are not *additional* assumptions (see Chapter 12). However, while, in theory, consistency of the true treatment effects in a given population must hold, there may be inconsistency in the evidence. Methods to assess evidence consistency are addressed in Chapter 7.

The consistency equations can also be seen as an example of the distinction between the $(S-1)$ *basic* parameters (Eddy et al., 1992) $d_{12}, d_{13}, d_{14}, \dots, d_{1S}$, the treatment effects relative to the reference treatment, on which prior distributions are placed, and the *functional* parameters, which are simply functions of the basic parameters and represent all the remaining contrasts. It is precisely this reduction in the number of dimensions, from the number of functions on which there are data to the number of basic parameters, that allows all data, whether directly informing basic or functional parameters, to be combined within a coherent, internally consistent, model. The exchangeability assumptions regarding the treatment effects $\delta_{i,12}$ and $\delta_{i,13}$ therefore make it possible to derive indirect comparisons of treatment 3 versus treatment 2, from trials of treatment 1 versus 2 and 1 versus 3, and also allow us to include trials of treatments 2 versus 3 in a coherent synthesis with the 1 versus 2 and 1 versus 3 trials.

For simplicity we will assume equal between-study variances in all models, that is, $\sigma_{12}^2 = \sigma_{13}^2 = \sigma_{23}^2 = \cdots = \sigma^2$, and this implies that the correlation between any two treatment contrasts in a multi-arm trial is 0.5 (Higgins and Whitehead, 1996). Although seemingly very strong, this is actually a reasonable assumption in most cases. If the patient populations are similar in all trials and the trial designs similar across comparisons, it seems reasonable to assume that between-study variability, beyond that explained by known effect modifiers (see Chapters 8 and 9), will be similar across comparisons. The assumption of a common, 'shared', between-study variance means that all studies contribute to its estimation, thereby reducing its uncertainty.

Heterogeneous variance models, where the between-study variances are allowed to be different for different comparisons (Lu and Ades, 2009), can be fitted but are more complex since the between-study variances are themselves subject to constraints implied by the consistency model. Failure to ensure these constraints are met may lead to invalid inferences (Lu and Ades, 2009), although they can be tricky to implement in practice.

To allow for comparisons of multiple treatments, we need to use a notation that distinguishes between arm k of trial i and the treatment compared in that arm, since not all studies will compare the same treatments. The trial-specific treatment effects of the treatment in arm k, relative to the treatment in arm 1 *in that trial*, δ_{ik}, are drawn from a common random effects distribution:

$$\delta_{ik} \sim \text{Normal}\left(d_{t_{i1},t_{ik}}, \sigma^2\right) \tag{2.8}$$

where $d_{t_{i1},t_{ik}}$ represents the mean effect of the treatment in arm k in trial i, t_{ik}, compared with the treatment in arm 1 of trial i, t_{i1}, and σ^2 represents the between-trial variability in treatment effects (heterogeneity). For trials that compare treatments 1 and 2, $d_{t_{i1},t_{ik}} = d_{12}$; for trials that compare treatments 2 and 3, $d_{t_{i1},t_{ik}} = d_{23}$; and so on. Note that now δ only has subscripts i for trial and k for arm, but it is always implied that δ_{ik} refers to the effect of the treatment in arm k compared with the **treatment in arm 1** in that trial, whatever that may be. We set $\delta_{i1} = 0$, as before. Again, the fixed effects model is a special case obtained by setting the between-study variance to zero, implying that $\delta_{ik} = d_{t_{i1},t_{ik}}$ for all i and k.

We write the consistency equations more generally as

$$d_{t_{i1}t_{ik}} = d_{1,t_{ik}} - d_{1,t_{i1}} \tag{2.9}$$

Note that only the basic parameters d_{1k}, $k = 2, \ldots, S$ are estimated, and they will be given non-informative prior distributions as before

$$d_{1k} \sim \text{Normal}\left(0, 100^2\right) \tag{2.10}$$

2.2.2.1 Incorporating Multi-Arm Trials

Suppose we have a number of multi-arm trials involving some or all of the treatments of interest, 1, 2, 3, 4, and so on. Among commonly suggested stratagems for dealing with such trials are combining all active arms into one, splitting the control group between all relevant experimental groups or ignoring all but two of the trial arms (Higgins and Green, 2008). None of these are satisfactory. In the NMA framework set out previously, multi-arm trials are naturally incorporated in the synthesis. No special considerations are required when fitting fixed effects models as long as the data are set out correctly (see Section 2.2.3.1).

The question of how to conduct a random effects meta-analysis of multi-arms trials has been considered in a Bayesian framework by Lu and Ades (2004), and in a frequentist framework by Lumley (2002) and Chootrakool and Shi (2008). Based on the same exchangeability assumptions described previously, a single multi-arm trial will estimate a vector of random effects $\boldsymbol{\delta}_i$. For example, a three-arm trial will produce two random effects and a four-arm trial three. However, these multiple random effects are correlated and this must be correctly accounted for. Assuming, as before, that the relative effects all have the same between-trial variance, we have

$$
\boldsymbol{\delta}_i = \begin{pmatrix} \delta_{i,12} \\ \vdots \\ \delta_{i,1a_i} \end{pmatrix} \sim \text{Normal}_{a_i - 1} \left(\begin{pmatrix} d_{t_{i1}t_{i2}} \\ \vdots \\ d_{t_{i1},t_{ia_i}} \end{pmatrix}, \begin{pmatrix} \sigma^2 & \sigma^2/2 & \cdots & \sigma^2/2 \\ \vdots & \vdots & \ddots & \vdots \\ \sigma^2/2 & \sigma^2/2 & \cdots & \sigma^2 \end{pmatrix} \right) \tag{2.11}
$$

where $\boldsymbol{\delta}_i$ is the vector of random effects, which follows a multivariate normal distribution, a_i represents the number of arms in trial i ($a_i = 2, 3, \ldots$) and $d_{t_{i1}t_{ik}}$ is as given in equation (2.9). Then the conditional univariate distribution (Raiffa and Schlaiffer, 2000) for the random effect of arm $k > 2$, given all arms from 2 to $k - 1$, is

$$
\delta_{i,1k} \mid \begin{pmatrix} \delta_{i,12} \\ \vdots \\ \delta_{i,1(k-1)} \end{pmatrix} \sim \text{Normal} \left(\left(d_{1,t_{ik}} - d_{1,t_{i1}} \right) + \frac{1}{k-1} \sum_{j=1}^{k-1} \left[\delta_{i,1j} - \left(d_{1,t_{ij}} - d_{1,t_{i1}} \right) \right], \right.
$$

$$
\left. \frac{k}{2(k-1)} \sigma^2 \right) \tag{2.12}
$$

Either the multivariate distribution in equation (2.11) or the conditional distributions in equation (2.12) must be used to estimate the random effects for each multi-arm study so that the between-arm correlations between parameters are taken into account. We will use the formulation in equation (2.12) as it allows for a more generic code that works for trials with any number of arms. This general formulation is no different from the model presented by Higgins and Whitehead (1996) and provides another interpretation of the exchangeability assumptions. It is indeed another way of deducing the consistency relations: we may consider a connected network of M trials involving S treatments to originate from M S-arm trials, but with some of the arms missing at random, conditional on the trial design and choice of treatments, at randomisation (Section 1.5).

The WinBUGS code presented in Section 2.2.3.1 exactly instantiates the theory behind NMA that relates it to pairwise meta-analysis. Therefore it will analyse pairwise meta-analysis, indirect comparisons and NMA (including combined direct and indirect evidence) with or without multi-arm trials without distinction.

2.2.3 Worked Example: Network Meta-Analysis

We now synthesise the extended thrombolytics dataset presented in Table 2.1, comprising 36 RCTs comparing seven treatments. The treatment network is given in Figure 2.1. We define SK as our reference or 'control' treatment, coded 1, and number the other treatments according to Figure 2.1.

As before, we define r_{ik} as the number of events (deaths), out of the total number of patients in each arm, n_{ik}, for arm k of trial i, and assume that the data generation process follows the binomial likelihood described in equation (2.3). The probabilities of an event p_{ik} are modelled on the logit scale as

$$\text{logit}\left(p_{ik}\right) = \mu_i + \delta_{ik} \tag{2.13}$$

where μ_i are trial-specific baselines representing the log odds of the outcome in the 'control' treatment (i.e. the treatment in arm 1) and δ_{ik} are the trial-specific LORs of an event on the treatment in arm k compared with the treatment in arm 1. Note that, in different trials, μ_i may represent the log odds of an event on different treatments. They are treated as nuisance parameters and given the (non-informative) prior distributions in equation (2.7). We set the relative effect of the treatment in arm 1 compared with itself to zero, $\delta_{i1} = 0$, so that the treatment effects δ_{ik} are only estimated when $k > 1$. For a random effects model, the trial-specific LORs come from the random effects distribution in equation (2.8).

For a fixed effects model, equation (2.13) is replaced by

$$\text{logit}(p_{ik}) = \mu_i + d_{t_{i1}t_{ik}} = \mu_i + d_{1,t_{ik}} - d_{1,t_{i1}} \tag{2.14}$$

where $d_{11} = 0$.

The basic parameters representing the pooled LOR of mortality for treatments 2, ..., 7 compared with treatment 1 (the reference treatment) are assigned the prior distributions in equation (2.10). The prior distribution for the between-trial heterogeneity standard deviation σ is chosen as

$$\sigma \sim \text{Uniform}(0, 2) \tag{2.15}$$

which is appropriately wide in this case (see Section 2.3.2).

2.2.3.1 WinBUGS Implementation

The implementation of the NMA model with fixed treatment effects is given as follows:

WinBUGS code for NMA: Binomial likelihood, logit link, fixed effect model.

The # symbol is used for comments – text after this symbol is ignored by WinBUGS. Note that WinBUGS specifies the normal distribution in terms of its mean and precision.

The full code with data and initial values is presented in the online file *Ch2_FE_Bi_logit.odc*.

```
# Binomial likelihood, logit link
# Fixed effect model
model{                      # *** PROGRAM STARTS
for(i in 1:ns){             # LOOP THROUGH STUDIES
    mu[i] ~ dnorm(0,.0001)  # vague priors for all trial baselines
    for (k in 1:na[i])  {   # LOOP THROUGH ARMS
      r[i,k] ~ dbin(p[i,k],n[i,k])      # binomial likelihood
      logit(p[i,k]) <- mu[i] + d[t[i,k]] - d[t[i,1]]      # model for
                                                          linear
                                                          predictor
    }
  }
d[1]<-0          # treatment effect is zero for reference treatment
for (k in 2:nt){  d[k] ~ dnorm(0,.0001) }   # vague priors for
                                              treatment effects
}                           # *** PROGRAM ENDS
```

Most of the lines in this fully generic (network) meta-analysis code are the same as in the previous code for pairwise meta-analysis with a fixed effect. The likelihood and prior distributions for mu[i] are the same, but now the loop covering all treatment arms needs to have k going from 1 to the total number of arms in that trial, which may be greater than 2, defined in na[i], loaded as data. The linear predictor now translates equation (2.14) into code where t[i,k] represents the treatment in arm k of trial i and d[k] $= d_{1k}$. As before d[1] is set to zero, and prior distributions for all other d[k] are specified, with nt representing the total number of treatments compared in the network, which is given as data. Note that no special consideration is needed for multi-arm trials in fixed effects models since no random effects are estimated, and therefore no within-trial correlation needs to be accounted for.

The data structure is presented in the succeeding text and has two components: a list specifying the number of studies ns and treatments nt (in this example ns = 36, nt = 7) and the main body of data. Both sets of data need to be loaded for the model to run (see Section 2.2.1.2). The main body of data is in a vector format, and we need to allow for a three-arm trial. Three column places are therefore required to specify the treatments, t, the number of events, r, and the number of patients, n, in each arm; 'NA' indicates that values are missing for a particular cell: r[,1] and n[,1] are the numerators and denominators for the first arm, r[,2] and n[,2] are the numerators and denominators for the second arm and r[,3] and n[,3] represent the numerators and denominators for the third arm of each trial, respectively. We specify t[,1], t[,2] and t[,3] as the treatment number identifiers for the first, second and third arms of each trial, respectively, that is, they identify the treatments compared in each arm to which r and n correspond. We also add a column with the number of arms in each trial, na[]. Only the first trial in this dataset has three arms; all other trials have two arms. Trial identifiers are added as comments, which are ignored by WinBUGS:

```
# Data (Thrombolytics example)
list(ns=36, nt=7)
na[] t[,1] t[,2] t[,3] r[,1]  n[,1]  r[,2]  n[,2]  r[,3] n[,3]  #ID       year
 3    1     3    4    1472   20251   652   10396   723   10374  #GUSTO-1   1993
 2    1     2    NA      3      65     3      64    NA     NA    #ECSG      1985
 2    1     2    NA     12     159     7     157    NA     NA    #TIMI-1    1987
 2    1     2    NA      7      85     4      86    NA     NA    #PAIMS     1989
 2    1     2    NA     10     135     5     135    NA     NA    #White     1989
 2    1     2    NA    887   10396   929   10372    NA     NA    #GISSI-2   1990
 2    1     2    NA      5      63     2      59    NA     NA    #Cherng    1992
 2    1     2    NA   1455   13780  1418   13746    NA     NA    #ISIS-3    1992
 2    1     2    NA      9     130     6     123    NA     NA    #CI        1993
 2    1     4    NA      4     107     6     109    NA     NA    #KAMIT     1991
 2    1     5    NA    285    3004   270    3006    NA     NA    #INJECT    1995
 2    1     7    NA     11     149     2     152    NA     NA    #Zijlstra  1993
 2    1     7    NA      1      50     3      50    NA     NA    #Riberio   1993
 2    1     7    NA      8      58     5      54    NA     NA    #Grinfeld  1996
 2    1     7    NA      1      53     1      47    NA     NA    #Zijlstra  1997
 2    1     7    NA      4      45     0      42    NA     NA    #Akhras    1997
 2    1     7    NA     14      99     7     101    NA     NA    #Widimsky  2000
 2    1     7    NA      9      41     3      46    NA     NA    #DeBoer    2002
 2    1     7    NA     42     421    29     429    NA     NA    #Widimsky  2002
 2    2     7    NA      2      44     3      46    NA     NA    #DeWood    1990
 2    2     7    NA     13     200     5     195    NA     NA    #Grines    1993
 2    2     7    NA      2      56     2      47    NA     NA    #Gibbons   1993
 2    3     5    NA     13     155     7     169    NA     NA    #RAPID-2   1996
 2    3     5    NA    356    4921   757   10138    NA     NA    #GUSTO-3   1997
```

2	3	6	NA	522	8488	523	8461	NA	NA	#ASSENT-2	1999
2	3	7	NA	3	55	1	55	NA	NA	#Ribichini	1996
2	3	7	NA	10	94	3	95	NA	NA	#Garcia	1997
2	3	7	NA	40	573	32	565	NA	NA	#GUSTO-2	1997
2	3	7	NA	5	75	5	75	NA	NA	#Vermeer	1999
2	3	7	NA	5	69	3	71	NA	NA	#Schomig	2000
2	3	7	NA	2	61	3	62	NA	NA	#LeMay	2001
2	3	7	NA	19	419	20	421	NA	NA	#Bonnefoy	2002
2	3	7	NA	59	782	52	790	NA	NA	#Andersen	2002
2	3	7	NA	5	81	2	81	NA	NA	#Kastrati	2002
2	3	7	NA	16	226	12	225	NA	NA	#Aversano	2002
2	3	7	NA	8	66	6	71	NA	NA	#Grines	2002

END

In setting up the data structure, the maximum number of columns needed to define each variable is the maximum number of arms in the trials included in the dataset, that is, the maximum number specified in column na.

An important feature of the code presented is the assumption that the treatments are always presented in ascending (numerical) order and that treatment 1 is taken as the reference treatment. So, for example, the first row of data has the columns in order so that data for treatments 1, 3 and 4 are presented in the first, second and third columns, respectively, rather than having, for example, data for treatment 1, then 4 and then 3. Although this is not very important for the simple NMA models presented in this chapter and in Chapters 3 and 4 (the code will work regardless), it is crucial to have the data set up in this way to explore inconsistency (Chapter 7) and, when extending the model to include covariates (Chapter 8) or bias models (Chapter 9), to ensure the correct relative effects are estimated and appropriate assumptions are implemented.

We will run three chains using the following initial values:

```
# Initial values
#chain 1
list(d=c( NA,  0,0,0,0,    0,0), mu=c(0,0,0,0,0,  0,0,0,0,0,   0,0,0,0,0,
0,0,0,0,0,    0,0,0,0,0,  0,0,0,0,0,    0,0,0,0,0,  0))
#chain 2
list(d=c( NA, -1,-1,-1,-1,  -1,-1),  mu=c(-3,-3,-3,-3,-3,  -3,-3,-3,-3,-3,
-3,-3,-3,-3,-3,    -3,-3,-3,-3,-3,  -3,-3,-3,-3,-3,   -3,-3,-3,-3,-3,
-3,-3,-3,-3,-3,   -3))
#chain 3
list(d=c( NA,  2,5,-3,1,    -7,4), mu=c(-3,5,-1,-3,7,   -3,-4,-3,-3,9,
-3,-3,-4,3,5,        -3, -2, 1, -3, -7,   -3,5,-1,-3,7,   -3,-4,-3,-3,0,
-3, 5,-1,-3,7,     -7))
```

Again note that d is a vector with seven components corresponding to the number of treatments being compared, but the initial value for the first component of vector d needs to be 'NA' since d[1] is set to zero and is therefore not estimated.

Running this model gives the following posterior summaries for the basic parameters, which represent the pooled LORs of mortality for all treatments compared with treatment 1 (SK). These are also summarised in Figure 2.4.

node	mean	sd	MC error	2.5%	median	97.5%	start	sample
d[2]	−0.003229	0.03035	1.362E-4	−0.06236	−0.003249	0.0564	20001	150000
d[3]	−0.1571	0.04349	2.77E-4	−0.2427	−0.1571	−0.07212	20001	150000
d[4]	−0.04324	0.04654	1.916E-4	−0.135	−0.04307	0.048	20001	150000
d[5]	−0.1105	0.06014	3.667E-4	−0.2288	−0.1106	0.007321	20001	150000
d[6]	−0.1518	0.0771	4.588E-4	−0.3028	−0.1517	−6.668E-4	20001	150000
d[7]	−0.4744	0.1006	4.5E-4	−0.6722	−0.474	−0.2775	20001	150000

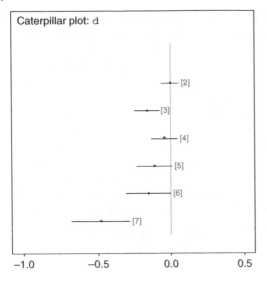

Figure 2.4 Extended thrombolytics example: caterpillar plot – from WinBUGS. Dots are posterior medians and lines represent 95% CrI for the log odds ratios of all treatments compared with SK, the reference treatment, from the fixed effects model. Numbers represent the treatment being compared (see Figure 2.1) and negative log odds ratios favour that treatment.

These LORs can be displayed in WinBUGS as a caterpillar plot, which can be obtained using the **Inference→Compare...** menu, by typing d in the '**node**' box and clicking '**caterpillar**'. Treatment 7, PTCA, provides the largest reduction in mortality with posterior median of the LOR of –0.47 and 95% CrI (–0.67, –0.28).

Additional code can be added before the last closing brace to estimate all the pairwise LORs and ORs (i.e. for all treatments compared with every other), to generate ranking statistics and to calculate the probability that each treatment is the best treatment as well as the probabilities that each treatment is ranked 1st, 2nd, 3rd, and so on:

```
# pairwise ORs and LORs for all possible pairwise
comparisons
for (c in 1:(nt-1)) {
  for (k in (c+1):nt)  {
      or[c,k] <- exp(d[k] - d[c])
      lor[c,k] <- (d[k]-d[c])
      }
  }
# ranking on relative scale
for (k in 1:nt) {
# rk[k] <- nt+1-rank(d[],k) # assumes events are "good"
  rk[k] <- rank(d[],k)  # assumes events are "bad"
```

```
best[k] <- equals(rk[k],1)  # calculate probability that
                                  treat k is best
for (h in 1:nt){ prob[h,k] <- equals(rk[k],h) }
   # calculate probability that treat k is h-th best
}
```

In addition, given an assumption about the absolute effect of one treatment (see Chapter 5), it is possible to produce estimates of absolute effects on all treatments; to express the treatment effect on other scales such as the risk difference (RD), defined as the difference in probabilities of an event on treatment k compared with the reference treatment, and the relative risk (RR), defined as the ratio of the probabilities of an outcome on treatment k compared with treatment 1; or to calculate the number needed to treat (NNT), defined as 1/RD (Hutton, 2000; Deeks, 2002; Higgins and Green, 2008). An advantage of the Bayesian MCMC is that appropriate distributions, and therefore credible intervals, are automatically generated for all these quantities. This is achieved in WinBUGS by adding the following code:

```
# Provide estimates of treatment effects T[k] on the
# natural (probability) scale, given a mean effect, meanA,
# for 'standard' treatment 1, with precision precA
A ~ dnorm(meanA,precA)
for (k in 1:nt) { logit(T[k]) <- A + d[k]   }
```

In this case, meanA and precA will be the assumed mean and precision of the log odds of mortality on treatment 1, respectively, and T[k] will represent the absolute probabilities of mortality on each treatment $k = 1, ..., 7$. Given the information on the absolute probabilities of mortality T[1], other quantities can be calculated using the following code:

```
# Provide estimates of number needed to treat NNT[k]
# Risk Difference RD[k] and Relative Risk RR[k],
# for each treatment relative to treatment 1
for (k in 2:nt) {
#    NNT[k] <- 1/(T[k] - T[1]) # assumes events are
                                     "good"
   NNT[k] <- 1/(T[1]- T[k])      # assumes events are
                                     "bad"
   RD[k] <- T[k] - T[1]
   RR[k] <- T[k]/T[1]
}
```

We will assume the absolute probability of mortality on SK is 8% with 95% CrI from 7 to 10%, which is approximately equivalent to assuming that the log odds of mortality on SK follows a normal distribution with a mean of −2.39 and precision of 11.9. For considerations on how to estimate or elicit these values, see Chapter 5. These values can be given as data or replaced in the aforementioned code. We will include them in the data by changing the list to read

```
list(ns=36, nt=7, meanA=-2.39, precA=11.9)
```

and loading that into WinBUGS.

The ORs of mortality for each treatment compared with every other can be obtained by monitoring node or. WinBUGS output for the fixed effects model is given in the succeeding text and summarised in Table 2.3 and Figure 2.5.

node	mean	sd	MC error	2.5%	median	97.5%	start	sample
or[1,2]	0.9972	0.03027	1.359E-4	0.9395	0.9968	1.058	20001	150000
or[1,3]	0.8554	0.0372	2.37E-4	0.7845	0.8547	0.9304	20001	150000
or[1,4]	0.9587	0.04463	1.838E-4	0.8737	0.9578	1.049	20001	150000
or[1,5]	0.897	0.05399	3.293E-4	0.7955	0.8953	1.007	20001	150000
or[1,6]	0.8617	0.06652	3.959E-4	0.7387	0.8592	0.9993	20001	150000
or[1,7]	0.6254	0.06305	2.823E-4	0.5106	0.6225	0.7577	20001	150000
or[2,3]	0.8586	0.04551	2.682E-4	0.7726	0.8573	0.9515	20001	150000
or[2,4]	0.9623	0.05353	2.266E-4	0.8613	0.9609	1.071	20001	150000
or[2,5]	0.9003	0.06062	3.591E-4	0.7876	0.8985	1.025	20001	150000
or[2,6]	0.8649	0.07175	4.158E-4	0.7323	0.8622	1.013	20001	150000
or[2,7]	0.6277	0.0657	2.987E-4	0.508	0.6245	0.7653	20001	150000
or[3,4]	1.122	0.06037	2.421E-4	1.008	1.121	1.245	20001	150000
or[3,5]	1.049	0.05817	2.914E-4	0.9411	1.048	1.169	20001	150000
or[3,6]	1.007	0.06435	2.826E-4	0.8875	1.005	1.139	20001	150000
or[3,7]	0.7316	0.07166	2.9E-4	0.6007	0.7284	0.8815	20001	150000
or[4,5]	0.9373	0.06649	3.24E-4	0.814	0.935	1.075	20001	150000
or[4,6]	0.9002	0.07511	3.575E-4	0.7619	0.8974	1.056	20001	150000
or[4,7]	0.6535	0.07034	2.867E-4	0.5259	0.6499	0.801	20001	150000
or[5,6]	0.9629	0.08145	3.988E-4	0.8129	0.9594	1.132	20001	150000
or[5,7]	0.6992	0.07725	3.395E-4	0.5598	0.6952	0.8615	20001	150000
or[6,7]	0.7293	0.08556	3.694E-4	0.5754	0.7245	0.911	20001	150000

We can see that PTCA is better at reducing mortality than all other treatments, while, for example, Acc t-PA is slightly better than SK + t-PA, reducing the odds of mortality by about 12% with 95% CrI from 1 to 25%.

Note however that the estimates for the OR of mortality on PTCA compared with Acc t-PA are slightly different from those in Table 2.2 and the 95% CrI is narrower. This is expected since we are now using all the evidence in the network (both direct and indirect) to estimate this relative treatment effect.

Table 2.3 Extended thrombolytics example: median and 95% CrI for the odds ratios and between-study standard deviation (heterogeneity) from fixed and random effects models.

		Fixed effect		Random effects	
X	Y	Median	95% CrI	Median	95% CrI
SK	t-PA	1.00	(0.94, 1.06)	0.98	(0.76, 1.10)
SK	Acc t-PA	0.85	(0.78, 0.93)	0.85	(0.69, 1.00)
SK	SK + t-PA	0.96	(0.87, 1.05)	0.96	(0.76, 1.23)
SK	r-PA	0.90	(0.80, 1.01)	0.89	(0.67, 1.06)
SK	TNK	0.86	(0.74, 1.00)	0.85	(0.60, 1.16)
SK	PTCA	0.62	(0.51, 0.76)	0.61	(0.47, 0.76)
t-PA	Acc t-PA	0.86	(0.77, 0.95)	0.87	(0.71, 1.16)
t-PA	SK + t-PA	0.96	(0.86, 1.07)	0.98	(0.79, 1.44)
t-PA	r-PA	0.90	(0.79, 1.03)	0.91	(0.71, 1.21)
t-PA	TNK	0.86	(0.73, 1.01)	0.87	(0.64, 1.33)
t-PA	PTCA	0.62	(0.51, 0.77)	0.63	(0.49, 0.84)
Acc t-PA	SK + t-PA	1.12	(1.01, 1.25)	1.13	(0.91, 1.50)
Acc t-PA	r-PA	1.05	(0.94, 1.17)	1.04	(0.83, 1.26)
Acc t-PA	TNK	1.01	(0.89, 1.14)	1.00	(0.77, 1.32)
Acc t-PA	PTCA	0.73	(0.60, 0.88)	0.72	(0.58, 0.88)
SK + t-PA	r-PA	0.94	(0.81, 1.08)	0.92	(0.65, 1.18)
SK + t-PA	TNK	0.90	(0.76, 1.06)	0.89	(0.60, 1.24)
SK + t-PA	PTCA	0.65	(0.53, 0.80)	0.64	(0.45, 0.83)
r-PA	TNK	0.96	(0.81, 1.13)	0.96	(0.70, 1.39)
r-PA	PTCA	0.70	(0.56, 0.86)	0.70	(0.53, 0.91)
TNK	PTCA	0.72	(0.58, 0.91)	0.72	(0.50, 0.99)
Heterogeneity (σ)		–	–	0.064	(0.003, 0.30)

Treatment definitions are given in Figure 2.1.

A plot similar to Figure 2.5 can be obtained in WinBUGS by doing a 'caterpillar' plot of the parameter lor, although note that these will be the LORs. See Exercise 2.1.

The probabilities that each treatment is ranked 1st, 2nd, ..., 7th are shown in Figure 2.6. Peaks in these figures indicate the most likely rank for that treatment, and the probabilities of each rank can be read on the vertical axis.

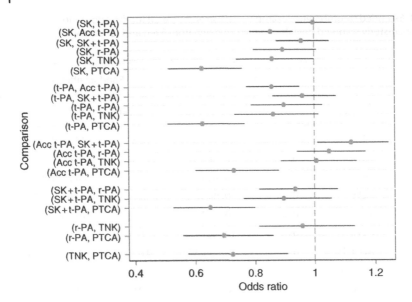

Figure 2.5 Extended thrombolytics example: summary forest plot – medians (dots) and 95% CrI (solid lines) for the ORs of all treatments compared with each other from the fixed effects model, plotted on a log scale. ORs < 1 favour the second treatment. Treatment definitions are given in Figure 2.1.

As expected, PTCA has a very high probability of being the best treatment, and so the probability that it is ranked 1st is close to 100%. Among the other treatments, Acc t-PA has probability of 42% being ranked 2nd and 48% of being ranked 3rd, and TNK also has about 43% probability of being ranked 2nd, suggesting these may be the best alternatives to PTCA, when it is not available. This agrees with the ORs in Figure 2.5, where apart from PTCA, Acc t-PA and TNK are the only other treatments that are favoured over the alternatives, and there seems to be no advantage of one over the other. The absolute probabilities, ranks and probabilities of being the best for all treatments are shown in Table 2.4.

The implementation of the random effects model accounting for the correlations in multi-arm trials is as follows:

WinBUGS code for NMA: Binomial likelihood, logit link, random effects model.
The # symbol is used for comments – text after this symbol is ignored by WinBUGS. Note that WinBUGS specifies the normal distribution in terms of its mean and precision.
The full code with data and initial values is presented in the online file *Ch2_RE_Bi_logit.odc*.

```
# Binomial likelihood, logit link
# Random effects model for multi-arm trials
model{                                                # *** PROGRAM STARTS
for(i in 1:ns){                                       # LOOP THROUGH STUDIES
  w[i,1] <- 0                                         # adjustment for multi-arm trials is zero for control
                                                      #   arm
  delta[i,1] <- 0                                     # treatment effect is zero for control arm
  mu[i] ~ dnorm(0,.0001)                              # vague priors for all trial baselines
  for (k in 1:na[i]) {                                # LOOP THROUGH ARMS
    r[i,k] ~ dbin(p[i,k],n[i,k])                      # binomial likelihood
    logit(p[i,k]) <- mu[i] + delta[i,k]              # model for linear predictor

  for (k in 2:na[i]) {                                # LOOP THROUGH ARMS
    delta[i,k] ~ dnorm(md[i,k],taud[i,k])            # trial-specific LOR distributions
    md[i,k] <- d[t[i,k]] - d[t[i,1]] + sw[i,k]       # mean of LOR distributions (multi-arm trial
                                                      #   correction)
    taud[i,k] <- tau *2*(k-1)/k                       # precision of LOR distributions (with multi-arm trial
                                                      #   correction)
    w[i,k] <- (delta[i,k] - d[t[i,k]] + d[t[i,1]])   # adjustment for multi-arm RCTs
    sw[i,k] <- sum(w[i,1:k-1])/(k-1)                  # cumulative adjustment for multi-arm trials
    }
  }
d[1] <- 0                                             # treatment effect is zero for reference treatment
for (k in 2:nt){  d[k] ~ dnorm(0,.0001) }            # vague priors for treatment effects
sd ~ dunif(0,2)                                       # vague prior for between-trial SD.
tau <- pow(sd,-2)                                     # between-trial precision = (1/between-trial variance)
  }                                                   # *** PROGRAM ENDS
```

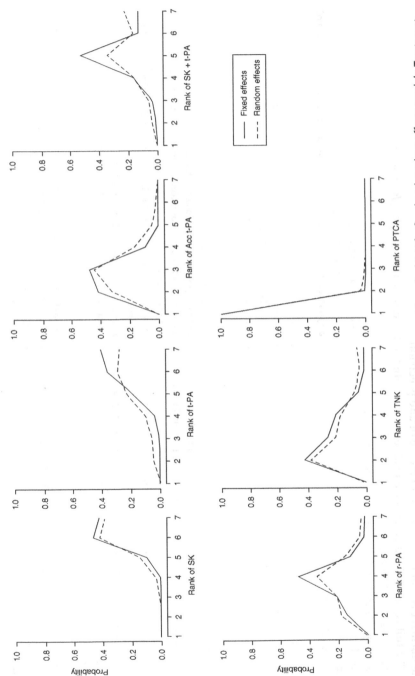

Figure 2.6 Extended thrombolytics example: probability that each treatment is ranked 1–7 for fixed and random effects models. Treatment definitions are given in Figure 2.1.

Table 2.4 Extended thrombolytics example: results from the fixed effects network meta-analysis model.

| Treatment | Probability of 35 day mortality | | Rank | | | |
	Median	95% CrI	Mean	Median	95% CrI	Pr(best)
SK	0.08	(0.05, 0.14)	6.31	6	(5, 7)	0.00
t-PA	0.08	(0.05, 0.14)	6.13	6	(4, 7)	0.00
Acc t-PA	0.07	(0.04, 0.12)	2.69	3	(2, 4)	0.00
SK + t-PA	0.08	(0.05, 0.13)	5.13	5	(3, 7)	0.00
r-PA	0.08	(0.04, 0.13)	3.75	4	(2, 6)	0.00
TNK	0.07	(0.04, 0.12)	3.00	3	(2, 6)	0.00
PTCA	0.05	(0.03, 0.09)	1.00	1	(1, 1)	1.00

Treatment definitions are given in Figure 2.1.

Readers should compare this code with the random effects code for pairwise meta-analysis and the fixed effects code for NMA and note the similarities. The main difference in this code is the implementation of the conditional univariate normal distributions for the trial-specific random effects given in equation (2.12).

We have coded the mean of the random effects normal distribution as md (WinBUGS does not allow algebraic expressions in distributions) and defined md as $d_{1,t_{ik}} - d_{1,t_{i1}} + sw_{ik}$ where sw_{ik} is the adjustment for the conditional normal distribution given as follows:

$$\delta_{i,k} \mid \delta_{i,2},...,\delta_{i,(k-1)}$$
$$\sim \text{Normal}\left((d_{t_k} - d_{t_1}) + \underbrace{\sum_{j=1}^{k-1} \underbrace{\frac{1}{(k-1)}(\delta_{i,j} - (d_{t_j} - d_{t_1}))}_{w_{i,j}}}_{sw_{i,k}}, \frac{k}{2(k-1)}\sigma^2 \right)$$

and taud is the adjusted precision for the conditional univariate normal, that is, the inverse of the aforementioned variance. We set w[i,1] to zero so that the code reduces to a simple univariate normal distribution (equation (2.8)) for two-arm trials.

Changes to the initial values are also required. Initial values need to be specified for the between-study heterogeneity sd, and WinBUGS can be left to generate initial values for the parameter delta, as before.

Results from the random effects model are given in Table 2.3. The posterior median of the between-study standard deviation is small compared with the

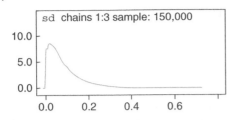

Figure 2.7 Extended thrombolytics example: posterior distribution of the between-study standard deviation – from WinBUGS.

size of the treatment effect of PTCA (Table 2.3), and its posterior distribution is concentrated around small values (Figure 2.7). We might therefore question whether a random effects model is really necessary here. Model comparison and model choice will be discussed in Chapter 3.

The posterior distribution of the between-study standard deviation (Figure 2.7) is far from the uniform distribution between zero and two that was set up as prior. This confirms that updating has taken place and that the prior distribution was not too restrictive (see Section 2.3.2 for more details). We should also note that the distribution is slightly different from that in Figure 2.3, since we are now using more evidence to estimate this parameter. In this case this has meant that the distribution is slightly narrower and concentrated around smaller values (see Section 2.3.2).

2.3   Technical Issues in Network Meta-Analysis

The use of the WinBUGS Bayesian MCMC software has advantages but it also requires some care. Users are strongly advised to acquire a good understanding of the Bayesian theory (Spiegelhalter et al., 2004) and to follow advice given in the WinBUGS manual and book (Lunn et al., 2013). Particular care must be taken in checking convergence, and we suggest that at least three chains are run, starting from widely different (yet sensible) initial values. The diagnostics recommended in the literature should be used to check convergence (Gelman, 1996; Brooks and Gelman, 1998). Users should *also* ensure that, after convergence, each chain is sampling from the same posterior distribution. Posteriors should be examined visually for spikes and unwanted peculiarities, and both the initial 'burn-in' and the posterior samples should be conservatively large (Lunn et al., 2013). The number of iterations used for both must be always reported.

Beyond these warnings, which apply to all Bayesian MCMC analyses, NMA models have particular properties that require careful examination.

2.3.1   Choice of Reference Treatment

While the likelihood is not altered by a change in which treatment is taken to be 'treatment 1', the choice of the reference treatment can sometimes affect the posterior estimates because prior distributions cannot be totally non-informative. However, for the vague prior distributions we suggest throughout for μ_i and d_{1k} (see Section 2.3.2), we expect the effect to be negligible. Choice should therefore be based on ease of interpretation, with placebo or a standard treatment usually taken as treatment 1. In larger networks, it is preferable to choose as treatment 1 a treatment that is in the 'centre' of the network. In other words, choose the treatment that has been trialled against the highest number of other treatments. The purpose of this is to reduce strong correlations that may otherwise be induced between mean treatment effects for pairs of treatments, which can slow convergence and make for inefficient sampling from the posterior distribution, although in theory if enough samples are collected, results can still be used for inference.

2.3.2 Choice of Prior Distributions

We recommend vague or flat prior distributions, such as Normal($0,100^2$), throughout for μ_i and d_{1k}. Informative prior distributions for relative effect measures would require special justification but can be easily implemented by specifying, for example, d[k]~dnorm(0,0.1) to represent a normal distribution with mean 0 and precision 0.1 (variance = 10) for the relative effects of all treatments compared with the reference. Normal prior distributions are appropriate when the treatment effects are defined on a continuous scale ranging from minus to plus infinity (such as the LORs). For treatment effects defined on scales with a different range, different prior distributions need to be considered – see Section 2.3.3.

It has become standard practice to also set vague prior distributions for the between-trial variances. For binomial with logit link models, the usual practice is to place a uniform prior distribution on the standard deviation, for example, $\sigma \sim$ Uniform(0,2). The upper limit of 2 represents a huge range of trial-specific treatment effects (Spiegelhalter et al., 2004, table 5.2). For example, if the median treatment effect was an OR of 1.5 (i.e. LOR of 0.405), then we would expect 95% of trials to have *true* ORs between 0.2 and 11 (calculated as exp(0.405 ± 2)). The posterior distribution of σ should always be inspected to ensure that it is sufficiently different from the prior distribution as this would otherwise indicate that the prior distribution is dominating the data and no posterior updating has taken place. This can be done in WinBUGS by plotting the posterior density of sd (see Figures 2.1 and 2.7) or by exporting the simulated values into other software and plotting them there.

An alternative approach, which was once popular but has since fallen out of favour, is to set a vague gamma prior distribution on the precision, for example, $1/\sigma^2 \sim$ Gamma(0.001,0.001). This approach gives a low prior weight to unfeasibly large σ on the logit scale but is actually quite informative when low values of σ are possible, and inferences are sensitive to the specific parameters of the gamma distribution chosen when data are sparse (Gelman, 2006). One specific disadvantage is that this puts considerable weight on values of σ near zero but, on the other hand, it rules out values of σ at zero. This may be desirable, because it is not uncommon, particularly when data are sparse, that MCMC sampling can 'get stuck' at $\sigma \approx 0$, leading to spikes in the posterior distribution of both σ and the treatment effect parameters d_{1k}. In these cases a gamma prior distribution may improve numerical stability and speed convergence. Half-normal prior distributions constrained to be positive can also be useful when data are sparse.

However they are formulated, there are major disadvantages in routinely using vague prior distributions for the between-study heterogeneity, although this has become a widely accepted practice. In the absence of large numbers of large trials for at least one comparison (at least four or five trials have been suggested as a minimum (Gelman, 2006)), the posterior distribution of σ will be poorly identified and likely to include values that, on reflection, are implausibly high or, possibly, implausibly low. Two further alternatives may be useful when there is insufficient data to adequately estimate the between-trial variation. The first is the use of external data (Higgins and Whitehead, 1996). If there is insufficient data in the meta-analysis, it may be reasonable to use an estimate for σ from a larger meta-analysis on the same trial outcome involving a similar treatment for a similar condition. The posterior distribution, or a posterior predictive distribution (Lunn et al., 2013) (see also Chapters 3 and 8), from such an analysis could be used to approximate an informative prior distribution. Alternatively, prior distributions derived from large numbers of meta-analyses can be used, with the appropriate prior distribution chosen depending on the outcome type and the type of treatments being compared (Turner et al., 2012, 2015b). Chapter 6 discusses issues caused by sparse data in more detail.

If there are no data on similar treatments and outcomes that can be used, an informative prior distribution can be elicited from a clinician who knows the field. This can be done by posing the question in this way: 'Suppose we accept that different trials, even if infinitely large, can produce different effect sizes. If the average effect was an OR of 1.8 [choose a plausible average], what do you think an extremely high and an extremely low effect would be, in a very large trial?' Based on the answer to this, it should be possible, by trial and error, to construct an informative gamma prior distribution for $1/\sigma^2$ or a normal prior distribution for σ, subject to $\sigma > 0$ (half-normal). For further discussion of prior distributions for variance parameters, see Lambert et al. (2005) and Spiegelhalter et al. (2004).

There can be little doubt that the vague prior distributions that are generally recommended for the heterogeneity parameter produce posterior distributions that are biased upwards. The extent of the bias is likely to be greater when the true variance is low and when there are few data: either few trials or small trials. However, this is also a problem when using frequentist estimators. Although we can be reassured that the bias tends to be conservative, ultimately it may be preferable to use informative prior distributions, perhaps tailored to particular outcomes and disease areas, based on studies of many hundreds of meta-analyses (Turner et al., 2012; Rhodes et al., 2015; Turner et al., 2015b). An easier approach might be to identify a large meta-analysis of other treatments for the same condition and using the same outcome measures and use the posterior distribution for the between-trial heterogeneity from this meta-analysis to inform the current analysis (Dakin et al., 2010).

Whichever prior distribution is chosen, it can be easily implemented in WinBUGS by changing the relevant lines of code. See Exercise 2.4.

2.3.3 Choice of Scale

The logit model presented in this chapter assumes linearity of effects on the log odds scale. It should be emphasised that it is important to use a scale in which effects are additive, as is required by the linear model (Deeks, 2002; Caldwell et al., 2012). Choice of scale can be guided by goodness of fit (see Chapter 3) or by lower between-study heterogeneity, but there is seldom enough data to make this choice reliably, and logical considerations may play a larger role (Caldwell et al., 2012). Quite distinct from choice of scale for modelling is the issue of how to report treatment effects. Thus, while one might assume linearity of effects on the log odds scale, the investigator, given information on the absolute effect of one treatment, is free to derive treatment effects with 95% CrI on other scales, such as RD, RR or NNT using the additional code provided in Section 2.2.3.1. Therefore, the most appropriate scale for each problem should be chosen based on statistical and logical considerations and **not** on reporting preferences.

Warn et al. (2002) suggest models for pairwise meta analysis of binomial data using the RD and log RR while still using the arm-based binomial likelihood in equation (2.3). The difficulty in modelling the probabilities on these scales is that while the LOR is unbounded, that is, it can take any value between plus and minus infinities, the RD must be between −1 and 1 and the log RR is bounded by the probability of an event, the risk, in the reference group, for each study. Thus the model must be constrained to ensure the probabilities of an event in each arm of each study remain between zero and one. To model binomial data on the RD scale, equation (2.13) is replaced by (Warn et al., 2002)

$$p_{i1} = \mu_i$$

$$p_{ik} = \mu_i + \min\left(\max\left(\delta_{ik}, -p_{i1}\right), 1 - p_{i1}\right), \quad k = 2, \ldots, na_i$$

which ensures that $0 \leq p_{ik} \leq 1$. The random effects distribution is the same as in equation (2.8), but the basic parameters are now the RD relative to treatment 1 which are bounded between −1 and 1. For a fixed effects model, we write

$$\delta_{ik} = d_{1,t_{ik}} - d_{1,t_{i1}}, \quad k = 2, \ldots, na_i \qquad (2.16)$$

The prior distributions in equation (2.10) are replaced with $d_{1k} \sim \text{Uniform}(-1,1)$. The μ_i now represent the probability of an event in arm 1 of trial i, which must also lie between zero and one, so they are given Uniform(0,1) prior distributions. The prior distribution for the between-study heterogeneity can be the same as presented in equation (2.15) or a suitable alternative (see Section 2.3.2).

To model binomial data on the log RR scale, equation (2.13) is replaced by (Warn et al., 2002)

$$\mu_i = \log(p_{i1})$$
$$\log(p_{ik}) = \mu_i + \min(\delta_{ik}, -\log(p_{i1})), \quad k = 2,\ldots,na_i$$

with $\delta_{i1} = 0$, which ensures that $0 \le p_{ik} \le 1$. The random effects distribution is the same as in equation (2.8), but the basic parameters are now the log RR relative to treatment 1. For a fixed effects model, we use equation (2.16). Warn et al. (2002) suggest that normal prior distributions with mean zero and variance of 10 are sufficiently vague for d_{1k}. The μ_i now represent the log risks in the control arm, that is, the log of the probability of an event in arm 1 of trial i. Prior distributions can be given by noting that $p_{i1} = \exp(\mu_i)$ is the probability of an event in arm 1 of trial i, which must lie between zero and one. Thus Uniform(0,1) prior distributions can be given to p_{i1}. The prior distribution for the between-study heterogeneity can be the same as presented in equation (2.15). For further details on the WinBUGS implementation of these models, see Warn et al. (2002) and Exercise 2.5.

Although attractive, these models can pose some computational and interpretation challenges when there are studies with zero events or with 100% events. In particular the RD model can cause numerical problems when the relative treatment effects are close to −1 or 1, which can happen by chance, particularly during the burn-in period. Some workarounds are possible, including using a slightly more restrictive prior distribution for the basic parameters, for example, uniform between −0.9999 and 0.9999, to avoid very extreme values being sampled (see Exercise 2.5). Warn et al. (2002) provide a thorough discussion of the issues and suggested workarounds.

2.3.4 Connected Networks

The models described assume that the network of treatments is connected, that is, there is a path leading from each treatment to every other. If this is not the case, steps should be taken to connect the network before inference on all relative effects can be made (see Section 1.6). However, the models presented here will still run for disconnected networks, and relative effects will be produced by WinBUGS. For disconnected treatments the relative effects produced will reflect only the prior distribution given (typically vague) as that is the only available information. Careful inspection of the output will immediately show this, as some parameters will have very large posterior variances, reflecting the variance in the prior distributions.

Network connectivity should always be checked before putting the data through the WinBUGS code. Particular care should be taken when there are studies with zero cells to ensure there is information on all contrasts – see Chapter 6 for further details. However, if this step is missed or performed incorrectly, inspecting the posterior distributions obtained for the basic parameters (d) will show these to be very wide and not updated from the prior distributions, highlighting the problem.

2.4   Advantages of a Bayesian Approach

Readers should note that implementation of NMA models with fixed or random effects, consistency equations and incorporating multi-arm correlations are straightforward in WinBUGS. Furthermore, the code presented in this chapter can be easily adapted to other data types (Chapter 4) and implemented to other models, such as those including a covariate or bias adjustment terms (Chapters 8 and 9).

An advantage of the Bayesian model with binomial likelihood is that exact arm-based likelihoods are used, and this means special precautions do not usually need to be taken in the case of the occasional trial with a zero cell count, since a binomial likelihood with a zero cell is allowed by WinBUGS (see Chapter 6). Chapters 4 and 10 extend this approach to other data types where, again, the arm-based likelihoods are used whenever possible. This is a major strength of the Bayesian MCMC approach, because some popular frequentist approaches for LORs have to add an arbitrary constant, usually 0.5, to cells in order to obtain non-infinite estimates of treatment effects and non-infinite variance, but in so doing they generate biased estimates of effect size (Sweeting et al., 2004; Bradburn et al., 2007). Simulation studies on fixed effect estimators (Sweeting et al., 2004) have also shown that Bayesian MCMC has performed well, ranks with Mantel–Haenszel (Mantel and Haenszel, 1959) and Exact method (Fleiss, 1994) estimators and is superior to the Peto method (Yusuf et al., 1985) and inverse-variance weighting in a wide range of situations.

However, in extreme cases where several trials have zero cells and many of the trials are small, the models we have recommended can be numerically unstable, either failing to converge or converging to a posterior distribution with very high standard deviation for some of the treatment effects. This is unlikely to happen with fixed effects models, and it can often be remedied in random effects models by using a (more) informative prior distribution on the variance parameter. The last resort, recognising the assumptions being made, is to put a random effects model on the trial-specific baselines μ_i as well as the relative treatment effects d_{1k}. See Chapter 5 for comments on this model and Chapter 6 for a thorough discussion of issues with sparse data and zero cells.

A specific problem arises in sparse networks, in which, for example, there is only one trial making the comparison X versus Y, and treatment Y only appears in this one trial. If the trial contains a zero cell, it may not be possible to estimate a treatment effect, and the output for that relative effect will resemble that of a disconnected network, that is, the prior distribution will not have been updated (Section 2.3.4). Trials with zero cells in *both* arms do not contribute evidence on the treatment effect and can be excluded, unless a model has been assumed for the trial-specific baselines. See Chapter 6 for further details.

Furthermore, in random effects models, the MCMC implementation proposed in this book automatically takes into account the uncertainty in the between-study heterogeneity parameter σ^2. While this is also possible using frequentist approaches (Zelen, 1971; Hardy and Thompson, 1996; Biggerstaff and Tweedie, 1997), it is rarely done in practice, and instead σ^2 is estimated and then assumed to be fixed and known for the purpose of computing the pooled treatment effects. By taking into account the uncertainty in σ^2 and propagating that to the uncertainty in the posterior distributions of the relative effects, the Bayesian implementation proposed here better reflects the overall uncertainty in the estimates, which is a crucial factor in decision-making (see Chapter 5).

2.5 Summary of Key Points and Further Reading

We have shown how to implement the two-treatment meta-analysis models with binary data in WinBUGS and that it is relatively simple to extend these models to meta-analysis of any number of trials on the same $S \geq 2$ comparators. In this framework, extension of the exchangeability assumption required for pairwise meta-analysis automatically delivers the assumption required for meta-analysis with more than two treatments, and vice versa.

In this chapter we have shown how to obtain the outputs generally required for reporting the results of an NMA. Different ways of summarising and presenting the results in tables and figures have been proposed (Chaimani et al., 2013a; Tan et al., 2014). For example, similar to the rank plots in Figure 2.6, the surface under the curve has been suggested as a summary (Salanti et al., 2011), and this can be calculated in WinBUGS. An analysis of the network geometry or of the relative weight contributions of each contrast to the pooled result have also been suggested as useful outputs, and readers are referred to the original references for more details (Salanti et al., 2008b; König et al., 2013).

The models with fixed and random effects have been implemented in WinBUGS using code that will work for any number of treatments and trials with any number of arms, when the outcome is binary. The code respects the randomisation in the original studies by modelling relative treatment effects and accounts for the correlation in trials with more than two arms in random effects models.

The issues of model fit and model choice are crucial to valid inference and decision-making and are explored in Chapter 3. Chapter 4 extends the model presented for binomial data with a logit link to the framework of generalised linear models, where the basic modelling framework stays the same, but the likelihood and link functions change to allow synthesis of different data types. Chapters 8 and 9 cover issues of heterogeneity, looking at meta-regression and treatment effects in subgroups and methods for bias adjustment, including publication bias and the so-called 'small-study bias'. In addition, Chapter 8 considers models where restrictions are placed on the relative treatment effects, for example, that they are exchangeable within class or additive. The synthesis of both direct and indirect evidence on the same contrast also raises questions about inconsistency between evidence sources (Chapter 1), which will be addressed in Chapter 7. Further considerations on the validity of the proposed approach to synthesis are made in Chapter 12.

Bayesian analysis is by no means a panacea: one area that clearly deserves more work is how to specify a 'vague' prior distribution for the variance parameter. Readers are referred to Section 2.3.2 and also to standard Bayesian textbooks (Bernardo and Smith, 1994; Gelman et al., 2004; Spiegelhalter et al., 2004) for general discussions of this issue.

2.6 Exercises

2.1 Using the extended thrombolytics example, run the WinBUGS code provided to obtain the values presented in Table 2.3.
   a) Draw a plot similar to Figure 2.5 in WinBUGS by plotting l or W What are the main differences between the two plots?
   b) Monitor the variable rk and plot its density for both the fixed and random effects models and compare with Figure 2.6.

2.2 Using the extended thrombolytics example, compare the OR for PTCA relative to Acc t-PA for the fixed effect pairwise meta-analysis (Table 2.2) with the corresponding OR in Table 2.3. Can you explain any differences in effect size and uncertainty?
   Compare also the OR in Table 2.3 with those for the reduced network presented in Chapter 1. What are the main differences?

2.3 Using the extended thrombolytics example, suppose there was a new study comparing treatment 3 (Acc t-PA) with a new (fictitious) treatment 8, where there were 5 deaths out of 32 patients randomised to treatment 3 and 8 deaths out of 35 patients randomised to treatment 8.
   a) How would the network plot in Figure 2.1 change with the inclusion of the new study?

b) How do you expect this study to influence the estimates of the relative effects for the other treatments? If a random effects model was fitted, how would you expect the between-study heterogeneity to change? Confirm your answers by running the random effects model with the new study included.

c) Now suppose there was a further study added comparing the new (fictitious) treatment 8 with treatment 7 (PTCA). Would this be expected to influence the other relative effects and the between-study heterogeneity (in a random effects model)?

2.4 *Table IV of Turner et al. (2015b) lists a series of lognormal prior distributions for the between-study variance, based on empirical data, for different outcome and comparison types. Using the extended thrombolytics example, fit the random effects network meta-analysis model using an informative prior distribution for the between-study heterogeneity, as suggested by Turner et al. (2015b).

a) Decide on the most appropriate distribution to consider given the outcome and comparison types presented in this example. (Hint: if there is more than one comparison type, use multiple prior distributions and compare the results.)

b) Use the WinBUGS Help for details on how to code a lognormal distribution and change the code provided to implement the new prior distribution(s). Note that the prior distributions suggested by Turner et al. (2015b) are for the between-study *variance*.

c) Inspect the posterior distribution of sd and compare with Figure 2.3. Note any changes in relative treatment effects compared with those in Table 2.3.

2.5 *Using the extended thrombolytics example, adapt the standard network meta-analysis code to fit fixed and random effects models using the risk difference and the log relative risk as the scale of the linear predictor as suggested by Warn et al. (2002) (Section 2.3.3). Take care to ensure appropriate prior distributions are given and that the initial values are within the bounds of the prior distributions. Note that for the random effects model on the risk difference, you may need to reduce the range of the uniform prior distributions for the basic parameters.

Compare the risk difference/relative risk estimates with those obtained by using the logit link and then transforming the ORs (Section 2.2.3.1). Compare also the probability that PTCA is the best treatment and the between-study heterogeneity in the random effects models.

3

Model Fit, Model Comparison and Outlier Detection

3.1 Introduction

An important part of any statistical analysis is an assessment of how well the predictions from a particular model fit with the observed data. If the model does not describe the data well, then any outputs based on predictions from that model, such as parameter estimates, uncertainty in parameter estimates, expected net benefit, optimal treatment strategy and the uncertainty in the optimal strategy, will be a poor reflection of the evidence base. In the context of comparative effectiveness research and health technology assessment, it is therefore essential that we aim to identify good fitting models, so that any judgements made as to the most effective or cost-effective intervention are a fair reflection of the available evidence.

Methods for assessing how well the predictions from a particular model fit the observed data are well established in the field of frequentist statistics (McCullagh and Nelder, 1989). Many of these ideas translate naturally into Bayesian inference (Spiegelhalter et al., 2002; Gelman et al., 2004) and can be computed using WinBUGS. Because a Bayesian analysis estimates a posterior distribution for the model parameters, model predictions for the observed data also have posterior distributions, and so too do measures of model fit. In this chapter, we describe the posterior mean residual deviance as a measure of global model fit. We then describe how the posterior mean deviances and the deviance information criteria (DIC) (Spiegelhalter et al., 2002) can be used to compare the fit of different models, for example, comparing fixed and random effects models (see Section 3.3) or models with different covariates or assumptions (see Chapters 7–9).

In both pairwise and network meta-analysis, evidence is pooled from multiple sources (studies). This opens up the potential for inconsistencies between the different data sources, where, for example, one study may find a strong treatment effect for a particular comparison, whereas the majority of studies

Network Meta-Analysis for Decision-Making, First Edition. Sofia Dias, A. E. Ades, Nicky J. Welton, Jeroen P. Jansen and Alexander J. Sutton.
© 2018 John Wiley & Sons Ltd. Published 2018 by John Wiley & Sons Ltd.
Companion website: www.wiley.com/go/dias/networkmeta-analysis

find no effect. If the evidence sources are consistent, then this strengthens confidence in any conclusions based on that synthesis. However, ignoring inconsistency in the evidence may lead to biased estimates being obtained and result in suboptimal decisions made based on the model outputs. Studies giving results that do not fit with the predictions from a model that otherwise fits well to the remaining data points are known as 'outliers' with respect to that model. We describe a predictive cross-validation approach to identify outlying studies in a network meta-analysis in Section 3.4.

Throughout this chapter we consider only the case where there is a binary outcome; however details and examples of models for other outcome types are given in Chapter 4, and the approach described here is generally applicable to all models.

3.2 Assessing Model Fit

Standard statistical measures of goodness of fit such as residuals, residual sum of squares and residual deviance (often called the likelihood ratio statistic) can all be calculated within a Bayesian framework to assess the fit of a model. However, instead of point estimates of these statistics, a Bayesian analysis delivers posterior distributions for them. These can be summarised in various ways, but typically the posterior mean of these distributions is reported. We will focus on the residual deviance, but other measures of fit can be calculated similarly.

3.2.1 Deviance

The deviance statistic measures the fit of the predictions made by a particular model to the observed data using the likelihood function. The likelihood function measures how 'likely' observed data are given a particular model and so is a natural measure to focus on to assess model fit – in fact frequentist statistics revolves around maximising the likelihood function. The deviance for a particular model, D_{model}, is defined as -2 times the log-likelihood, $Loglik_{model}$, for a given model:

$$D_{model} = -2 Loglik_{model}$$

For a given model and observed data, the larger the likelihood, then the closer the model fit. Similarly, the larger the log of the likelihood, $Loglik_{model}$, the closer the model fit. Multiplying by -2 reverses this, so the smaller the deviance, D_{model}, the closer the model fit. D_{model} therefore measures how far the model predictions *deviate* from the observed data. The deviance is simply a function of model parameters that can be written down and calculated at each iteration of an MCMC simulation. Posterior summaries for D_{model} can then be obtained, including the posterior mean, $\bar{D}_{model}$. The smaller the posterior mean deviance statistic, $\bar{D}_{model}$, then the better the model fit.

3.2.2 Residual Deviance

The smaller the deviance statistic, $\bar{D}_{model}$, then the better the model fit, but how small is small? The disadvantage of using $\bar{D}_{model}$ is that there is no clear answer to this question. Instead, we define the residual deviance, D_{res}, which helps us gauge how good the model fit is, by providing a reference point. The residual deviance is equal to the deviance for a given model, D_{model}, minus the deviance for a saturated model (McCullagh and Nelder, 1989), D_{sat}:

$$D_{res} = D_{model} - D_{sat}$$

A saturated model is one where all of the predictions from the model are equal to the observed data values. For a binomial likelihood with r_{ik} events out of n_{ik} patients randomised to arm k of trial i, it can be confirmed that the residual deviance is calculated as

$$
\begin{aligned}
D_{res} &= \sum_i \sum_k dev_{ik} \\
&= \sum_i \sum_k 2\left(r_{ik} \log\left(\frac{r_{ik}}{\hat{r}_{ik}} \right) + (n_{ik} - r_{ik})\log\left(\frac{n_{ik} - r_{ik}}{n_{ik} - \hat{r}_{ik}} \right) \right)
\end{aligned}
\tag{3.1}
$$

where dev_{ik} is the residual deviance contribution for each data point and $\hat{r}_{ik} = n_{ik} p_{ik}$ is the expected (predicted) number of events in each trial arm, based on the current model. D_{res} is computed at each iteration of the MCMC simulation and then summarised by the posterior mean residual deviance, $\bar{D}_{res}$.

For a model that fits the data well, we would expect the individual contributions to the residual deviance to have a roughly chi-squared distribution with degrees of freedom equal to 1. This is exact if the observed data have a normal likelihood (Dempster, 1997) (see also Chapter 4). It follows that if we sum over N unconstrained data points, then we would expect the residual deviance to have a roughly chi-squared distribution with degrees of freedom equal to N. On this basis, we would expect the posterior mean of the residual deviance, $\bar{D}_{res}$, to be close to the number of unconstrained data points, N, if the model predictions are a good fit to the data (Spiegelhalter et al., 2002). If $\bar{D}_{res}$ is much greater than N, this suggests some lack of fit, and if $\bar{D}_{res}$ is much smaller than N, this suggests the model may 'over-fit' the data. We can examine the posterior means of the contributions to the residual deviance, dev_{ik}, to identify individual data points that are contributing most to $\bar{D}_{res}$. This can help to identify particular observations that appear to be 'outliers' with respect to the model fitted.

3.2.3 Zero Counts*

Note that the formula for the residual deviance for binomial data, equation (3.1), can be numerically unstable to compute when there are zero counts for either the number of observed events, r_{ik}, or the number of non-events, $(r_{ik} - n_{ik})$. This is because it depends on the terms

$$r_{ik} \log\left(\frac{r_{ik}}{\hat{r}_{ik}}\right) \quad \text{and} \quad (n_{ik} - r_{ik})\log\left(\frac{n_{ik} - r_{ik}}{n_{ik} - \hat{r}_{ik}}\right)$$

which, if each term in the products is computed separately, involve taking the log of 0. However, the product $r \log(r/\hat{r})$ tends to 0 in the limit as r tends to 0, and so this term should be set to zero when $r = 0$. The code presented in this book is adapted to enforce this limiting value, taking advantage of the way WinBUGS performs the computation (the term will be zero, without ever needing to compute the logarithm (Lunn et al., 2013)). Calculating this in other packages, for example, in R, the computation of equation (3.1) can give numerical errors, or inflated deviance contributions for those data points with zero counts, and care should be taken to enforce its limiting value of zero. We discuss issues with zero counts and sparse data further in Chapter 6.

3.2.4 Worked Example: Full Thrombolytic Treatments Network

To illustrate, we return to the network of RCTs comparing thrombolytic treatments for patients following acute myocardial infarction, presented in Section 2.2. The network presented in Figure 2.1 is based on a set of treatments defined in two comprehensive systematic reviews (Boland et al., 2003; Keeley et al., 2003b). However these reviews excluded trials involving either anistreplase (ASPAC) or urokinase (UK) because these treatments were no longer available in the United Kingdom. Just because treatments are not part of our 'decision set' of treatments does not mean they cannot provide useful evidence, indirectly, on the comparisons that *are* of interest, in a network meta-analysis. If we include these two additional treatments, then we obtain the evidence network displayed in Figure 3.1, which comprises 50 trials (including 2 three-arm studies) comparing nine thrombolytic treatments (Dias et al., 2010b). The complete dataset can be found in the file *Ch3_thromboFE_criticism.odc*.

3.2.4.1 Posterior Mean Deviance, $\bar{D}_{model}$

The deviance, D_{model}, can be monitored in two ways in WinBUGS. There is always a system node called `deviance` that will monitor the relevant deviance for a given likelihood. The `deviance` node can be set in the same way as any other node in the model by going to **Inference→Samples** and typing in 'deviance'. After convergence is deemed to be satisfactory, a further set of

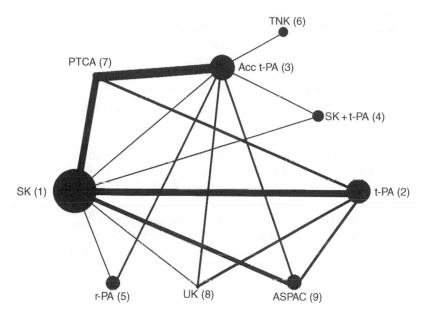

Figure 3.1 Full thrombolytic treatments network: lines connecting two treatments indicate that a comparison between these treatments (in one or more RCTs) has been made: streptokinase (SK), tissue plasminogen activator (t-PA), accelerated t-PA (Acc t-PA), reteplase (r-PA), tenecteplase (TNK), percutaneous transluminal coronary angioplasty (PTCA), urokinase (UK) and anistreplase (ASPAC). The width of the lines reflects the number of studies providing evidence on that comparison, the size of the circles is proportional to the number of patients randomised to that treatment, and the numbers by the treatment names are the treatment codes used in the modelling. There are 2 three-arm trials: SK versus Acc t-PA versus SK + t-PA and SK versus t-PA versus ASPAC.

samples can be generated, on which to base inference. Click on the '**stats**' button (making sure to set '**beg**' so that burn-in samples are discarded) to obtain posterior summaries for D_{model}. Doing this for the fixed effects model given in Section 2.2.3 using the full thrombolytic evidence network in Figure 3.1, we get

node	mean	sd	MC error	2.5%	median	97.5%	start	sample
deviance	535.5	10.89	0.03035	516.1	534.9	558.7	20001	150000

and the posterior mean is $\bar{D}_{model} = 535.5$.

Alternatively, the posterior mean of the deviance, $\bar{D}_{model}$, can be obtained from the DIC tool. After convergence, set the DIC tool: **Inference→DIC→set**. You can think of the DIC tool as an additional node that you monitor, but it must only be set after convergence has been achieved. Then, after updating further sufficient samples, click on '**DIC**'. The posterior mean of the deviance, $\bar{D}_{model}$, is given under the heading '**Dbar**'. The row labelled 'r' indicates that the model fit statistics given for that row correspond to the input data with the

Dbar = post.mean of –2logL; Dhat = –2LogL at post.mean of stochastic nodes				
	Dbar	Dhat	pD	DIC
r	535.557	477.864	57.693	593.250
total	535.557	477.864	57.693	593.250

Figure 3.2 Full thrombolytic network example: output from the DIC tool for the network meta-analysis – from WinBUGS.

name 'r'. The row labelled 'total' indicates that the model fit statistics for that row are summed over all input data. In our example all of our data is held in the matrix r [,], so the figures in the row 'total' are identical to those in the row 'r' (Figure 3.2).

It can be seen that both methods give exactly the same result: $\bar{D}_{model} = 535.5$. As described in Section 3.2.2, this figure isn't very helpful for assessing the fit of a single model. For that we need to compute the posterior mean residual deviance, $\bar{D}_{res}$.

3.2.4.2 Posterior Mean Residual Deviance, $\bar{D}_{res}$

There is no preset node for the residual deviance, and it is not included in the DIC tool. Instead we need to write out the relevant formula for D_{res} for a given likelihood in WinBUGS. For a binomial likelihood the formula is as given in equation (3.1), and the lines of code that need to be added to the core fixed effect network meta-analysis model presented in Chapter 2 are highlighted in bold.

WinBUGS code for NMA: Binomial likelihood, logit link, fixed effect model - code for residual deviance added to the core fixed effect network meta-analysis model.

New code highlighted in bold and larger font. The full code with data and initial values is presented in the online file *Ch3_throm-boFE_criticism.odc*.

```
# Binomial likelihood, logit link
# Fixed effects model
model{
for(i in 1:ns){                                              # *** PROGRAM STARTS
    mu[i] ~ dnorm(0,.0001)                                   # LOOP THROUGH STUDIES
    for (k in 1:na[i]) {                                     # vague priors for all trial baselines
        r[i,k] ~ dbin(p[i,k],n[i,k])                         # LOOP THROUGH ARMS
        logit(p[i,k]) <- mu[i] + d[t[i,k]] - d[t[i,1]]       # binomial likelihood
                                                             # model for linear predictor
        rhat[i,k] <- p[i,k] * n[i,k]                         # expected value of the numerators
        dev[i,k] <- 2 * (r[i,k] * (log(r[i,k])-log(rhat[i,k]))  #Deviance contribution
             + (n[i,k]-r[i,k]) * (log(n[i,k]-r[i,k]) - log(n[i,k]-rhat[i,k])))

    }
    resdev[i] <- sum(dev[i,1:na[i]])                         # summed residual deviance contribution for this
                                                             #   trial
    }
totresdev <- sum(resdev[])                                   # Total Residual Deviance
d[1]<-0                                                       # treatment effect is zero for reference treatment
for (k in 2:nt){  d[k] ~ dnorm(0,.0001) }                    # vague priors for treatment effects
}                                                            # *** PROGRAM ENDS
```

The WinBUGS code firstly computes $\hat{r}_{ik} = n_{ik} p_{ik}$, the expected (predicted) number of events in each trial arm, in node rhat[i,k], followed by dev[i,k], the contribution of each data point to the residual deviance (equation (3.1)). The contribution of each study to the residual deviance, resdev[i], is the sum of the dev[i,k] over all na[i] arms in study i. The total residual deviance, totresdev, is the sum of the resdev[i] over all studies. Note that the code in bold can be added to any model with a binomial likelihood, taking care to include the lines of code within the appropriate loops.

Setting and monitoring the node totresdev provides posterior summaries for the residual deviance:

node	mean	sd	MC error	2.5%	median	97.5%	start	sample
totresdev	105.9	10.89	0.03035	86.5	105.2	129.1	20001	150000

The posterior mean of the residual deviance is $\bar{D}_{res} = 105.9$. In the full thrombolytic network dataset, there are 48 two-arm trials and 2 three-arm trials, making $(48 \times 2) + (2 \times 3) = 102$ data points in total. $\bar{D}_{res} = 105.9$ is greater than 102 by 3.9, suggesting there may be some lack of fit of the predictions from this network meta-analysis model to the observed data points.

A visual representation of the contribution of individual studies to the residual deviance can be obtained by monitoring the node resdev and using either a caterpillar plot (see Section 2.2.3) or a box plot (Figure 3.3). The box plot can be obtained in the same way as the caterpillar plot in WinBUGS: **Inference→Compare….** In the '**node**' box type 'resdev', in the '**beg**' box type the first iteration post burn-in (in our example 20001), then click on '**boxplot**'. Various display options can be set by right-clicking on the graph and selecting **Properties…** (see **Help→User manual→WinBUGS Graphics** for details).

Inspection of Figure 3.3 shows that studies 44 and 45 in particular stand out as having a large contribution to the residual deviance. These studies are both two-arm studies comparing treatments Acc t-PA versus ASPAC and are in fact the only studies that make this comparison. One possible explanation for the observed lack of fit is that there is inconsistency between the indirect and direct evidence on Acc t-PA versus ASPAC. We will discuss this in more detail in Chapter 7.

3.3 Model Comparison

A model selection process systematically compares the fit of a set of potential models. For example, we can simplify a random effects model by reducing it to a fixed effects model (model contraction), or we could complicate it further by adding in covariates in a meta-regression (model expansion; see Chapter 8). Both the deviance and residual deviance statistics can be used to compare the fit of different models, as long as the data and distribution chosen for the likelihood are the same across all models fitted. In fact, because the two measures

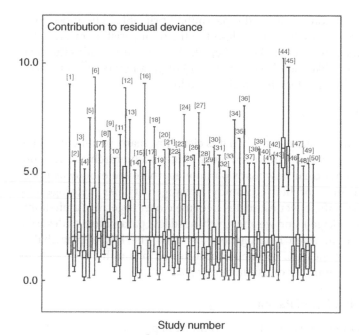

Figure 3.3 Full thrombolytics example: box plot of the contribution of each study to the residual deviance – from WinBUGS. The boxes represent the interquartile range, the line across the box represents the median, and the whiskers represent the 95% credible intervals. Numbers above the lines represent the study. The horizontal line indicates a contribution to residual deviance of 2, which is expected from a two-arm trial. For the 2 three-arm trials (studies 1 and 6), we would expect a contribution of 3.

differ only by a constant term, D_{sat}, which doesn't depend on the model fitted, when we look at the difference in either measure between two models, we get exactly the same result, that is,

$$D_{res1} - D_{res2} = \left(D_{model1} - D_{sat}\right) - \left(D_{model2} - D_{sat}\right) = D_{model1} - D_{model2} \quad (3.2)$$

Model fit isn't the only consideration to make when selecting a model – we may wish to take into consideration other contextual factors (e.g. certain covariates may be considered to be important to include). In general, the more parameters we include (i.e. the more complicated the model gets), the better the model fit will be. The model that gives the smallest deviance will be one with the same number of parameters as there are unconstrained data points, that is, the saturated model where each data point has its own prediction (akin to just considering each study separately and not performing any meta-analysis). However, such a model is of no use for prediction purposes and cannot be used in a decision analysis (unless making a decision for each study population separately). Instead, we would like to select a model that remains as simple as possible while still fitting well to the observed data. We wish to make a trade-off between model fit and model complexity – the parsimony principle. The most commonly used measure to resolve this trade-off in the frequentist literature is the Akaike information criterion (AIC) (Akaike, 1974), which simply adds the deviance at the maximum likelihood estimate (MLE) of parameters, θ^{MLE}, to twice the total number of parameters, p, in a given model:

$$AIC = D\left(\theta^{MLE}\right) + 2p$$

The AIC can easily be calculated when it is clear how many parameters there are (we can just count them). However, when there is a hierarchical (random effects) model, it is not clear how many parameters there are exactly as it will depend on the degree of between-study heterogeneity. As heterogeneity decreases, study-specific effect estimates become closer together (a phenomenon known as 'shrinkage'), and we effectively need less parameters to describe study differences than when heterogeneity is high. The DIC (Spiegelhalter et al., 2002) extends the AIC to handle non-nested hierarchical models by defining the effective number of parameters, p_D.

3.3.1 Effective Number of Parameters, p_D

For a fixed effects model, it is clear that the number of parameters is equal to the number of study-specific baselines (μ_i), which is equal to the number of studies included, plus $(S-1)$ basic parameters d_k, $k = 2,...,S$. In the full thrombolytic network, there are 50 studies and $S = 9$ treatments, giving 58 parameters in total. At the other extreme, we can fit an independent effects model where each study estimates a treatment effect parameter completely independent of the other studies. In this case the number of parameters is equal to the number of study-specific baselines (μ_i) plus the number of study-specific treatment effects (δ_{ik}). In the full thrombolytic network, there are 50 study-specific baselines and 52 study treatment effects (allowing for the 2 three-arm trials), giving 102 parameters in total (equal to the number of data points).

The fixed effects model is a special case of a random effects model when the between-study standard deviation, σ, equals zero (see also Chapter 2), whereas the independent effects model is equivalent to a random effects model as $\sigma \to \infty$. In practice, a random effects model will estimate a value of σ somewhere between 0 and ∞, and the number of parameters will lie somewhere between the minimum value (fixed effects model) and the maximum (independent effects model), depending on the extent of heterogeneity. In the full thrombolytic treatments network, if there is very little heterogeneity (i.e. σ close to 0), the number of parameters will be close to 58, and as the heterogeneity increases ($\sigma \to \infty$), the number of parameters increases towards an upper limit of 102. This is quantified by the effective number of parameters (Spiegelhalter et al., 2002), p_D, which is defined as

$$p_D = \overline{D}_{model} - D(\hat{\theta}) \tag{3.3}$$

The effective number of parameters is the posterior mean deviance for a given model, $\bar{D}_{model}$, minus the deviance calculated at some 'plug-in' estimate for the parameters, $\hat{\theta}$. For binomial data, the likelihood function depends on $\hat{r}_{ik} = n_{ik} p_{ik}$ (equation (3.1)), where p_{ik} is a function of model parameters θ. However, there are several possible options on parameter estimates to 'plug in', and different choices reflect a different 'focus' for model predictions (Lunn et al., 2013). The DIC tool in WinBUGS automatically uses the posterior mean of the parameters that are given prior distributions, to plug into the equation (3.3), that is, $\hat{\theta} = \bar{\theta}$. In the DIC tool output (Figure 3.2) $\bar{D}_{model}$ and $D(\hat{\theta})$ are labelled 'Dbar' and 'Dhat', respectively.

For a binomial likelihood, the posterior means of the parameters that are given prior distributions will be fed through the logistic regression model to obtain an estimate for p_{ik} to plug into the deviance formula (equation (3.1)). Note that this plug-in estimate will be different from the posterior mean for p_{ik} due to the non-linearity of the model. Unfortunately, in such cases where the relationship between the model predictions and the model parameters that are given prior distributions is highly non-linear and/or non-normal, using the posterior mean of the parameters that are given prior distributions can be problematic and can even cause p_D to be negative. In such cases (and arguably in general if the 'focus' of predictions is the observed data), it is more appropriate to use the posterior mean of the model predictions for the *observed data* as the plug-in and calculate p_D externally to WinBUGS (e.g. in R or Microsoft Excel). For binomial data, the plug-in would be the posterior mean of $\hat{r}_{ik} = n_{ik} p_{ik}$, which can be obtained in WinBUGS by monitoring the node rhat and then using the posterior means of the rhat's in the deviance formula (equation (3.1)) to obtain Dhat for use in equation (3.3). We illustrate these ideas in the worked example (Section 3.3.3).

3.3.2 Deviance Information Criterion (*DIC*)

The DIC (Spiegelhalter et al., 2002, 2007) is equal to the posterior mean deviance, $\bar{D}_{model}$, plus the effective number of parameters, p_D:

$$DIC = \bar{D}_{model} + p_D \tag{3.4}$$

Lower values of the DIC suggest a more parsimonious model. It has been suggested that differences in DIC over five are important (MRC Biostatistics Unit, 2015b), whereas if there are only small differences (<3) in DIC there is probably little to choose between two models – although one should still check robustness of conclusions to choice of model.

Either the deviance or the residual deviance can be used in equations (3.3) and (3.4). Although they lead to different numerical values for the DIC, they only differ by a constant (equation (3.2)), so that when comparing two different models the, difference in DIC will be the same whether the deviance or residual deviance is used.

3.3.2.1 *Leverage Plots

Another way to conceptualise p_D is as the sum of the leverage of each individual data point. Leverage is a statistical measure used to assess the influence that each data point has on the model parameters. The leverage for each data point, $leverage_{ik}$, is calculated as the posterior mean of the residual deviance, dev_{ik}, minus the deviance at the posterior mean of the parameters, that is, the contribution of each data point to equation (3.3), and as described in Section 3.3.1, it often makes most sense to use the posterior mean of the fitted values for the plug-in estimates in equation (3.3). For a binomial likelihood, letting $\tilde{r}_{ik}$ be the posterior mean of $\hat{r}_{ik}$ and $\overline{dev}_{ik}$ the posterior mean of dev_{ik},

$$p_D = \sum_i \sum_k leverage_{ik} = \sum_i \sum_k \left[\overline{dev}_{ik} - \widetilde{dev}_{ik} \right]$$

where $\widetilde{dev}_{ik}$ is the posterior mean of the deviance calculated by replacing $\hat{r}_{ik}$ with $\tilde{r}_{ik}$ in equation (3.1).

Leverage plots (Spiegelhalter et al., 2002) may be used to identify influential and/or poorly fitting observations and can be used to check how each point is affecting the overall model fit and DIC. Leverage plots show each data point's contribution to p_D ($leverage_{ik}$) plotted against its contribution to $\bar{D}_{res}$, $\overline{dev}_{ik}$. It is useful to display these summaries in a plot of $leverage_{ik}$ versus w_{ik} for each data point, where $w_{ik} = \pm\sqrt{\overline{dev}_{ik}}$, with sign given by the sign of $\left(\hat{r}_{ik} - r_{ik} \right)$ to indicate whether the data are over- or underestimated by the model. Curves of the form $x^2 + y = c$, $c = 1, 2, 3, ...$, where x represents w_{ik} and y represents the leverage, are marked on the plots, and points lying on such parabolas each contribute an amount c to the DIC (Spiegelhalter et al., 2002). Points that lie outside the line with $c = 3$ can generally be identified as contributing to the model's poor fit. Points with a high leverage are influential, which means that they have a strong influence on the model parameters that generate their fitted values.

3.3.3 Worked Example: Full Thrombolytic Treatments Network

We will use $\bar{D}_{res}$, $\bar{D}_{model}$, p_D, σ and DIC to compare the fit of fixed and random effects models for the full thrombolytic treatments network. These summary measures are obtained in exactly the same way as for the fixed effects model; we just have to be careful to place the code for the residual deviance within the right loops, as shown in the WinBUGS code as follows.

WinBUGS code for NMA: Binomial likelihood, logit link, random effects model – code for residual deviance added to the core random effects network meta-analysis model.
New code highlighted in bold. The full code with data and initial values is presented in the online file *Ch3_thromboRE_criticism.odc.*

```
# Binomial likelihood, logit link
# Fixed effects model
model{                                                  # *** PROGRAM STARTS
for(i in 1:ns){                                         # LOOP THROUGH STUDIES
    mu[i] ~ dnorm(0,.0001)                              # vague priors for all trial baselines
    for (k in 1:na[i]) {                                # LOOP THROUGH ARMS
        r[i,k] ~ dbin(p[i,k],n[i,k])                    # binomial likelihood
        logit(p[i,k]) <- mu[i] + d[t[i,k]] - d[t[i,1]]  # model for linear predictor
        rhat[i,k] <- p[i,k] * n[i,k]                    # expected value of the numerators
        dev[i,k] <- 2 * (r[i,k] * (log(r[i,k])-log(rhat[i,k]))  #Deviance contribution
           + (n[i,k]-r[i,k]) * (log(n[i,k]-r[i,k]) - log(n[i,k]-rhat[i,k])))
    }
    resdev[i] <- sum(dev[i,1:na[i]])                    # summed residual deviance contribution
                                                        #    for this trial
}
totresdev <- sum(resdev[])                              # Total Residual Deviance
d[1]<-0                                                 # treatment effect is zero for reference
                                                        #    treatment
for (k in 2:nt){  d[k] ~ dnorm(0,.0001) }               # vague priors for treatment effects
}                                                       # *** PROGRAM ENDS
```

Table 3.1 Full thrombolytics example: $\bar{D}_{res}$, $\bar{D}_{model}$, p_D, σ and *DIC* for both the fixed and random effects models.

Summary	Fixed effects model	Random effects model
Posterior mean residual deviances[a], $\bar{D}_{res}$	105.9	102.8
Posterior mean deviances, $\bar{D}_{model}$	535.5	532.4
Effective number of parameters, p_D	57.6	61.7
Between-study standard deviation, σ: posterior median (95% credible interval)		0.077 (0.01, 0.29)
Deviance information criteria, *DIC*	593.2	594.1

All results are based on 50,000 iterations on three chains after a burn-in of 20,000.
[a] Compare to 102 data points.

Table 3.1 shows $\bar{D}_{res}$, $\bar{D}_{model}$, p_D, σ and *DIC* for both the fixed and random effects models. The random effects model gives a posterior mean residual deviance of $\bar{D}_{res} = 102.8$, which is close to the number of unconstrained data points (102), suggesting that the model gives predictions that fit well to the data. This is very often the case in random effects models because the between-study standard deviation parameter can increase to account for heterogeneity between the results of different studies. Note that the difference between $\bar{D}_{res}$ for the random effects and fixed effects models is the same as the difference between $\bar{D}_{model}$ for the random effects and fixed effects models. This is always true. Here the difference is 3.1, suggesting a modest improvement in fit from using a random effects model. The effective number of parameters for the fixed effects model is approximately $p_D = 58$ as we expected (see Section 3.3.1), whereas for the random effects model, it is $p_D = 61.7$, which is much closer to 58 (fixed effects model) than to 102 (independent effects). This suggests that there is very little heterogeneity in this network of evidence, which can also be seen in the estimate of $\sigma = 0.08$, which is close to 0. The *DIC* is a sum of $\bar{D}_{model}$ and p_D and is very similar for both models (*DIC* = 593.2 for the fixed effects model and *DIC* = 594.1 for the random effects model). This suggests that although the random effects model fits better, this comes at a cost of added complexity (p_D), and if we balance both fit and complexity, there is little to choose between the two models. In such cases we typically use the simpler model, since it is easier to interpret, unless there is a good intuitive clinical rationale otherwise (e.g. heterogeneity may be expected, but insufficient studies are available to detect it). Figure 3.4 shows the estimated log odds ratios for each treatment relative to treatment 1 (SK) for both fixed effects (solid) and random effects (dashed) models. The estimated treatment effects are comparable for the two models, although uncertainty is wider in the estimates from

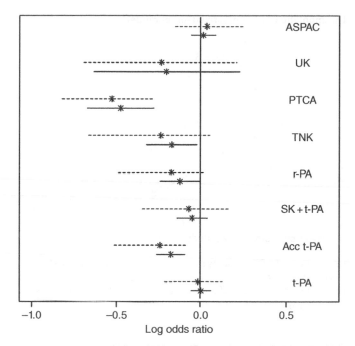

Figure 3.4 Full thrombolytics example: posterior mean (asterisks) and 95% credible intervals displayed for log odds ratios of each treatment compared to treatment 1 (SK) for both fixed effects (solid) and random effects (dashed) models. The vertical line indicates the line of no effect (log odds ratio = 0).

the random effects model (which is typically the case in random effects models due to uncertainty in the between-study variance parameter). The overall conclusion that PTCA (treatment 7) is the most effective treatment is robust to the choice of model. This robustness also supports our choice of the fixed effects model for these data.

The leverage plots for the fixed and random effects models are displayed in Figures 3.5 and 3.6, respectively. The Excel file to produce these plots can be found in the file *Ch3_leverageplots.xlsx*. Because the residuals from all fitted models comprised over- and underestimates, it is expected that the leverage plots are roughly symmetrical, as seen in Figures 3.5 and 3.6. For the fixed effects model, there are two points on the left and two points on the right that lie outside of the curve of DIC contribution equal to 3. These points correspond to each arm of trials 44 and 45, which we already identified as having a large contribution to the residual deviance (Figure 3.3). Although these points contribute a large amount to the residual deviance, they do not have an especially high contribution to p_D (leverage). The leverage plot for random effects model is very similar to that for the fixed effects model; however the points for

Thrombolytic treatments, fixed effect model

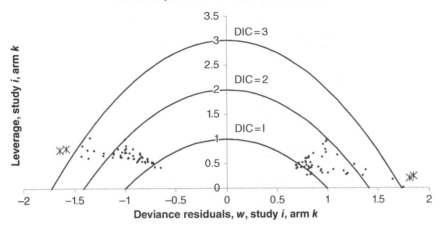

Figure 3.5 Full thrombolytics example, fixed effects model: plot of leverage, *leverage*$_{ik}$, versus posterior mean deviance residual, w_{ik}, for each data point, with curves of the form $x^2 + y = c$ representing a contribution to the DIC of $c = 1, 2, 3$ as indicated. Studies 44 and 45 are indicated with a star plotting symbol.

Thrombolytic treatments, random effects model

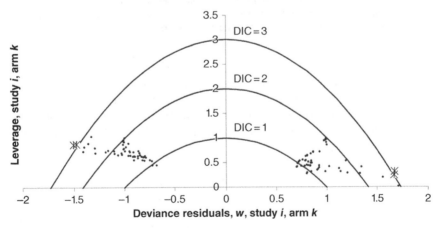

Figure 3.6 Full thrombolytics example, random effects model: plot of leverage, *leverage*$_{ik}$, versus posterior mean deviance residual, w_{ik}, for each data point, with curves of the form $x^2 + y = c$ representing a contribution to the DIC of $c = 1, 2, 3$ as indicated. Studies 44 and 45 are indicated with a star plotting symbol.

trials 44 and 45 have lower residual deviance and lie close to the DIC contribution of three curve.

WinBUGS will calculate p_D and the posterior mean of the deviance for the current model $\bar{D}$, but will not output the contributions of the individual data points to the calculations. Therefore users wishing to produce leverage plots such as those in Figures 3.5 and 3.6 need to calculate the contributions of individual studies to $\bar{D}_{res}$ and to the leverage themselves. The latter needs to be calculated outside WinBUGS, for example, in R or Microsoft Excel. The p_D and therefore the DIC, calculated in the way we suggest, are not precisely the same as that calculated in WinBUGS. This is because WinBUGS calculates the fit at the mean value of the parameter values, while we propose the fit at the mean value of the fitted values. The latter is more stable in highly non-linear models, and we recommend users to calculate it themselves externally to WinBUGS. Calculating p_D externally for the full thrombolytics example gives 56.7 and 60.4 for the fixed and random effects models, respectively, which are very similar to the values given by WinBUGS (Table 3.1). See also Section 3.3.1 and the worked example file: *Ch3_leverageplots.xlsx*.

3.4 Outlier Detection in Network Meta-Analysis

In this section, the focus is not on the overall level of variation in trial results, but on one or two specific trials that seem to have results that are particularly different from the others. Outlier detection is closely related to heterogeneity. A single outlying trial may impact greatly the measure of heterogeneity. Conversely, a high level of heterogeneity makes it difficult to detect a true outlier. The existence of a potential outlier may be detected through lack of fit identified using global model fit statistics (as described in Section 3.3), if a particular study makes a large contribution to the deviance. In this section we describe methods to further explore whether individual study results are 'outliers'. We begin by briefly summarising methods for outlier detection in pairwise meta-analysis before going on to extend to network meta-analysis using a worked example of treatments to reduce the incidence of febrile neutropenia, an adverse event during chemotherapy (Madan et al., 2011).

3.4.1 Outlier Detection in Pairwise Meta-Analysis

A much-analysed meta-analysis of magnesium versus placebo for patients with acute myocardial infarction is a good example to illustrate the main ideas of outlier detection. The key feature of this meta-analysis is that prior to 1995 there were a lot of small trials that showed magnesium to be effective in preventing mortality, leading to a publication entitled 'An effective, safe, simple, and inexpensive intervention' (Yusuf et al., 1993). However, in 1995 results from the Fourth International Study of Infarct Survival (ISIS-4) mega-trial were published that, based on very large numbers, showed no effect of

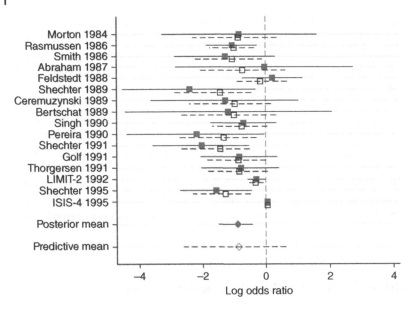

Figure 3.7 Magnesium example: crude log odds ratios with 95% CI (filled squares, solid lines); posterior mean with 95% CrI of the trial-specific log odds ratios, 'shrunken' estimates (open squares, dashed lines); posterior mean with 95% CrI of the posterior (filled diamond, solid line) and predictive distribution (open diamond, dashed line) of the pooled treatment effect, obtained from a random effects model including all the trials.

magnesium (ISIS-4 Collaborative Group, 1995). Figure 3.7 shows a forest plot with the crude log odds ratios calculated from the data and the 'shrunken' estimates from a random effects model (i.e. the trial-specific treatment effects, assumed to be exchangeable) (Sterne et al., 2001).

The first point to note from Figure 3.7 is that the ISIS-4 study gives a relative treatment effect estimate that is somewhat different from the estimates from the other trials. In particular neither the crude 95% confidence interval (CI) nor the 'shrunken' 95% CrI for this trial overlap with the 95% CrI for the mean treatment effect (Figure 3.7). Investigators might wonder whether this trial is an 'outlier' in some sense. The second point to note is that there is a substantial heterogeneity, with the posterior median of the between-study standard deviation of 0.68 with 95% CrI (0.35, 1.30), which is comparable in size with the mean treatment effect of −0.89 with 95% CrI (−1.49, −0.41) on the log odds ratio scale. This high degree of heterogeneity can also be seen by the very wide 95% predictive interval for the treatment effect (labelled 'predictive mean' in Figure 3.7). The predictive distribution can be interpreted as the treatment effect we would expect to see in a 'new' (or different) trial population that is exchangeable with the trials included in the meta-analysis. The predictive distribution reflects two different sources of uncertainty: uncertainty in the random effects mean and variance parameters and uncertainty as to where in the random effects distribution the 'new' population will lie. In a pairwise meta-analysis, the predictive distribution is obtained from WinBUGS by monitoring δ_{new}:

$$\delta_{new} \sim \text{Normal}\left(d_{12}, \sigma^2\right) \tag{3.5}$$

where d and σ are drawn from their posterior distributions.

The appropriate tool for examination of single trials in a meta-analysis is predictive cross-validation (DuMouchel, 1996; Marshall and Spiegelhalter, 2003) based on a 'leave one out' approach. The procedure is to remove the trial from the synthesis and compare the observed treatment effect with the treatment effect that we would predict for the omitted trial *based on an analysis of the remaining trials*. For the magnesium meta-analysis, this is achieved by the following steps (full details can be found in (Dias et al., 2011a; Welton et al., 2012)):

Step 1: Fit random effects model excluding trial under investigation (i.e. ISIS-4).
Step 2: Obtain a predicted relative effect in the 'new' trial population (i.e. ISIS-4).
Step 3: Obtain a predicted absolute effect in the magnesium arm of ISIS-4 by combining an estimate of the absolute effect in the placebo arm with the predicted relative effect.
Step 4: Obtain a prediction for the number of events in the magnesium arm of ISIS-4 that accounts for sampling error using the predicted absolute effect for the magnesium arm of ISIS-4, the likelihood and the known sample size.
Step 5: Compare the observed number of events with the predicted number of events in the magnesium arm of ISIS-4 to obtain a Bayesian p-value that measures the probability that we predict results as or more extreme as those observed in ISIS-4.

The result is a p-value of 0.056, indicating that a trial with a result as extreme as ISIS-4 would be unlikely, but still possible, given our model for the remaining data. In examining these results, however, one must take into account the effective number of tests that could be undertaken. In carrying out cross-validation for ISIS-4, we have picked the most extreme of 16 trials, so there is an implication that $n = 16$ tests could be performed and the test on ISIS-4 would give the most extreme result (i.e. have the smallest p-value). To correctly interpret the significance of the observed p-value, we need to compare it with its expected value, which is $1/(n+1) = 0.059$, the value of the nth uniform order statistic. The observed p-value therefore suggests that ISIS-4 is not necessarily incompatible with a random effects model fitted to the remaining data. This can also be seen in Figure 3.8, which presents the 'shrunken' estimates ($\delta_{i2}, i = 1, ..., 15$), random

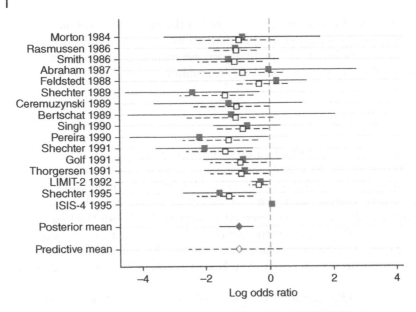

Figure 3.8 Magnesium example: crude log odds ratios with 95% CI (filled squares, solid lines); posterior mean with 95% CrI of the trial-specific log odds ratios, 'shrunken' estimates (open squares, dashed lines); posterior mean with 95% CrI of the posterior (filled diamond, solid line) and predictive distribution (open diamond, dashed line) of the pooled treatment effect, obtained from a random effects model excluding the ISIS-4 trial.

effects mean and predictive distribution mean treatment effect for a random effects meta-analysis *excluding* the ISIS-4 trial, together with the observed log odds ratio and CI for ISIS-4 for comparison. It can be seen that the observed log odds ratio from the ISIS-4 trial although well outside the CrI for the posterior mean still lies within the bounds of the CrI for the predictive mean treatment effect, which is the basis for predictive cross-validation.

This is a statistical result only: it is impossible to deduce whether ISIS-4 is a deviant result or whether the other trials are. This particular meta-analysis has been discussed repeatedly (Egger and Davey-Smith, 1995; Higgins and Spiegelhalter, 2002), and current opinion is that ISIS-4 is in fact the 'correct' result (Li et al., 2007).

The magnesium dataset holds several important messages about random effects models in decision making. First, we should note that the random effects models both with and without the ISIS-4 trial give a good fit to the observed data. This is because a random effects model can generally fit any random distribution of effects, simply by increasing the between-study standard deviation parameter estimate. Second, it illustrates the weakness in basing inference on the random effects mean. Within the entire ensemble of trials, whether including or even excluding itself, ISIS-4 is not particularly remarkable. It is, however, markedly different from the random effects mean. The presence of a random effect acknowledges variation between results that cannot be explained, and making a decision on the mean of such a distribution implies basing decisions on a parameter without any real-world definition. In the ISIS-4 example, the random effects mean is a summary from a model *in which the different sources of evidence are in an unexplained conflict.* A decision based on the predictive distribution, on the other hand, is compatible with all the data.

3.4.2 Predictive Cross-Validation for Network Meta-Analysis

We now show how predictive cross-validation can be extended to network meta-analysis. We illustrate the method using a synthesis of evidence on three treatments to reduce the incidence of febrile neutropenia, an adverse event during chemotherapy (Madan et al., 2011). We will take 'No Treatment', coded 1, as the reference for the analysis. The three treatments of interest – filgrastim, pegfilgrastim and lenograstim – are coded 2–4. All treatments are given to patients undergoing chemotherapy. Table 3.2 shows the number of patients with febrile neutropenia, r_{lk}, out of all included patients, n_{lk}, and the treatments compared, t_{lk}, in each arm of the included trials ($i = 1, ..., 25$; $k = 1, 2$). The network diagram is presented in Figure 3.9.

Figure 3.10a shows a forest plot with the crude log odds ratios calculated from the data and the 'shrunken' estimates (i.e. the trial-specific treatment effects, assumed to be exchangeable) for the trials comparing treatments 1 and 3, along with the posterior and predictive effects of treatment 1 compared with 3, from a random effects model including all the trials in Table 3.2. Although the random effects network meta-analysis fits the data well (posterior mean of the residual deviance is 49.6, compared with 50 data points), trial 25 has an estimated trial-specific log odds ratio that is somewhat different from the other trials and may be contributing to the high estimated heterogeneity in this network (posterior median of $\sigma = 0.42$ with 95% CrI (0.20, 0.73)). Is this study an outlier?

As for pairwise meta-analysis the appropriate tool for examination of single trials is predictive cross-validation (DuMouchel, 1996; Marshall and Spiegelhalter, 2003). The procedure is to remove the trial from the network meta-analysis and compare the observed treatment effect to the predicted effect that we would expect *based on an analysis of the remaining trials.* Predictive cross-validation requires a series of steps described using the febrile neutropenia example to illustrate:

Step 1: Fit random effects model excluding trial under investigation.
The first step in predictive cross-validation is to fit the random effects meta-analysis model excluding the trial under investigation (trial 25 in febrile neutropenia example). It is convenient if the trial under investigation is the highest numbered in the dataset, as it can easily be excluded by changing the number

Table 3.2 Adverse events in chemotherapy example: number of adverse events r_{ik}, out of the total number of patients receiving chemotherapy n_{ik}, in arms 1 and 2 of 25 trials for the four treatments t_{ik}. Data from Madan et al. (2011).

	Treatments		Number of events		Number of patients	
	Arm 1	Arm 2	Arm 1	Arm 2	Arm 1	Arm 2
Study ID	t_{i1}	t_{i2}	r_{i1}	r_{i2}	n_{i1}	n_{i2}
1	2	3	15	10	75	77
2	2	3	27	14	147	149
3	2	3	2	5	25	46
4	2	3	6	6	31	29
5	2	3	1	0	13	14
6	1	2	26	34	72	276
7	1	2	17	9	39	41
8	1	2	15	4	72	77
9	1	2	86	72	192	197
10	1	2	52	34	104	101
11	1	2	62	40	125	125
12	1	2	27	16	85	90
13	1	2	80	38	104	95
14	1	2	34	17	64	65
15	1	2	38	25	130	129
16	1	4	18	5	28	23
17	1	4	42	36	59	61
18	1	4	15	5	26	22
19	1	4	62	52	80	82
20	1	4	14	5	43	43
21	1	3	27	11	73	73
22	1	3	34	14	343	343
23	1	3	5	4	29	30
24	1	3	10	3	118	123
25	1	3	78	6	465	463

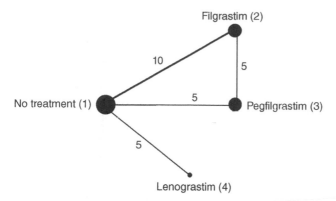

Figure 3.9 Adverse events in chemotherapy example: treatment network. Lines connecting two treatments indicate that a comparison between these treatments has been made. The width of the lines is proportional to the number of studies providing evidence on that comparison (also marked on the lines), the size of the circles is proportional to the number of patients randomised to that treatment, and the numbers by the treatment names are the treatment codes used in the modelling.

of studies to loop over from ns=25 to ns=24. The only other change required is to reduce the number of initial values for mu by 1.

Figure 3.10b shows the same results as in Figure 3.10a, but from the random effects network meta-analysis model omitting trial 25. It can be seen that the 95% CI for the crude log odds ratio for trial 25 does not overlap with the mean log odds ratio estimated from the network meta-analysis excluding trial 25 (labelled 'mean effect'). This suggests that trial 25 may be an outlier, but is this a real result, or just due to chance?

Step 2: Obtain predicted treatment effects in the 'new' trial population.
The next step is to make a prediction for the treatment effect, δ_{new}, we would expect to see in the excluded trial using the predictive distribution based on the other trials. The predictive distribution is obtained by making random draws from the random effects distribution. In pairwise meta-analysis this is obtained using equation (3.5). However, in network meta-analysis, the predictive distribution needs to be multivariate normal to account for the fact that we are making predictions for multiple treatments. So, for a network with S treatments, the predictive distribution for the $S-1$ treatment effects relative to treatment 1 (the basic parameters; see Section 2.2.2) is given by

$$\delta^{new} = \begin{pmatrix} \delta_{12}^{new} \\ \vdots \\ \delta_{1S}^{new} \end{pmatrix} \sim \text{Normal}_{S-1} \left(\begin{pmatrix} d_{12} \\ \vdots \\ d_{1S} \end{pmatrix}, \begin{pmatrix} \sigma^2 & \sigma^2/2 & \cdots & \sigma^2/2 \\ \vdots & \vdots & \ddots & \vdots \\ \sigma^2/2 & \sigma^2/2 & \cdots & \sigma^2 \end{pmatrix} \right) \quad (3.6)$$

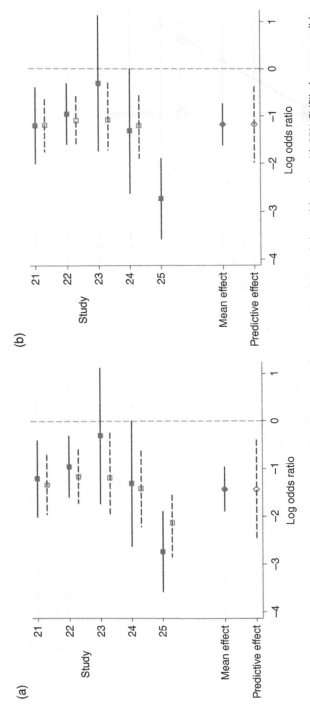

Figure 3.10 Adverse events in chemotherapy example – comparison of treatment 1 and 3: crude log odds ratios with 95% CI (filled squares, solid lines); posterior mean with 95% CrI of the trial-specific log odds ratios, 'shrunken' estimates (open squares, dashed lines); posterior mean with 95% CrI of the posterior (filled diamond, solid line) and predictive distribution (open diamond, dashed line) of the pooled treatment effect for a random effects model (a) including all the trials and (b) excluding trial 25 (cross-validation model).

where d and σ are sampled from the posterior distributions (given the data). Equation (3.6) can be rewritten as a series of conditional univariate normal distributions (Raiffa and Schlaifer, 1967):

$$\delta_{1k}^{new} \left| \begin{pmatrix} \delta_{12}^{new} \\ \vdots \\ \delta_{1(k-1)}^{new} \end{pmatrix} \sim \text{Normal} \left(d_{1k} + \frac{1}{(k-1)} \sum_{j=1}^{k-1} \left(\delta_{1j}^{new} - d_{1j} \right), \frac{k}{2(k-1)} \sigma^2 \right) \quad (3.7)$$

Equation (3.7) is more convenient for coding in WinBUGS.

Note that predictive distributions for any pair of treatments can be obtained from the consistency equations (Section 2.2.2):

$$\delta_{XY}^{new} = \delta_{1Y}^{new} - \delta_{1X}^{new}$$

thus ensuring that the correlations between the predictive treatment effects are carried through correctly to all treatment contrasts.

Step 3: Obtain predicted absolute effects for all treatments by combining an esti-
mate of the absolute effect in the reference arm in the trial under investigation
with the predicted relative effects (step 2).

For a binomial likelihood the absolute effect on the reference arm is the logit of the probability of mortality on arm 1 of the excluded trial, p_1^{new}, which could be estimated from the observed proportion of events on arm 1 of the excluded trial. In the febrile neutropenia example, this would give $p_1^{new} = 78/465 = 0.168$. However, this does not convey our uncertainty about this probability. Instead we can assume that the probability of an event in arm 1 of a study like trial 25 has a beta distribution:

$$p_1^{new} \sim \text{Beta}(a, b) \quad (3.8)$$

where $a = r_{25,1} = 78$, the number of events in arm 1 of trial 25, and $b = n_{25,1} - r_{25,1} = 387$, the number of non-events in arm 1 of trial 25. This distribution is shown in Figure 3.11, together with the observed proportion from trial 25.

The predictive absolute probability of an event on arm k of a study like trial 25, p_k^{new}, can then be obtained by combining the predicted absolute probability of event on arm 1 with the relevant predicted relative effects:

$$\text{logit}\left(p_k^{new} \right) = \text{logit}\left(p_1^{new} \right) + \delta_{1,t_{25,k}}^{new} - \delta_{1,t_{25,1}}^{new}$$

where the indexes for δ^{new} ensure that the correct treatment comparison for arm k versus arm 1 of trial 25 is made.

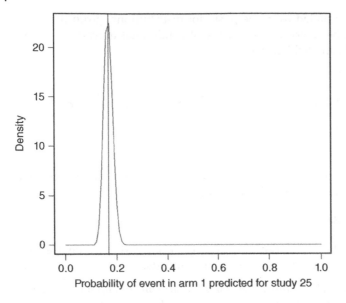

Figure 3.11 Adverse events in chemotherapy example: beta distribution representing the probability of an event in arm 1 predicted for a study like trial 25, equation (3.8). The solid line indicates the observed proportion of events in arm 1 of trial 25, $78/465 = 0.168$.

Step 4: Obtain a prediction for the number of events in non-reference arms of the trial under investigation using the predicted absolute effects of treatments (step 3) and the known sample size.

The predictive absolute effects for treatment k, p_k^{new}, are predictions from an *infinitely sized* new study. However, the omitted study has a (known) finite sample size on each arm, and an assumed likelihood, so there will be an additional element of uncertainty in the observed data from the trial that is due to sampling error. For binomial data, the predicted number of events, r_k^{new}, on arm k of a future trial of the same size and design as trial 25 ($k = 2$ in this case) is

$$r_k^{new} \sim \text{Binomial}\left(p_k^{new}, n_{25,k}\right)$$

Step 5: Compare predictions with observed data.

The predicted number of events, r_k^{new}, can be compared with the observed number of events on arm k of trial 25 to obtain a Bayesian p-value: the probability of obtaining a value as extreme as that observed in trial 25, that is, $\Pr(r_k^{new} > r_{25,k})$. Within a Bayesian MCMC framework, this is done by setting up a variable that, at each iteration, takes the value 1 if $r_k^{new} > r_{25,k}$ and is 0 otherwise. By averaging over a large number of iterations, this variable gives the desired probability.

Care needs to be taken in the interpretation of these Bayesian p-values. We have chosen to perform cross-validation on this particular trial because it looked like a possible outlier from the results. There are 25 trials that could have been excluded and therefore 25 corresponding Bayesian p-values that could have been computed. To correctly interpret the significance of the observed p-value, we need to compare it with its expected value, which is $1/(M+1)$, the value of the Mth uniform order statistic, where M is the number of trials.

The result from the predictive cross-validation is a p-value of 0.004, indicating that a trial with results as extreme as trial 25 would be very unlikely, given our model for the remaining data (convergence was achieved after 60,000 burn-in iterations and results are based on 100,000 samples from three independent chains). Taking into account that we could have conducted $M = 25$ such tests, we compare the observed p-value with its expected value, which is $1/(M+1) = 0.038$, the value of the Mth uniform order statistic. The observed p-value is substantially less than this, indicating that trial 25 may be an 'outlier'. This can also be seen in Figure 3.10b, which presents the 'shrunken' estimates mean and predictive treatment effects for the trials comparing treatments 1 and 3, along with the posterior and predictive effects of treatment 1 compared with 3, from a random effects model excluding trial 25 (but including the observed log odds ratio and CI for this trial). The 95% CI for the observed log odds ratio from trial 25 is $(-3.57, -1.89)$, which is well outside the 95% CrI for the posterior mean $(-1.61, -0.74)$ and only marginally within the bounds of the 95% CrI for the predictive mean treatment effect $(-1.98, -0.38)$, which is the basis for predictive cross-validation. The posterior median for the between-trial heterogeneity for the random effects network meta-analysis excluding trial 25 is 0.29 with 95% CrI $(0.05, 0.58)$, smaller than for the model with the full data.

This is a statistical result only: it is impossible to deduce whether trial 25 is a deviant result or whether the other trials are, based on the statistical analysis alone. In such circumstances we would advise checking the trial design and patient population in all included trials for potential reasons why a study has been identified as an outlier.

3.4.3 Note on Multi-Arm Trials

When there are multi-arm trials in a network, there are $(S - 1)$ study-specific relative effect parameters to estimate, and the predicted outcomes in this study will be a multivariate prediction, r^{new}. It is not strictly correct to compute cross-validation p-values for each arm independently, as this does not allow for the fact that the outcomes will be correlated. For a multi-arm study to be considered an 'outlier', the most straightforward approach is to compute the probability that *at least one* of the predicted outcomes is greater than the observed outcome to obtain a cross-validation p-value that accounts for within-study correlations. So, for a three-arm trial of treatments t_1 versus t_2 versus t_3, we compute $\Pr\left(r_{t_2}^{new} > r_{i,2} \text{ or } r_{t_3}^{new} > r_{i,3}\right)$.

3.4.4 WinBUGS Code: Predictive Cross-Validation for Network Meta-Analysis

The WinBUGS code for predictive cross-validation in a network meta-analysis is given as follows, highlighted in bold. Note that this code is completely general for data with a binomial likelihood and can be used for predictive cross-validation in networks with or without multi-arm trials and in pairwise meta-analysis.

WinBUGS Code for predictive cross-validation in NMA with multi-arm trials: Binomial likelihood, logit link, random effects model. The full code with data and initial values is presented in the online file *Ch3_Crossval.odc*.

```
# Binomial likelihood, logit link, network meta-analysis (multi-arm trials)
# Random effects model with Predictive Cross-validation
model{
for(i in 1:ns){                                                      # *** PROGRAM STARTS
                                                                     # LOOP THROUGH STUDIES
    w[i,1] <- 0                                                      # adjustment for multi-arm trials is zero for control arm
    delta[i,1] <- 0                                                  # treatment effect is zero for control arm
    mu[i] ~ dnorm(0,.0001)                                           # vague priors for all trial baselines
    for (k in 1:na[i]) {                                             # LOOP THROUGH ARMS
        r[i,k] ~ dbin(p[i,k],n[i,k])                                 # binomial likelihood
        logit(p[i,k]) <- mu[i] + delta[i,k]                          # model for linear predictor
        rhat[i,k] <- p[i,k] * n[i,k]                                 # expected value of the numerators
        dev[i,k] <- 2 * (r[i,k] * (log(r[i,k])-log(rhat[i,k]))       # Deviance contribution
          + (n[i,k]-r[i,k]) * (log(n[i,k]-r[i,k]) - log(n[i,k]-rhat[i,k])))
    }
    resdev[i] <- sum(dev[i,1:na[i]])                                 # summed residual deviance contribution for this trial
    for (k in 2:na[i]) {                                             # LOOP THROUGH ARMS
        delta[i,k] ~ dnorm(md[i,k],taud[i,k])                        # trial-specific LOR distributions
        md[i,k] <-  d[t[i,k]] - d[t[i,1]] + sw[i,k]                  # mean of LOR distributions (with multi-arm trial
                                                                     # correction)
        taud[i,k] <- tau *2*(k-1)/k                                  # precision of LOR distributions (with multi-
                                                                     # arm trial correction)
        w[i,k] <- (delta[i,k] - d[t[i,k]] + d[t[i,1]])               # adjustment for multi-arm RCTs
        sw[i,k] <- sum(w[i,1:k-1])/(k-1)                             # cumulative adjustment for multi-arm trials
    }
}
```

```
totresdev <- sum(resdev[])                                        # Total Residual Deviance
d[1]<-0                                                           # treatment effect is zero for reference treatment
for (k in 2:nt){ d[k] ~ dnorm(0,.0001) }                         # vague priors for treatment effects
sd ~ dunif(0,5)                                                   # vague prior for between-trial SD
tau <- pow(sd,-2)                                                 # between-trial precision = (1/between-trial
                                                                  #   variance)

# predictive distribution for future trial is multivariate normal
delta.new[1] <- 0                                                 # treatment effect is zero for reference treatment
w.new[1] <- 0                                                     # adjustment for conditional mean is zero for ref.
                                                                  #   treat.

for (k in 2:nt) {                                                 # LOOP THROUGH TREATMENTS
    delta.new[k] ~ dnorm(m.new[k],tau.new[k])                    # conditional distribution of each delta.new
    m.new[k] <- d[k] + sw.new[k]                                  # conditional mean of delta.new
    tau.new[k] <- tau *2*(k-1)/k                                  # conditional precision of delta.new
    w.new[k] <- delta.new[k] - d[k]                               # adjustment for conditional mean
    sw.new[k] <- sum(w.new[1:k-1])/(k-1)                          # cumulative adjustment for cond. mean
}

p.base ~ dbeta(a,b)                                               # draw baseline (control group) effect
a <- r[ns+1,1]                                                    # no. of events in control group
b <- n[ns+1,1]-r[ns+1,1]                                          # no of non-events in control group
for (k in 2:na[ns+1]) {                                           # LOOP THROUGH ARMS
# predictive prob of event for each treatment arm of the new trial
    logit(p.new[k]) <- logit(p.base) + (delta.new[t[ns+1,k]] - delta.new[t[ns+1,1]])
# draw predicted number of events for each arm of the new trial
    r.new[k] ~ dbin(p.new[k], n[ns+1,k])
# Bayesian p-value: probability of obtaining a value as extreme as the
# value observed (r[ns+1,2]), given the model and the remaining data
# extreme value "smaller"
    p.cross[k] <- step(r[ns+1,2] - r.new[k]) - 0.5*equals(r.new[k],r[ns+1,2])
}
}                                                                 # *** PROGRAM ENDS
```

When we are considering a multi-arm trial to be a possible outlier, the following code can be added before the closing brace to find the probability of predicting a value as extreme as the value observed on *at least one of* arms 2, 3, and so on

```
s <- na[ns+1]-1
p.cross.multi <- ranked(p.cross[2:na[ns+1]],s)
```

The following code can be added before the closing brace to predict all pairwise log odds ratios and odds ratios in a new trial.

```
# pairwise ORs and LORs for all possible pairwise
# comparisons, if nt>2
for (c in 1:(nt-1)) {
    for (k in (c+1):nt)   {
       lor.new[c,k] <- delta.new[k]- delta.new[c]
       or.new[c,k] <- exp(lor.new[c,k])
       }
 }
```

3.5   Summary and Further Reading

It is important to check model fit and select adequately fitting models, so that any judgements made as to the most effective or cost-effective intervention are a fair reflection of the available evidence. The posterior mean of the residual deviance can be used to assess model fit (smaller values indicate a better fit, and we would expect a value close to the number of unconstrained data points in an adequately fitting model). The posterior mean deviances and the DIC can be used to compare models (smaller values indicate a better fitting/more parsimonious model, respectively). Leverage plots can indicate individual data points that have a poor fit and/or exert a high influence on the fitted model parameters. Predictive cross-validation can be used to explore whether individual studies are outliers; however this can be computationally expensive if there are many studies to compute cross-validation *p*-values for.

Model choice should be based on contextual factors as well as on statistical measures of model fit and adequacy. For example, if inclusion criteria are strict, then fixed effects models are plausible; however where populations and interventions are heterogeneous, we may not consider a fixed effects model plausible, even if model fit statistics indicate it to be the most parsimonious model. This is especially the case in situations where the evidence available to inform a network meta-analysis is sparse, so that there is very little power to detect heterogeneity if it exists. In such cases, informative prior distributions may be used on the between-study variance parameter (Turner et al., 2012, 2015b) as discussed in Section 2.3.2.

We have assumed a binomial likelihood in this chapter; however all methods can be extended to other types of outcome measures. The only change required in the WinBUGS code is the likelihood, the scale on which relative effects are modelled (link), and any measures of fit and prediction that rely on the likelihood. The prior distribution for the between-study standard deviation parameter needs to be sufficiently wide, given the scale of the relative effects. We give details and examples in Chapter 4.

The books by Gelman et al. (2004), Spiegelhalter et al. (2004) and Lunn et al. (2013) are excellent texts on Bayesian model criticism and model selection. For further information specifically on the DIC and p_D measures, including frequently asked questions, see the DIC web page (MRC Biostatistics Unit, 2015b) and also a retrospective reflection on the DIC (Spiegelhalter et al., 2014).

Predictive cross-validation (DuMouchel, 1996; Stern and Cressie, 2000) can be computationally expensive when there are many studies to remove, one at a time. Approximate cross-validation measures have been proposed that can be computed using the full dataset, including posterior predictive p-values (Stern and Cressie, 2000; Gelman et al., 2004) and mixed predictive p-values (Marshall and Spiegelhalter, 2003), although these tend to be conservative. Examples of predictive cross-validation and mixed predictive checks applied to pairwise meta-analysis and to meta-analysis of single arms (institutional rankings) are given in Welton et al. (2012) and Lunn et al. (2013), respectively. If predictive cross-validation (or mixed prediction) is performed for every trial in the dataset, then the resulting p-values can be ordered and plotted against the relevant uniform order statistics to gauge whether any of the p-values are unusually small or large, allowing for repeated tests (see Welton et al. (2012) for an illustration).

If all (or most) of the trials providing information on a particular comparison are outliers compared with the rest of the evidence, then the direct evidence on that particular comparison is said to be *inconsistent* with the remaining evidence. We describe methods for the detection of inconsistency in Chapter 7.

3.6 Exercises

3.1 For the febrile neutropenia example, fit fixed and random effects models, to the complete dataset (i.e. all 25 trials). The WinBUGS code can be found in the file *Ch3_Ex1_febneutro.odc*. Complete the following table:

Summary	Fixed effects model	Random effects model
Posterior mean residual deviance, $\bar{D}_{res}$		
Posterior mean deviance, $\bar{D}_{model}$		
Effective number of parameters, p_D		
Deviance information criteria, DIC		
Between-study standard deviation, σ: posterior mean (95% credible interval)		
Log odds ratio for filgrastim vs no treatment: posterior mean (95% credible interval)		
Log odds ratio for pegfilgrastim vs no treatment: posterior mean (95% credible interval)		
Log odds ratio for lenograstim vs no treatment: posterior mean (95% credible interval)		

(a) Comment on the overall fit of the models to the observed data. Which model do you prefer, and why?

(b) *Use a leverage plot to explore the fit and influence of individual data points.

3.2 For the full thrombolytic treatments example, fit a random effects model and use predictive cross-validation for study 45. Is there evidence that study 45 is an outlier? You can use the code provided for the febrile neutropenia example (*Ch3_Crossval.odc*), but replace the data with the thrombolysis data.

Hints: Reorder the studies to make study 45 the last study in the dataset. Think about how many studies to loop over, ns. Think about how many initial values are required.

3.3 In the thrombolysis example, study 1 is a three-arm study comparing treatments 1 versus 3 versus 4, and it is the only study making these comparisons. Fit a random effects model and use predictive cross-validation for study 1 (exactly as for Exercise 3.2, but omitting study 1 instead of study 45). Is there evidence that the comparisons in study 1 are outliers?

3.4 *Using the same model and data as for Exercise 3.2, add in code to make predictions for the number of events in arm 2 of study 45 for varying sample sizes for arm 2: $N = 50$, 100, 150, 200 and 250. Compute the predicted odds ratio (assuming arm 1 remains the same: 3/138). So, if r.new.N holds the predicted number of events for N on arm 2, the odds ratio is (r.new.N/(N − r.new.N))/(3/135). How does this change as the sample size on arm 2, N, increases?

4

Generalised Linear Models

4.1   A Unified Framework for Evidence Synthesis

Chapter 2 describes fixed and random effects models that can be used for pairwise and network meta-analysis based on aggregate data from randomised controlled trials (RCTs) reporting a binary outcome. The models are formulated within the familiar framework of generalised linear models (GLM) (McCullagh and Nelder, 1989) and applied to data with a specific likelihood (binomial) that are pooled on the log odds ratio scale using a logit link function. Chapter 3 describes methods for model critique and comparison for the same type of data.

In order to cover the variety of outcomes reported in trials and the range of data transformations required to achieve approximate linearity, we now extend this framework to allow synthesis of data with normal, binomial, Poisson and multinomial likelihoods, with identity, logit, log, complementary log–log, and probit link functions, based on the common core fixed or random effects models for the linear predictor.

We describe a general modular approach where different likelihoods and link functions may be employed, but the synthesis (or network meta-analysis model), which occurs at the level of the linear predictor, takes the exact same form in every case. Thus we present a single model for pairwise meta-analysis, indirect comparisons and network meta-analysis (mixed treatment comparisons) with or without multi-arm trials for any combination of likelihood and (appropriate) link function (McCullagh and Nelder, 1989), implemented in a Bayesian MCMC context and supported by WinBUGS code.

The common GLM framework can, of course, be applied in either frequentist (McCullagh and Nelder, 1989) or Bayesian contexts (Ntzoufras, 2009). However, the use of WinBUGS and MCMC dovetails with the GLM approach, as it allows us to take full advantage of the modularity implied by GLMs and implicit in the WinBUGS code provided. Thus we are able to present a unified

Network Meta-Analysis for Decision-Making, First Edition. Sofia Dias, A. E. Ades, Nicky J. Welton, Jeroen P. Jansen and Alexander J. Sutton.
© 2018 John Wiley & Sons Ltd. Published 2018 by John Wiley & Sons Ltd.
Companion website: www.wiley.com/go/dias/networkmeta-analysis

approach to fitting fixed and random effects models for pairwise or network meta-analysis as well as critiquing each model and choosing between them (Chapter 3). In later chapters we extend this framework to inconsistency checking (Chapter 7), meta-regression (Chapter 8) and bias adjustment (Chapter 9), as well as more complex data structures (Chapters 10 and 11).

The modular nature of WinBUGS and GLM also allows different trial reporting formats to be accommodated within the same synthesis with an appropriate likelihood through 'shared parameter models', thus allowing all data informing the relative treatment effects of interest to be synthesised in a single model (Section 4.6).

4.2 The Generic Network Meta-Analysis Models

We now extend the approaches in Chapters 2 and 3 to models with an arbitrary link function for data with a generic likelihood. The essential idea is that the basic apparatus of the meta-analysis and model checking remains the same, but the *likelihood* and the *link function* can change to reflect the nature of the data (continuous, rate, categorical, etc.) and the sampling process that generated it (normal, Poisson, multinomial, etc.). In GLM theory (McCullagh and Nelder, 1989), a likelihood is defined in terms of some unknown parameters γ, while a link function, $g(\cdot)$, maps these parameters onto the plus/minus infinity range, where the treatment effects can be assumed to be additive. Our meta-analysis model is now written as a GLM taking the form

$$g(\gamma) = \theta_{ik} = \mu_i + \delta_{ik} \tag{4.1}$$

where g is an appropriate link function (e.g. the logit link; Chapter 2) and θ_{ik} is the linear predictor, usually a continuous measure of the treatment effect in arm k of trial i (e.g. the log odds). This linear predictor is simply a regression model with $S-1$ treatment effect parameters for any S-treatment network.

As before, μ_i are the trial-specific effects of the treatment in arm 1 of trial i, treated as unrelated nuisance parameters, and the δ_{ik} are the trial-specific treatment effects of the treatment in arm k relative to the treatment in arm 1 in that trial, where $\delta_{i1} = 0$ and for $k > 1$

$$\delta_{ik} \sim \text{Normal}\left(d_{t_{i1},t_{ik}}, \sigma^2\right) \tag{4.2}$$

with $d_{t_{i1}t_{ik}} = d_{1,t_{ik}} - d_{1,t_{i1}}$ defining the consistency equations as in Chapter 2.
The fixed effects model is obtained by replacing equation (4.1) with

$$\theta_{ik} = \mu_i + d_{t_{i1}t_{ik}} = \mu_i + d_{1,t_{ik}} - d_{1,t_{i1}} \tag{4.3}$$

Readers will note that the model for the linear predictor given in equations (4.1) and (4.3) is exactly the same as the model defined in Section 2.2. Therefore, the correlation in random effects from multi-arm trials is accounted for in the same way (Section 2.2.2). Prior distributions are also defined in the same way, but we need to pay careful attention to the specification of the prior distribution for the between-trial standard deviation parameter to ensure it is appropriately wide for the scale being used (see Section 2.3.2).

The implementation of the generic network meta-analysis model with fixed treatment effects follows the following structure, where the likelihood, link function and residual deviance formula would be replaced with appropriate code to reflect the model being used.

Generic WinBUGS code for NMA: fixed effect model. The likelihood, link function and residual deviance formula need to be specified in each case.

```
# Fixed effect model for multi-arm trials
model{                                   # *** PROGRAM STARTS
for(i in 1:ns){                          #  LOOP THROUGH STUDIES
  mu[i] ~ dnorm(0,.0001)                 # vague priors for all
                                         trial baselines

  for (k in 1:na[i]) {                   # LOOP THROUGH ARMS
    likelihood                           # define likelihood
    # model for linear predictor
    link function <- mu[i] + d[t[i,k]] - d[t[i,1]]
    dev[i,k] <- residual deviance        # Deviance contribution
                formula
  }
  resdev[i] <- sum(dev[i,1:na[i]])       # summed residual deviance
                                         contribution for this trial
}
totresdev <- sum(resdev[])               # Total Residual Deviance
d[1]<-0                                  # treatment effect is zero
                                         for reference treatment
for (k in 2:nt){  d[k] ~ dnorm(0,.0001) }  # vague priors for treatment
                                         effects
}                                        # *** PROGRAM ENDS
```

To obtain the code for a fixed effects model with binomial data and a logit link, described in Chapters 2 and 3, we would need to replace only the high-lighted text and add an extra line, within the i and k loops, for the model predictions (rhat[i,k]) to be calculated.

So, for each outcome being analysed, the likelihood needs to be coded as binomial, Poisson, and so on, as appropriate. The linear predictor translates equation (4.3) and an appropriate link needs to be coded. As before d[1] is set to zero and prior distributions for all other d[k] are specified, with nt representing the total number of treatments compared in the network.

The generic implementation of the random effects model accounting for the correlations in multi-arm trials is shown as follows.

Generic WinBUGS code for NMA: random effects model. The likelihood, link function, residual deviance formula and upper bound for the uniform prior distribution need to be specified in each case.

```
# Random effects model for multi-arm trials
model{
for(i in 1:ns){                                      # *** PROGRAM STARTS
w[i,1] <- 0                                          # LOOP THROUGH STUDIES
delta[i,1] <- 0                                      # adjustment for multi-arm trials is zero for control arm
mu[i] ~ dnorm(0,.0001)                               # treatment effect is zero for control arm
for (k in 1:na[i]) {                                 # vague priors for all trial baselines
                                                     # LOOP THROUGH ARMS
    likelihood                                       # define likelihood
    link function <- mu[i] + delta[i,k]              # model for linear predictor
    dev[i,k] <- residual deviance formula            # Deviance contribution
}
resdev[i] <- sum(dev[i,1:na[i]])                     # summed residual deviance contribution for this trial
for (k in 2:na[i]) {                                 # LOOP THROUGH ARMS
    delta[i,k] ~ dnorm(md[i,k],taud[i,k])            # trial-specific random effects distributions
    md[i,k] <- d[t[i,k]] - d[t[i,1]] + sw[i,k]       # mean of LOR distributions (with multi-arm trial
                                                     # correction)
    taud[i,k] <- tau*2*(k-1)/k                       # precision of LOR distributions (with multi-arm trial
                                                     # correction)
    w[i,k] <- (delta[i,k] - d[t[i,k]] + d[t[i,1]])   # adjustment for multi-arm RCTs
    sw[i,k] <- sum(w[i,1:k-1])/(k-1)                 # cumulative adjustment for multi-arm trials
}
}
totresdev <- sum(resdev[])                           # Total Residual Deviance
d[1]<-0                                              # treatment effect is zero for reference treatment
for (k in 2:nt){  d[k] ~ dnorm(0,.0001) }            # vague priors for treatment effects
sd ~ dunif(0, appropriate upper bound)               # vague prior for between-trial SD
tau <- pow(sd,-2)                                    # between-trial precision = (1/between-trial variance)
}                                                    # *** PROGRAM ENDS
```

To obtain the code for binomial data with a logit link with random treatment effects, presented in Chapters 2 and 3, we would need to make the same changes as for the fixed effects model, as well as specifying an upper bound for the vague prior distribution for the between-study heterogeneity. The linear predictor now translates equation (4.1). An appropriate link needs to be coded; an upper bound for the prior distribution of sd that is considered large, given the plausible values of the outcome, needs to be specified; and initial values need to be specified for the between-study heterogeneity sd. For most examples, WinBUGS can be left to generate initial values for the parameter delta; however these can be specified by the user, if required.

To get the posterior summaries of the parameters of interest for inference, the nodes d and sd (in the random effects model only) need to be monitored. To obtain the posterior means of the parameters required to assess model fit and model comparison, dev, totresdev and the DIC (from the WinBUGS DIC tool) need to be monitored.

In addition, code to obtain all pairwise relative and absolute effects of each treatment will be similar to the code presented in Section 2.2.3.1, but we will need to consider the scale of measurement and link being used. The code presented for ranking interventions and calculating the probability of each intervention being at each rank is exactly the same for all models (Section 2.2.3.1), but in each case we need to ensure rank 1 is the best, depending on whether higher values of the outcome are desirable or undesirable.

The next sections give examples of different types of outcome data commonly available from trials and the GLM required to analyse them. In each case, the basic synthesis model remains the same (equations (4.1) or (4.3)), but the likelihood and link function may change. Box 4.1 details the steps required

---

Box 4.1 List of Steps to Obtain the Appropriate Network Meta-Analysis Code from the Generic Code

Steps to follow:

1) Decide whether to fit fixed or random effects models and choose the appropriate generic code.
2) Decide on the likelihood that best describes the data generating process. Replace the likelihood and corresponding deviance calculation formulae in the generic code (see Table 4.1).
3) Decide which link function to use based on which scale the relative effects are most likely to be additive and replace in the generic code (see Table 4.1).
4) If using a random effects model, choose an appropriate upper bound for the prior distribution for the between-study standard deviation, or choose an appropriate informative prior distribution. Replace this in the generic code for random effects.
5) Add any extra code required to obtain all relative effects in an appropriate scale, absolute effects or other summaries of efficacy such as the ranks or NNT for all treatments.

Table 4.1 Commonly used likelihood, link functions, inverse link functions and formulae for the residual deviance.

Likelihood	Link	Link function $\theta = g(\gamma)$	Inverse link function $\gamma = g^{-1}(\theta)$	Model prediction	Residual deviance
Normal $y_{ik} \sim \text{Normal}(\bar{y}_{ik}, se_{ik}^2)$ se_{ik} assumed known	Identity	γ	θ	$\bar{y}_{ik}$	$\sum_i \sum_k \left(\dfrac{(y_{ik} - \bar{y}_{ik})^2}{se_{ik}^2} \right)$
	Log	$\ln(\gamma)$	$\exp(\theta)$		
Multivariate normal $y_{i,1:k} \sim \text{Normal}_k\left(\bar{y}_{i,1:k}, \Sigma_{(k \times k)} \right)$ Σ assumed known	Identity	γ	θ	$\bar{y}_{i,1:k}$	$\sum_i (y_{i,1:k} - \bar{y}_{i,1:k})^T \, \Sigma^{-1} \, (y_{i,1:k} - \bar{y}_{i,1:k})$
	Log	$\ln(\gamma)$	$\exp(\theta)$		
Binomial $r_{ik} \sim \text{Binomial}(p_{ik}, n_{ik})$	Logit	$\ln\left(\dfrac{\gamma}{(1-\gamma)} \right)$	$\dfrac{\exp(\theta)}{1+\exp(\theta)}$	$\hat{r}_{ik} = n_{ik} p_{ik}$	$\sum_i \sum_k 2 \left(r_{ik} \log\left(\dfrac{r_{ik}}{\hat{r}_{ik}} \right) + (n_{ik} - r_{ik}) \log\left(\dfrac{n_{ik} - r_{ik}}{n_{ik} - \hat{r}_{ik}} \right) \right)$
	Complementary log–log (cloglog)	$\ln\{-\ln(1-\gamma)\}$	$1 - \exp\{-\exp(\theta)\}$		
	Probit	$\Phi^{-1}(\gamma)$	$\Phi(\theta)$		
Multinomial $r_{i,k,1:J} \sim \text{Multinomial}(p_{i,1:J}, n_{ik})$	Logit	$\ln\left(\dfrac{\gamma}{(1-\gamma)} \right)$	$\dfrac{\exp(\theta)}{1+\exp(\theta)}$	$\hat{r}_{ikj} = n_{ik} p_{ikj}$	$\sum_i \sum_k 2 \left(\sum_j r_{ikj} \log\left(\dfrac{r_{ikj}}{\hat{r}_{ikj}} \right) \right)$
	Complementary log–log (cloglog)	$\ln\{-\ln(1-\gamma)\}$	$1 - \exp\{-\exp(\theta)\}$		
	Probit	$\Phi^{-1}(\gamma)$	$\Phi(\theta)$		
Poisson $r_{ik} \sim \text{Poisson}(\lambda_{ik} E_{ik})$	Log	$\ln(\gamma)$	$\exp(\theta)$	$\check{r}_{ik} = \lambda_{ik} E_{ik}$	$\sum_i \sum_k 2 \left((\hat{r}_{ik} - r_{ik}) + r_{ik} \log\left(\dfrac{r_{ik}}{\hat{r}_{ik}} \right) \right)$

to obtain appropriate code for the problem at hand, which will be followed in the next sections. Table 4.1 has details of the most commonly used likelihoods, link and inverse link functions along with formulae for the residual deviance and the predicted values needed to calculate p_D (Chapter 3).

4.3   Univariate Arm-Based Likelihoods

The preferred form of summary data for meta-analysis is arm-based summaries, which allow an exact likelihood to be defined for each arm of each trial and avoid the need to add 0.5 or other values to zero event cells, for count data. This is the data format used in Chapters 2 and 3. In this section we expand the generic model to other data types while still assuming that we have outcome data available for each arm of each study. Models for data available only as relative effects are described in Section 4.4.

4.3.1   Rate Data: Poisson Likelihood and Log Link

When the data available for the RCTs included in the meta-analysis are in the form of counts over a certain time period (which may be different for each trial), a Poisson likelihood and a log link can be used. Examples would be the number of deaths, or the number of patients in whom a device failed, for a given exposure time, usually given as the number of *person-years* at risk. For patients who do not reach the end event, the time at risk is the same as their follow-up time, but for those who do, it is the time from the start of the trial to the event, so the method allows for censored observations.

Defining r_{ik} as the number of events occurring during the trial follow-up period, E_{ik} as the exposure time (e.g. person-years) and λ_{ik} as the rate at which events occur, in arm k of trial i, we can write the likelihood as

$$r_{ik} \sim \text{Poisson}\left(\lambda_{ik} E_{ik}\right) \tag{4.4}$$

The parameter of interest in the likelihood is the hazard, the rate at which the events occur in each trial arm. Since this can only take positive values, it is modelled using the log link so that the linear predictor in equation (4.1) is on the log rate scale, and we write

$$\theta_{ik} = \log\left(\lambda_{ik}\right) = \mu_i + \delta_{ik} \tag{4.5}$$

for a random effects model, with $\delta_{i1} = 0$ and random effects distribution given in equation (4.2). We set $\theta_{ik} = \log(\lambda_{ik})$ in equation (4.3) for a fixed effects model.

A key assumption of this model is that in each arm of each trial the hazard is constant over the follow-up period. This can only be the case in homogeneous populations where all patients have the same hazard rate. In populations with constant but heterogeneous rates, the average hazard must necessarily decrease over time, as those with higher hazard rates tend to reach their endpoints earlier and exit from the risk set. A model with piecewise constant hazards may need to be used in this case (see Chapters 10 and 11).

This model is also useful for certain *repeated* event data. Examples would be the number of accidents, or the number of hypoglycaemic events in patients with diabetes (National Clinical Guideline Centre, 2015), where each individual may have more than one event. Here one would model the total number of events in each arm, which may need to be calculated as the average number of events multiplied by the number of patients in each arm. The Poisson model can also be used for observations repeated in space rather than time, for example, the number of teeth requiring fillings (Dias et al., 2010c) (see also Chapter 9). Using the Poisson model for repeated event data makes the additional assumption that the events are independent, so that, for example, an accident (or tooth filling) is no more likely in an individual who has already had an accident than in one who has not. Other examples include the analysis of the rate of stroke and the rate of major or fatal bleeding episodes per 1000 person-years for different stroke prevention treatments (Cooper et al., 2006) and a review of asthma self-management education in children that looked at school absences and emergency room visits (Guevara et al., 2004).

4.3.1.1 WinBUGS Implementation

The code to fit this model is obtained by specifying the likelihood as Poisson in the generic code. This can be done by including the following lines to define the likelihood:

```
r[i,k] ~ dpois(theta[i,k])        # Poisson likelihood
theta[i,k] <- lambda[i,k]*E[i,k]  # event rate * exposure
```

Note that an extra line has been added to define the Poisson parameter as the product of the rate (lambda) and the exposure (E) since WinBUGS does not allow mathematical operations within the definition of distributions. The log link in a fixed effects model is specified as log(lambda[i,k]) and the code to calculate each data points' contribution to the residual deviance (Table 4.1) is

```
dev[i,k] <- 2*((theta[i,k]-r[i,k]) + r[i,k]*log(r[i,k]/theta[i,k]))
                                       #Deviance contribution
```

To calculate the effective number of parameters, p_D, outside of WinBUGS (as recommended in Chapter 3), the node theta needs to be monitored and used to replace $\hat{r}_{ik}$ in the residual deviance formula (Table 4.1).

The implementation of the random effects model accounting for the correlations in multi-arm trials makes these same changes to the likelihood, link function and deviance formula in the generic random effects code, but the upper bound for the between-trial standard deviation needs to be specified. This value will depend on the specific example being considered, although note that it is on a log rate scale.

The code to monitor all the $S(S-1)/2$ log hazard/rate ratios and hazard/rate ratios is given in Chapter 2, but to avoid confusion when interpreting the output, we recommend replacing `or` with `hr` in the code. The code to generate absolute effects (rates) for each treatment is now `log(T[k]) <- A + d[k]`.

A further variable that may be required for cost-effectiveness modelling might be the proportion of patients that would be expected to have an event, after a follow-up of, say, x months, under each treatment. If rates are per year, time is $x/12$, and the following code can be added before the last closing brace, with x replaced by the appropriate value or given as data:

```
for (k in 1:nt) { pt[k] <- 1-exp(-T[k]*x/12) }
```

The variable `pt` will hold the required proportions for each treatment, given the assumed log rate of an event for the reference treatment.

4.3.1.2 Example: Dietary Fat

In a Cochrane review of RCTs to assess the effect of change in dietary fats on total and cardiovascular mortality (Hooper et al., 2000), data were extracted as the number of events per person-years observed (Table 4.2). Only two generic interventions were of interest: a reduced fat dietary intervention ('diet') and a non-reduced fat diet ('control'). Most of the trials compared only one type of diet with the control. However, trial 2 compared two types of reduced fat diets against control (for more details see Hooper et al. (2000)). For the purpose of this example, we consider the two different types of diet as the same intervention (treatment 2), but keep the treatment arms separately, so that in a random effects model, this trial will provide two correlated estimates of the trial-specific treatment effect δ_{ik} ($k=2, 3$) with identical means and in the fixed effects model, both arms will contribute to the estimate of the common treatment effect d_{12}.

The data are assumed to have a Poisson likelihood and modelled with a log link. The WinBUGS code is changed as detailed earlier, and the upper bound for the prior distribution for the between-study standard deviation is chosen as 5, which is large enough, given the plausible values of the log rate of mortality in this example. Thus the code is changed to `sd ~ dunif(0,5)`.

The data structure has two components: a list specifying the number of treatments `nt`, the number of studies `ns` and the mean and precision for the distribution of log rates on the reference treatment `meanA` and `precA`. Both data components need to be loaded into WinBUGS for the program to run. Since there is a three-arm study, three columns are required in the main body of data to specify the treatments `t[,]`, the exposure times `E[,]` and the number of events `r[,]` in each arm. The number of arms in each study is in column `na[]`, as before.

```
# Data (Dietary fat example)
list(ns=10, nt=2 , meanA=-3, precA=1.77)

t[,1]  t[,2]  t[,3]  E[,1]   E[,2]   E[,3]  r[,1]  r[,2]  r[,3]  na[]  #ID
1      2      NA     1917    1925    NA     113    111    NA     2     #2  DART
1      2      2      43.6    41.3    38     1      5      3      3     #10 London Corn /Olive
1      2      NA     393.5   373.9   NA     24     20     NA     2     #11 London Low Fat
1      2      NA     4715    4823    NA     248    269    NA     2     #14 Minnesota Coronary
1      2      NA     715     751     NA     31     28     NA     2     #15 MRC Soya
1      2      NA     885     895     NA     65     48     NA     2     #18 Oslo Diet-Heart
1      2      NA     87.8    91      NA     3      1      NA     2     #22 STARS
1      2      NA     1011    939     NA     28     39     NA     2     #23 Sydney Diet-Heart
1      2      NA     1544    1588    NA     177    174    NA     2     #26 Veterans Administration
1      2      NA     125     123     NA     2      1      NA     2     #27 Veterans Diet & Skin CA

END
```

Table 4.2 Dietary fat example: study names and treatment codes for the 10 included studies and person-years and total mortality observed in each study. Data from Hooper et al. (2000).

Study name and ID	Treatment			Person-years observed			Total mortality			Number randomised		
	Control t[,1]	Diet t[,2]	Diet 2 t[,3]	Control E[,1]	Diet E[,2]	Diet 2 E[,3]	Control r[,1]	Diet r[,2]	Diet 2 r[,3]	Control n[,1]	Diet n[,2]	Diet 2 n[,3]
1. DART	1	2		1917	1925		113	111		1015	1018	
2. London Corn/Olive	1	2	2	43.6	41.3	38	1	5	3	26	28	26
3. London Low Fat	1	2		393.5	373.9		24	20		129	123	
4. Minnesota Coronary	1	2		4715	4823		248	269		4516	4516	
5. MRC Soya	1	2		715	751		31	28		194	199	
6. Oslo Diet-Heart	1	2		885	895		65	48		206	206	
7. STARS	1	2		87.8	91		3	1		30	30	
8. Sydney Diet-Heart	1	2		1011	939		28	39		237	221	
9. Veterans Administration	1	2		1544	1588		177	174		422	424	
10. Veterans Diet & Skin CA	1	2		125	123		2	1		67	66	

No particular changes are required for the structure of the initial values, apart from accounting for the number of treatments and trials included in this particular example.

The full code with data and initial values for the fixed and random effects models for this example are presented in the online files *Ch4_FE_Po_log.odc* and *Ch4_RE_Po_log.odc*, respectively.

4.3.1.3 Results: Dietary Fat

The results from the two models (three chains: 100,000 iterations after a burn-in of 20,000 for the fixed effects model and a burn-in of 100,000 for the random effects model) are compared in Table 4.3, where p_D was calculated outside WinBUGS. Results given by the WinBUGS DIC tool are presented in Figure 4.1,

Table 4.3 Dietary fat example: posterior median and 95% CrI for both fixed and random effects models for the treatment effect d_{12}, absolute effects of the control diet (T_1) and the reduced fat diet (T_2) for a log rate of mortality on the control diet with mean -3 and precision 1.77; heterogeneity standard deviation, σ; and model fit statistics.

	Fixed effects model		Random effects model	
	Median	95% CrI	Median	95% CrI
d_{12}	-0.01	$(-0.11, 0.10)$	-0.01	$(-0.19, 0.16)$
T_1	0.05	$(0.01, 0.22)$	0.05	$(0.01, 0.22)$
T_2	0.05	$(0.01, 0.22)$	0.05	$(0.01, 0.22)$
σ	$-$	$-$	0.10	$(0.00, 0.43)$
$\bar{D}_{res}{}^a$	22.32		21.45	
p_D	11.00		13.55	
DIC	33.32		35.00	

[a] Compare to 21 data points.

Fixed Effects model
Dbar = post.mean of −2logL; Dhat = −2LogL at post.mean of stochastic nodes

	Dbar	Dhat	pD	DIC
r	125.016	114.160	10.855	135.871
total	125.016	114.160	10.855	135.871

Random Effects model
Dbar = post.mean of −2logL; Dhat = −2LogL at post.mean of stochastic nodes

	Dbar	Dhat	pD	DIC
r	124.165	110.820	13.345	137.510
total	124.165	110.820	13.345	137.510

Figure 4.1 Dietary fat example: WinBUGS DIC tool output for the fixed and random effects models.

Figure 4.2 Dietary fat example: posterior distribution of the between-study standard deviation – from WinBUGS.

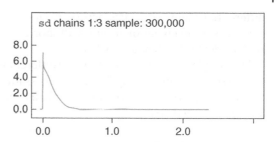

and in this example, the differences in p_D calculated using the different methods are very small. The random and fixed effects models are indistinguishable in terms of model fit, and both appear to fit the data well with $\bar{D}_{res}$ close to 21, the number of data points. We would therefore prefer the fixed effects model as it is simpler and more interpretable.

Inspecting the posterior distribution for the between-study heterogeneity (Figure 4.2), we can see that it has been updated from the Uniform(0,5) prior distribution and also that the density is concentrated around small values, which lends support to the choice of the fixed effects model. The posterior median of the pooled log rate of a reduced fat diet compared with the control diet is −0.01 in the fixed effects model with 95% CrI (−0.11, 0.10), suggesting no difference in the number of cardiovascular mortalities in each group. The posterior medians of the absolute rates of mortality for each treatment (and their 95% CrIs), having assumed that the log rate of mortality on the control diet has mean −3 and precision 1.77, are the same on the control and reduced fat diets (Table 4.3).

4.3.2 Rate Data: Binomial Likelihood and Cloglog Link

In some meta-analyses, each included trial reports the proportion of patients reaching an endpoint at a specified follow-up time, but the trials do not all have the same follow-up time. Defining r_{ik} as the number of events in arm k of trial i, at follow-up time f_i (measured in days, weeks, etc.), out of n_{ik} patients randomised to each trial arm, the likelihood generating the data is binomial. However, using a logit model as described in Chapter 2 implies one of the following assumptions: that all patients who reach the endpoint do so by some specific follow-up time, and further follow-up would make no difference, or that the proportional *odds* assumption holds, which implies a complex form for the hazard rates (Collett, 1994). If longer follow-up results in more events, as is common with, for example, all-cause mortality or relapse, the standard logit model is hard to interpret. Holzhauer (2017) stated that a logit model would be reasonable only if dropout times are independent of event times and follow the same distribution for all treatment groups in a trial. In addition, differential follow-up times across groups within a trial will bias treatment comparisons in favour of groups with shorter follow-up (for undesirable events such as death) (Holzhauer, 2017).

The simplest way to account for the different length of follow-up in each trial is to assume an underlying Poisson process for each trial arm, with a constant event rate λ_{ik}, so that T_{ik}, the time until an event occurs in arm k of trial i, has an exponential distribution:

$$T_{ik} \sim \text{Exponential}(\lambda_{ik})$$

The probability that there are no events by time f_{ik} in arm k of trial i, the survival function (see also Chapter 10), can be written as

$$\Pr(T_{ik} > f_{ik}) = \exp(-\lambda_{ik} f_{ik})$$

Then, for each trial i, p_{ik}, the probability of an event in arm k of trial i after follow-up time f_{ik}, can be written as

$$p_{ik} = 1 - \Pr(T_{ik} > f_{ik}) = 1 - \exp(-\lambda_{ik} f_{ik}) \qquad (4.6)$$

which is time dependent.

We model the event rate λ_{ik} on the log scale, taking into account the different follow-up times f_{ik}. Since equation (4.6) is a non-linear function of $\log(\lambda_{ik})$, the complementary log–log (cloglog) link function (Prentice and Gloeckler, 1978) is used to obtain a GLM for $\log(\lambda_{ik})$ (see Table 4.1). This gives

$$\theta_{ik} = \text{cloglog}(p_{ik}) = \log(f_{ik}) + \log(\lambda_{ik}) \qquad (4.7)$$

and $\log(\lambda_{ik})$ is modelled as in equation (4.5) for a random effects model:

$$\theta_{ik} = \text{cloglog}(p_{ik}) = \log(f_{ik}) + \mu_i + \delta_{ik}$$

with the treatment effects δ_{ik} representing the log hazard ratios of the treatment in arm k compared with the treatment in arm 1 in trial i, and we would make similar changes to implement a fixed effects model.

Note, however, that the binomial likelihood is only appropriate for data where participants have at most one event (e.g. death). A different likelihood (e.g. Poisson) would need to be used when multiple events are allowed for each participant.

The assumptions made in this model are that the hazard ratios are constant over the entire duration of follow-up. Nonetheless, this assumption may be preferable to assuming that the follow-up time makes no difference to the number of events. The clinical plausibility of these assumptions should be discussed and supported by citing relevant literature or by examination of evidence of changes in outcome rates over the follow-up period in the included trials. However, it should be noted that if follow-up for all arms in a trial is the same and events occur at a very low rate (but not necessary constant), then a logit analysis will produce similar results to a time-to-event analysis (Holzhauer, 2017).

When the constant hazards assumption is not reasonable, but further follow-up time is believed to result in more events, extensions are available that allow for time-varying rates. One approach is to adopt piecewise constant hazards. These models can be fitted if there are data reported at multiple follow-up times within the same study (Lu et al., 2007; Stettler et al., 2007). See Chapter 11 for further details.

An alternative is to fit a Weibull model, which involves an additional 'shape' parameter α:

$$\Pr\left(T_{ik} > f_{ik}\right) = \exp\left[\left(-\lambda_{ik} f_{ik}\right)^{\alpha}\right]$$

which leads to

$$\theta_{ik} = \text{cloglog}\left(p_{ik}\right) = \alpha\left(\log\left(f_{ik}\right) + \mu_i + \delta_{i,bk} I_{\{k \neq 1\}}\right)$$

Although no longer a GLM, since a non-linear predictor is used, these extensions lead to major liberalisation of modelling, but require more data. The additional Weibull parameter, for example, can only be adequately identified if there are data on a wide range of follow-up times and if investigators are content to assume the *same* shape parameter for all treatments.

4.3.2.1 WinBUGS Implementation

The code to fit the model in equation (4.7) is obtained by specifying the likelihood as binomial in the generic code and the link as cloglog. For a fixed effects model the cloglog link is specified as

```
cloglog(p[i,k]) <- log(time[i,k]) + mu[i] + d[t[i,k]] - d[t[i,1]]
                                        # model for linear predictor
```

Note that a term to calculate the log of time for each arm of each study has been added and time will need to be given as data for each arm of each trial. In some cases, when time takes large values (e.g. 52 weeks), WinBUGS may produce a numerical error when calculating the log of time. This can be solved by calculating this for each study outside of WinBUGS and feeding it in as data so that we would type `lntime[i,k]` instead of `log(time[i,k])` where `lntime` now contains the log of the follow-up time for each arm of each study. In most cases all arms of a given study will have the same follow-up time; thus `time[i,k]` can be replaced with `time[i]` in the code. However in this case the `time[i]` term will cancel when calculating the relative treatment effects, so it can usually be omitted from the code.

Truncation has been suggested to avoid numerical errors when the probabilities get very close to zero or one (Ntzoufras, 2009). This can be achieved using the following code in place of the linear predictor. This sets `cloglog(p[i,k])` to be `-xi1` or `xi2` whenever the linear predictor is outside these bounds.

```
# cloglog truncated to avoid arithmetic overflow when close
# to 0 or 1; see Ntzoufras(2009, Chapter 7)
    eta[i,k] <- log(time[i,k]) + mu[i] + d[t[i,k]] - d[t[i,1]]
    cloglog(p[i,k]) <- eta[i,k]*(1-step(-xi1-eta[i,k]))
     *(1-step(eta[i,k]-xi2))
     -xi1*step(-xi1-eta[i,k])+ xi2*step(eta[i,k]-xi2)
```

We further add code setting these bounds to be 15 and 3, effectively setting the probabilities to zero or one, respectively, when the linear predictor is calculated outside these bounds. This prevents numerical overflow in cases when the probabilities get very small (rare events) or very large (common events).

```
# cloglog truncation values
xi1 <- 15
xi2 <- 3
```

Note however that setting these bounds effectively restricts the values of the probabilities that can be estimated, which corresponds to having a (weak) model on the arm-level parameters. This can have a considerable impact on the estimates of the relative effects in some situations. We therefore recommend that these constraints are not added unless absolutely necessary and that sensitivity to the choice of truncation values is assessed. The simulated values of the posterior distribution should also be checked to assess the impact of truncation by, for example, looking for extreme values in the WinBUGS history plots.

The code to calculate each data point's contribution to the residual deviance and their contribution to p_D is that of the binomial distribution, so the deviance code to be added is as described in Chapter 3.

The implementation of the random effects model accounting for the correlations in multi-arm trials makes these same changes to the generic random effects code, but we now need to also specify the upper bound for the between-trial standard deviation. This value will depend on the specific example being considered, although note that it is on a log hazard scale, so very large values will usually be unlikely.

The absolute probabilities of an event on each treatment will now depend on the time point at which they are evaluated, timeA (given as data), so the code that needs to be added is

```
for (k in 1:nt) { cloglog(T[k]) <- log(timeA) + A + d[k] }
```

4.3.2.2 Example: Diabetes

We illustrate this model with a network meta-analysis to assess the incidence of diabetes in RCTs of antihypertensive drugs (Elliott and Meyer, 2007). The outcome is new cases of diabetes observed over the trial duration period (measured in years) for six different drugs: diuretic (treatment 1), placebo (treatment 2), β-blocker (treatment 3), calcium channel blockers (CCB) (treatment 4), angiotensin-converting enzyme (ACE) inhibitors (treatment 5) and angiotensin receptor blockers (ARB) (treatment 6). The reference treatment chosen was diuretic, as recommended in this field – for more details see Elliott and Meyer (2007). The data are presented in Table 4.4 and the network diagram in Figure 4.3.

The data are assumed to have a binomial likelihood and the model uses the cloglog link. The upper bound for the between-study standard deviation is chosen as 5, which is large enough on this scale.

The data structure is given as follows. The main body of data is in the same format as the binomial likelihood with logit link in Chapter 2, with an additional vector containing the follow-up time for each study, time[] (which is the same for all arms within a study). Treatments are ordered numerically so that the treatment in arm 2 has a higher code than the treatment in arm 1 and the treatment in arm 3 has a higher code than the treatment in arm 2 (Chapter 2).

We will calculate the absolute probabilities of an event at 3 years for all treatments, assuming the cloglog of the probability of an event on diuretic is approximately normal with mean −4.2 and precision 1.11.

```
# Data (Diabetes example)
list(ns=22, nt=6, meanA=-4.2, precA=1.11, timeA=3)

time[] t[,1] r[,1]  n[,1] t[,2] r[,2]  n[,2] t[,3] r[,3] n[,3] na[]  #Study
5.8    1     43     1081  2     34     2213  3     37    1102  3     #MRC-E 38
4.7    1     29     416   2     20     424   NA    NA    NA    2     #EWPH 32
3      1     140    1631  2     118    1578  NA    NA    NA    2     #SHEP 42
3.8    1     75     3272  3     86     3297  NA    NA    NA    2     #HAPPHY33
4      1     302    6766  4     154    3954  5     119   4096  3     #ALLHAT 26
3      1     176    2511  4     136    2508  NA    NA    NA    2     #INSIGHT35
4.1    1     200    2826  5     138    2800  NA    NA    NA    2     #ANBP-2 18
1      1     8      196   6     1      196   NA    NA    NA    2     #ALPINE 27
3.3    2     154    4870  4     177    4841  NA    NA    NA    2     #FEVER 20
3      2     489    2646  5     449    2623  NA    NA    NA    2     #DREAM 31
4.5    2     155    2883  5     102    2837  NA    NA    NA    2     #HOPE 34
4.8    2     399    3472  5     335    3432  NA    NA    NA    2     #PEACE 40
3.1    2     202    2721  6     163    2715  NA    NA    NA    2     #CHARM 30
3.7    2     115    2175  6     93     2167  NA    NA    NA    2     #SCOPE 41
3.8    3     70     405   4     32     202   5     45    410   3     #AASK 25
4      3     97     1960  4     95     1965  5     93    1970  3     #STOP-2 43
5.5    3     799    7040  4     567    7072  NA    NA    NA    2     #ASCOT 28
4.5    3     251    5059  4     216    5095  NA    NA    NA    2     #NORDIL 39
4      3     665    8078  4     569    8098  NA    NA    NA    2     #INVEST 36
6.1    3     380    5230  5     337    5183  NA    NA    NA    2     #CAPPP 29
4.8    3     320    3979  6     242    4020  NA    NA    NA    2     #LIFE 37
4.2    4     845    5074  6     690    5087  NA    NA    NA    2     #VALUE 44
END
```

Table 4.4 Diabetes example: study names, follow-up time in years, treatments compared, total number of new cases of diabetes and number of patients in each trial arm, where diuretic=treatment 1, placebo=treatment 2, β-blocker=treatment 3, CCB=treatment 4, ACE inhibitor=treatment 5 and ARB=treatment 6. Data from Elliott & Meyer (2007).

Study ID	Study follow-up (in years) Time[]	Treatment Arm 1 t[,1]	Arm 2 t[,2]	Arm 3 t[,3]	New cases of diabetes Arm 1 r[,1]	Arm 2 r[,2]	Arm 3 r[,3]	Total number of patients Arm 1 n[,1]	Arm 2 n[,2]	Arm 3 n[,3]
1. MRC-E	5.8	1	2	3	43	34	37	1081	2213	1102
2. EWPH	4.7	1	2		29	20		416	424	
3. SHEP	3	1	2		140	118		1631	1578	
4. HAPPHY	3.8	1	3		75	86		3272	3297	
5. ALLHAT	4	1	4	5	302	154	119	6766	3954	4096
6. INSIGHT	3	1	4		176	136		2511	2508	
7. ANBP-2	4.1	1	5		200	138		2826	2800	
8. ALPINE	1	1	6		8	1		196	196	
9. FEVER	3.3	2	4		154	177		4870	4841	
10. DREAM	3	2	5		489	449		2646	2623	
11. HOPE	4.5	2	5		155	102		2883	2837	
12. PEACE	4.8	2	5		399	335		3472	3432	
13. CHARM	3.1	2	6		202	163		2721	2715	
14. SCOPE	3.7	2	6		115	93		2175	2167	
15. AASK	3.8	3	4	5	70	32	45	405	202	410
16. STOP-2	4	3	4	5	97	95	93	1960	1965	1970
17. ASCOT	5.5	3	4		799	567		7040	7072	
18. NORDIL	4.5	3	4		251	216		5059	5095	
19. INVEST	4	3	4		665	569		8078	8098	
20. CAPPP	6.1	3	5		380	337		5230	5183	
21. LIFE	4.8	3	6		320	242		3979	4020	
22. VALUE	4.2	4	6		845	690		5074	5087	

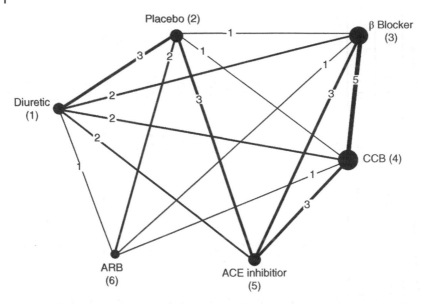

Figure 4.3 Diabetes network: each circle represents a treatment, and connecting lines indicate pairs of treatments that have been directly compared in randomised trials. The numbers on the lines indicate the numbers of trials making that comparison, and the numbers by the treatment names are the treatment codes used in the modelling. Line thickness is proportional to the number of trials making that comparison, and the width of the circles is proportional to the number of patients randomised to that treatment.

No particular changes are required for the structure of the initial values in the fixed effects model, apart from accounting for the number of treatments and trials included in this particular example. However, due to the structure of the model, numerical errors may occur when allowing WinBUGS to automatically generate initial values for `delta` in the random effects model. Thus in this case it is advisable for the user to initialise `delta`.

The full code with data and initial values for the fixed and random effects models for this example are presented in the online files *Ch4_FE_Bi_cloglog. odc* and *Ch4_RE_Bi_cloglog.odc*, respectively. The code to add truncation is included although it is commented out (using the # symbol) as it is not required for this particular example. Users should replace the simple link function with this code if required for their particular example, taking care to perform the checks described in Section 4.3.2.1.

4.3.2.3 Results: Diabetes
Results from fixed and random effects models (three chains: 100,000 iterations after a burn-in of 50,000) are presented in Table 4.5 where p_D was calculated outside WinBUGS (Chapter 3). The DIC tool in WinBUGS gave $p_D = 26.97$ and

Table 4.5 Diabetes example: posterior median and 95% CrI for both fixed and random effects models for the treatment effects of placebo (d_{12}), β-blocker (d_{13}), CCB (d_{14}), ACE inhibitor (d_{15}) and ARB (d_{16}) relative to diuretic; absolute effects of diuretic (T_1) placebo (T_2), β-blocker (T_3), CCB (T_4), ACE inhibitor (T_5) and ARB (T_6); heterogeneity parameter σ and model fit statistics.

	Fixed effects model		Random effects model	
	Median	95% CrI	Median	95% CrI
d_{12}	−0.25	(−0.36, −0.14)	−0.29	(−0.47, −0.12)
d_{13}	−0.06	(−0.17, 0.05)	−0.07	(−0.25, 0.10)
d_{14}	−0.25	(−0.36, −0.15)	−0.24	(−0.41, −0.07)
d_{15}	−0.36	(−0.46, −0.25)	−0.40	(−0.58, −0.24)
d_{16}	−0.45	(−0.58, −0.33)	−0.47	(−0.70, −0.26)
T_1	0.044	(0.01, 0.25)	0.044	(0.01, 0.25)
T_2	0.034	(0.01, 0.20)	0.033	(0.01, 0.20)
T_3	0.042	(0.01, 0.24)	0.041	(0.01, 0.24)
T_4	0.034	(0.01, 0.20)	0.035	(0.01, 0.20)
T_5	0.031	(0.00, 0.18)	0.030	(0.00, 0.18)
T_6	0.028	(0.00, 0.17)	0.028	(0.00, 0.17)
σ			0.12	(0.06, 0.23)
$\bar{D}_{res}{}^a$	78.25		53.55	
p_D	27.00		38.06	
DIC	105.25		91.61	

[a] Compare to 48 data points.

38.00 for the fixed and random effects models, respectively. The fixed effects model has a very poor fit (the posterior mean of the residual deviance is 78 compared with 48 data points), and the random effects model has a smaller DIC (difference in DIC is 13.6), so this should be preferred.

The posterior median of the pooled treatment effects, on the log hazard scale, of treatments 2–6 relative to the reference treatment show a beneficial effect of all the treatment with the exception of treatment 3 (Table 4.5 and Figure 4.4).

The posterior distribution of the between-study heterogeneity has median 0.12 and 95% CrI (0.05, 0.23) that can be considered moderate, when compared to the size of the log hazard ratios, and excludes values close to zero (Figure 4.5).

The posterior medians of the absolute probabilities of developing diabetes after a period of 3 years, assuming that the cloglog of the probability of developing diabetes on diuretic has mean −4.2 and precision 1.11, on each of the treatments are between 3 and 4% (Table 4.5).

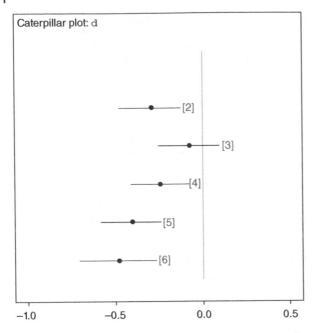

Caterpillar plot: d

Figure 4.4 Diabetes example: caterpillar plot showing log hazard ratios of all treatments compared to the reference for the random effects model – from WinBUGS. The treatment being compared is indicated in square brackets; negative values favour this treatment.

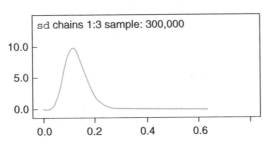

sd chains 1:3 sample: 300,000

Figure 4.5 Diabetes example: posterior distribution of the between-study standard deviation – from WinBUGS.

4.3.3 Continuous Data: Normal Likelihood and Identity Link

With continuous outcome data, meta-analysis is usually based on the sample means, y_{ik}, with standard errors, se_{ik}, assumed known. As long as the sample sizes are not too small, the central limit theorem allows us to assume that, even in cases where the underlying data are skewed, the sample means are approximately normally distributed, with likelihood

$$y_{ik} \sim \text{Normal}\left(\theta_{ik}, se_{ik}^2\right) \tag{4.8}$$

The parameter of interest is the mean, θ_{ik}, of this continuous measure that is unconstrained on the real line and so no transformation is needed. The identity link is used (Table 4.1), and for a random effects model, the linear model can be written on the natural scale (with no link function; equation (4.1)).

The trial-specific random effects δ_{ik} represent the mean differences between the treatment in arm k and the treatment in arm 1 of trial i. Similarly, a fixed effects model can be obtained using equation (4.3).

4.3.3.1 Before/After Studies: Change from Baseline Measures
In cases where the original trial outcome is continuous and measured at baseline (i.e. at the start of treatment or at randomisation) and at a pre-specified follow-up point, the most common method is to base the meta-analysis on the mean change from baseline for each patient and an appropriate measure of uncertainty (e.g. the variance or standard error) that takes into account any within-patient correlation. It should be noted that the most efficient and least biased statistic to use is the mean of the final reading, having adjusted for the baseline values via regression/ANCOVA, and these should be the preferred outcome measures when available (Higgins and Green, 2008).

The likelihood for the mean change from baseline in arm k of trial i, with change variance, can be assumed normal as in equation (4.8).

However, in practice many studies fail to report an adequate measure of the uncertainty for the before/after difference in outcome and instead report the mean and variance (or other measure of uncertainty) at baseline and at follow-up separately. While the mean change from baseline can be easily calculated, to calculate the variance for such trials, information on the within-patient correlation is required, but seldom available. It may be possible to obtain information on the correlation from a review of similar trials using the same outcome measures, or else a reasonable value, often 0.5 (which is considered conservative) or 0.7 (Frison and Pocock, 1992), can be used alongside sensitivity analyses (Follmann et al., 1992; Higgins and Green, 2008). A more sophisticated approach, which takes into account the uncertainty in the correlation, is to use whatever information is available within the dataset, from trials that report both the before/after variances and the change variance, and possibly external trials as well, to obtain an evidence-based prior distribution for the correlation, or even to estimate the correlation and the treatment effect simultaneously within the same analysis (Abrams et al., 2005). Alternatively, assuming trials reporting change from baseline and trial reporting follow-up measures are estimating the same mean difference between treatments, we can simply pool change from baseline and follow-up data (Higgins and Green, 2008; Senn et al., 2013) – the absolute values will be quite different, but their differences will not. In any case, the network meta-analysis model is the same whether data are means at follow-up or change from baseline.

4.3.3.2 Standardised Mean Differences
Often the scale chosen to pool the data will be the standardised mean differences (SMD) (Cohen, 1969; Hedges and Olkin, 1985). The main role of the SMD is to facilitate combining results from trials that have reported outcomes

measured on different continuous scales. The idea is that the different scales are measuring essentially the same quantity and that results can be placed on a common scale if the mean difference between two arms in each trial is divided by its standard deviation. The best known SMD measures are Cohen's d (Cohen, 1969; Hedges and Olkin, 1985) and Hedges' adjusted g (Hedges and Olkin, 1985), which differ only in how the pooled standard deviation is defined and the fact that Hedges' g is adjusted for small sample bias. The Cochrane Collaboration recommends the use of Hedges' g while noting that interpretation of the overall intervention effect is difficult (Higgins and Green, 2008). It recommends re-expressing the pooled SMD in terms of effect sizes as small, medium or large (according to some rules of thumb), transforming the pooled SMD into an odds ratio, or re-expressing the SMD in the units of one or more of the original measurement instruments (Higgins and Green, 2008), although it is conceded none of these manoeuvres mitigates the drawbacks of using this measure, which has been robustly criticised (Rothman et al., 2012). Alternatives are suggested in Section 11.7.

If arm-level data are available for each study and a standardising constant is defined for each study, the model can be adapted to give a pooled SMD by writing the likelihood as

$$y_{ik} \sim \text{Normal}\left(\varphi_{ik}, se_{ik}^2\right)$$

with $\varphi_{ik} = \theta_{ik} \times s_i$ and s_i the standardising constant for each study. The parameter of interest is the SMD, θ_{ik}, which is modelled directly, as before.

Often the data y_{ik} will already be transformed to SMD. To model such data see Section 4.4.1.

SMDs are also sometimes used for non-continuous outcomes. For example, in a review of topical fluoride therapies to reduce caries in children and adolescents, the outcomes were the number of new caries observed, but the mean number of caries in each trial arm was modelled as SMD (Salanti et al., 2009) – see Exercise 4.2. Where possible, it is preferable to use the appropriate GLM, in this case a Poisson likelihood and log link, as this is likely to reduce heterogeneity (Dias et al., 2010c) – see also Chapter 9 where this example is explored in more detail.

4.3.3.3 WinBUGS Implementation
The code is obtained by specifying the likelihood as normal in the generic code and the link as identity (i.e. no function). The likelihood is defined as

```
var[i,k] <- pow(se[i,k],2)          # calculate variances
prec[i,k] <- 1/var[i,k]             # set precisions
y[i,k] ~ dnorm(theta[i,k],prec[i,k]) # normal likelihood
```

where we have added two extra rows of code to calculate the trial-specific variances and precisions, since we assume the data will be given as means and their standard errors. If data were already given as variances or precisions, the corresponding lines of code should be deleted.

The linear predictor is specified simply as `theta[i,k]`, since the means are modelled directly, and the code for the residual deviance contribution is

```
dev[i,k] <- (y[i,k]-theta[i,k])*(y[i,k]-theta[i,k])*prec[i,k]
                                          #Deviance contribution
```

To calculate p_D outside WinBUGS, the node `theta` should be monitored and replaced into the residual deviance formula in Table 4.1.

To implement the random effects model, the upper bound for the between-trial standard deviation needs to be specified. This value will depend on the specific example being considered and the plausible values for the mean differences. Care should be taken to ensure the prior distribution is not overly constraining the possible values of the heterogeneity. Inspection of the posterior distribution of `sd` will provide an indication of whether the prior distribution was suitably wide (see Exercise 4.1).

The same code can be added to monitor all the $S(S-1)/2$ mean differences as before, but now we do not need to exponentiate the relative effects. To avoid confusion when interpreting the output, we recommend replacing `lor` with `diff` in the code presented in Chapter 2 (the line to calculate `or` is not needed).

The code to generate absolute effects (i.e. the means) for all treatments is `T[k] <- A + d[k]`.

4.3.3.4 Example: Parkinson's

The data presented in Table 4.6 are the mean reduction in time off work (in hours) in patients given dopamine agonists as adjunct therapy in Parkinson's disease (Franchini et al., 2012). The data available are the mean, standard deviation and number of patients in each trial arm for seven studies of five different drugs: placebo, coded 1, and five active drugs, namely, pramipexole, ropinirole, bromocriptine and cabergoline, coded 2–5. The network diagram is presented in Figure 4.6.

The data are assumed to follow a normal likelihood and the WinBUGS code is changed as detailed in Section 4.3.3.3. For the random effects model, the upper bound for the prior distribution of the between-study standard deviation is chosen as 5, which is large in comparison to the expected mean differences in time off work (Table 4.6).

Values for the mean time off work on placebo and its precision are assumed to be −0.73 and 21, respectively, and the absolute effects for all treatments are calculated based on this.

Table 4.6 Parkinson's example: mean off-time reduction (*y*) with its standard deviation (sd) and total number of patients in each trial arm (*n*); treatment differences (diff) and standard error of the differences (se(diff)); where treatment 1 is a placebo and treatments 2–5 are active drugs. Data from Franchini et al. (2012).

Study	Treatment	y	sd	n	diff	se(diff)
1	1	−1.22	3.7	54	−0.31	0.668
	3	−1.53	4.28	95		
2	1	−0.7	3.7	172	−1.7	0.383
	2	−2.4	3.4	173		
3	1	−0.3	4.4	76		
	2	−2.6	4.3	71	−2.3	0.718
	4	−1.2	4.3	81	−0.9	0.695
4	3	−0.24	3	128	−0.35	0.442
	4	−0.59	3	72		
5	3	−0.73	3	80	0.55	0.555
	4	−0.18	3	46		
6	4	−2.2	2.31	137	−0.3	0.274
	5	−2.5	2.18	131		
7	4	−1.8	2.48	154	−0.3	0.320
	5	−2.1	2.99	143		

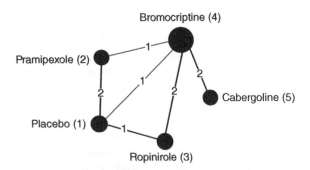

Figure 4.6 Parkinson's network: each circle represents a treatment, and connecting lines indicate pairs of treatments that have been directly compared in randomised trials. The numbers on the lines indicate the numbers of trials making that comparison, and the numbers by the treatment names are the treatment codes used in the modelling. Line thickness is proportional to the number of trials making that comparison, and the width of the circles is proportional to the number of patients randomised to that treatment. Adapted from Franchini et al. 2012.

To set up the data, note that the maximum number of arms is three, so three columns are needed for the treatment indicators, t [,]; the continuous outcomes, y [,]; and their standard errors, se [,].

```
# Data (Parkinson's example)
list(ns=7, nt=5 , meanA=-0.73, precA=21)
```

t[,1]	t[,2]	t[,3]	y[,1]	y[,2]	y[,3]	se[,1]	se[,2]	se[,3]	na[]
1	3	NA	-1.22	-1.53	NA	0.504	0.439	NA	2
1	2	NA	-0.7	-2.4	NA	0.282	0.258	NA	2
1	2	4	-0.3	-2.6	-1.2	0.505	0.510	0.478	3
3	4	NA	0.24	-0.59	NA	0.265	0.354	NA	2
3	4	NA	-0.73	-0.18	NA	0.335	0.442	NA	2
4	5	NA	-2.2	-2.5	NA	0.197	0.190	NA	2
4	5	NA	-1.8	-2.1	NA	0.200	0.250	NA	2
END									

No particular changes are required for the structure of the initial values, apart from accounting for the number of treatments and trials included in this particular example.

The full code with data and initial values for the fixed and random effects models for this example are presented in the online files *Ch4_FE_Normal_id.odc* and *Ch4_RE_Normal_id.odc*, respectively.

4.3.3.5 Results: Parkinson's

Results (based on three chains: 100,000 iterations after a burn-in of 50,000) are presented in Table 4.7. In this case p_D calculated externally or in WinBUGS are exactly the same since we have a normal likelihood and a linear model (Chapter 3). The random and fixed effects models both fit the data well, and since the random effects model has a higher DIC (due to having a higher effective number of parameters), the fixed effects model should be preferred. In addition, the posterior distribution of the between-study standard deviation (Figure 4.7) is concentrated around small values and peaked at values close to zero, lending further support to the choice of the fixed effects model.

The difference in mean of symptoms for each of the treatments compared with placebo and the absolute mean reduction in symptoms are given in Table 4.7 and Figure 4.8.

Treatment 2 (pramipexole) provides the largest reduction in time off work compared with placebo with posterior median of the mean difference −1.81 and 95% CrI (−2.46, −1.16). Other treatments appear to show no evidence of benefit compared with placebo.

Table 4.7 Parkinson's example: posterior median and 95% CrI for both fixed and random effects models for the treatment effects of treatments 2–5 (d_{12}–d_{15}) relative to placebo, absolute effects of placebo (T_1) and treatments 2–5 (T_2–T_5), heterogeneity parameter σ and model fit statistics for the data presented as arm-level means and standard errors.

	Fixed effects model		Random effects model	
	Median	95% CrI	Median	95% CrI
d_{12}	−1.81	(−2.46, −1.16)	−1.84	(−2.91, −0.85)
d_{13}	−0.47	(−1.43, 0.49)	−0.50	(−1.78, 0.75)
d_{14}	−0.52	(−1.46, 0.43)	−0.53	(−1.77, 0.71)
d_{15}	−0.82	(−1.84, 0.22)	−0.83	(−2.35, 0.69)
T_1	−0.73	(−1.16, −0.30)	−0.73	(−1.16, −0.30)
T_2	−2.54	(−3.32, −1.76)	−2.57	(−3.72, −1.50)
T_3	−1.20	(−2.25, −0.15)	−1.23	(−2.57, 0.10)
T_4	−1.25	(−2.28, −0.21)	−1.26	(−2.57, 0.05)
T_5	−1.55	(−2.66, −0.43)	−1.56	(−3.14, 0.02)
σ	−	−	0.28	(0.01, 1.55)
$\bar{D}_{res}{}^a$	13.3		13.6	
p_D	11.0		12.4	
DIC	24.3		26.0	

^{a} Compare to 15 data points.

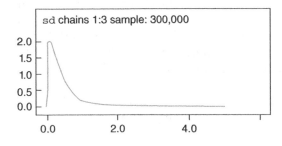

Figure 4.7 Parkinson's example: posterior distribution of the between-study standard deviation – from WinBUGS.

4.4 Contrast-Based Likelihoods

Trial results are sometimes only available as overall, trial-based summary measures, for example, as mean differences or SMD between treatments, log odds ratios, log relative risks (risk ratios), log hazard ratios, risk differences or some other trial summary statistic and its sample variance. In such cases, we can often assume a normal distribution for the continuous measure of

Figure 4.8 Parkinson's example: caterpillar plot showing posterior medians and 95% CrI of the mean differences of all treatments compared to the reference for the fixed effects model. The treatment being compared is indicated in square brackets; negative values favour this treatment – from WinBUGS.

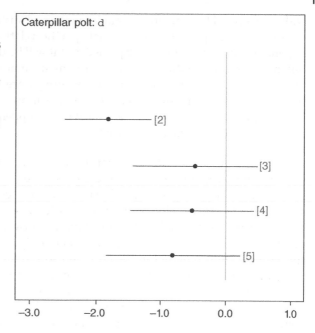

treatment effect of arm k relative to arm 1 in trial i. Although we might expect the normality assumption to hold for mean difference measures, we might doubt that even approximate normality will hold for log odds ratio, log hazard ratio or log relative risks measures, particularly when the sample sizes or number of events are small. In addition it often required to add 0.5 to the events and non-events in trial with zero cells, which can lead to biased estimates (see Chapter 6). Nevertheless, the assumption of (approximate) normality is likely to be reasonable in most cases and is, in fact, the standard assumption in most frequentist meta-analyses.

4.4.1 Continuous Data: Treatment Differences

Let y_{ik} define the continuous measure of treatment effect of arm k relative to arm 1 in trial i, with variance V_{ik}, $(k \geq 2)$ such that

$$y_{ik} \sim \text{Normal}(\delta_{ik}, V_{ik}) \tag{4.9}$$

The parameters of interest are the trial-specific mean treatment effects δ_{ik} that the data inform directly. An identity link is used, and no trial-specific effects of the control treatment need to be estimated, so the trial-specific baselines are eliminated and the linear predictor is reduced to

$$\theta_{ik} = \delta_{ik}$$

with δ_{ik} assumed to come from a random effects distribution or to be fixed, as before. The interpretation of the trial-specific relative effects δ_{ik} will differ depending on the measure being pooled, but it will be the relative effect (mean difference, log odds ratio, etc.) of the treatment in arm k compared with the treatment in arm 1 of trial i. The case where y_{ik} are log odds ratios and an inverse-variance weighting is applied, with variance based on the normal theory approximation, remains a mainstay in applied meta-analytic studies even when arm-level data are available.

4.4.1.1 Multi-Arm Trials with Treatment Differences (Trial-Based Summaries)

When results from multi-arm trials are presented as (continuous) treatment differences relative to the control arm (arm 1), a correlation between the treatment differences is induced, since all differences are taken relative to the same control arm. Unlike the correlations between the random effects parameters described in Chapter 2 that applies only to random effects models, this correlation is inherent in the *data* and so requires an adjustment to the *likelihood* (i.e. for both fixed and random effects models).

A trial with $a_i > 2$ arms produces $a_i - 1$ treatment differences that are correlated. The covariance between differences taken with respect to the same control arm is equal to the observed variance for the common control arm (Franchini et al., 2012) on the appropriate scale. So, for example, in a three-arm trial comparing treatments A, B and C, letting y_{AB} and y_{AC} represent the treatment differences (e.g. mean differences, log odds ratios, etc.) of treatments B and C relative to treatment A, we know that

$$Var(y_{AB} - y_{AC}) = Var(y_{AB}) + Var(y_{AC}) - 2Cov(y_{AB}, y_{AC}) \qquad (4.10)$$

and

$$\begin{aligned} Var(y_{AB}) &= Var(y_A) + Var(y_B) \\ Var(y_{AC}) &= Var(y_A) + Var(y_C) \end{aligned} \qquad (4.11)$$

with y_A, y_B and y_C representing the original measurements on each arm of the trial (e.g. mean, log odds, etc.), because, in an RCT, the measurements in each trial arm are independent. By successive replacement of the expressions in equation (4.11) into equation (4.10), we can see that $Cov(y_{AB}, y_{AC}) = Var(y_A)$.

So, the likelihood for a trial i with a_i arms is multivariate normal:

$$\begin{pmatrix} y_{i2} \\ y_{i3} \\ \vdots \\ y_{i,a_i} \end{pmatrix} \sim \text{Normal}_{a_i - 1} \left(\begin{pmatrix} \theta_{i2} \\ \theta_{i3} \\ \vdots \\ \theta_{i,a_i} \end{pmatrix}, \begin{bmatrix} V_{i2} & se_{i1}^2 & \cdots & se_{i1}^2 \\ se_{i1}^2 & V_{i3} & \cdots & se_{i1}^2 \\ \vdots & \vdots & \ddots & \vdots \\ se_{i1}^2 & se_{i1}^2 & \cdots & V_{i,a_i} \end{bmatrix} \right)$$

where the diagonal elements in the variance–covariance matrix represent the variances of the treatment differences and the off-diagonal elements represent the observed variance in the control arm in trial i, denoted by se_{i1}^2 (see Section 4.3.3). For example, when the treatment differences are given as log odds ratios, se_{i1}^2 is the variance of the log odds for arm 1 of trial i. When an SMD is used, we have $Cov(SMD_{AB}, SMD_{AC}) = Var(y_A)/s^2$ where s represents the standardising constant used for that trial and y_A is the original (unstandardised) mean.

When this variance is not reported, we may be able to calculate it if enough data are provided on all treatment contrasts. If a three-arm trial comparing treatments A, B and C reports the relative effects y_{AB}, y_{AC} and y_{BC} with their variances, the variance in the control arm A of that trial can be calculated as

$$Var(y_A) = \frac{Var(y_{AB}) + Var(y_{AC}) - Var(y_{BC})}{2} \qquad (4.12)$$

using an extension of equation (4.10).

If this is not available, an approximation should be made, perhaps based on the variances of the differences (Tierney et al., 2007; Woods et al., 2010), or if the value of the control variance is available only for some of the included trials, that information can be used to estimate the parameters of a distribution for the control variance (assumed to be common). This estimated distribution can then be used to predict the variance of the control arm in the trials where it is missing. This method has been used, in a slightly different context, to predict missing variances (Dakin et al., 2011). Riley (2009) provides a review of methods to impute unknown within-study correlations within the context of multivariate meta-analysis. These methods can also be applied to network meta-analysis with multi-arm trials.

However, note that when the covariance is approximated or estimated from external sources, care should be taken to ensure that the resulting variance–covariance matrix is still positive definite (i.e. invertible) as otherwise any inferences would be invalid (and WinBUGS model will not run!).

4.4.1.2 *WinBUGS Implementation

The code to fit this model has different components for the likelihood and deviance definitions that depend on the number of treatment arms in the included studies. This requires users to set up the data file with all two-arm trials first, then three-arm trials, then – if any are present – four-arm trials and so on.

For two-arm studies, the code is obtained by specifying the normal likelihood in the generic code as given in equation (4.9). This automatically specifies the link as identity (i.e. no function). We also need to take care to only loop over studies with two arms and hence with data on only one relative effect.

Assuming studies with only two arms are in the first `ns2` rows of data, the likelihood and residual deviance contribution are defined as

```
for(i in 1:ns2){                        # LOOP THROUGH 2-ARM STUDIES
  y[i,2] ~ dnorm(delta[i,2],prec[i,2]) # normal likelihood for 2-arm trials
  resdev[i] <- (y[i,2]-delta[i,2])*(y[i,2]-delta[i,2])*prec[i,2]
                                        #Deviance contribution for trial i
}
```

Assuming studies with three arms are given in the next `ns3` rows of data, the likelihood and residual deviance contribution for those data are defined as

```
for(i in (ns2+1):(ns2+ns3)) {          # LOOP THROUGH THREE-ARM STUDIES
    for (k in 1:(na[i]-1)) {           # set variance-covariance matrix
      for (j in 1:(na[i]-1)) {
        Sigma[i,j,k] <- V[i]*(1-equals(j,k)) + var[i,k+1]*equals(j,k)
        }
      }
    Omega[i,1:(na[i]-1),1:(na[i]-1)] <- inverse(Sigma[i,,])   #Precision
                                                              matrix
# multivariate normal likelihood for 3-arm trials
  y[i,2:na[i]] ~ dmnorm(delta[i,2:na[i]],Omega[i,1:(na[i]-1),
                                               1:(na[i]-1)])
#Deviance contribution for trial i
  for (k in 1:(na[i]-1)){  # multiply vector & matrix
    ydiff[i,k]<- y[i,(k+1)] - delta[i,(k+1)]
    z[i,k]<- inprod2(Omega[i,k,1:(na[i]-1)], ydiff[i,1:(na[i]-1)])
    }
  resdev[i]<- inprod2(ydiff[i,1:(na[i]-1)], z[i,1:(na[i]-1)])
}
```

The matrix `Sigma` will hold the observed variances of the relative effects in its diagonal and their covariance, given in the data as `V`, in the off-diagonal. This is achieved using the `equals` function in WinBUGS that returns a value of 1 when its two elements are equal and a zero when they are not. Careful working through the loops and code will show that this gives the correct variance–covariance matrix for each trial. `Omega` is then the precision matrix, taken as the inverse of `Sigma`. The code to obtain the deviance contributions for each data point simply translates the formula in Table 4.1 into WinBUGS code, requiring the use of a function to multiply vectors. The loop enclosing these lines of code only runs through the three-arm trials.

Finally we add two rows of code to calculate the trial-specific variances and precisions, since we assume the data will be given as means and their standard errors. This code will apply to trials with any number of arms, and so, if we only have two- and three-arm trials, we write

```
for(i in 1:(ns2+ns3)){              #   LOOP THROUGH ALL STUDIES
  for (k in 2:na[i]) {              #   LOOP THROUGH ARMS
    var[i,k] <- pow(se[i,k],2)      # calculate variances
    prec[i,k] <- 1/var[i,k]         # set precisions
    delta[i,k] <-  d[t[i,k]] - d[t[i,1]]
  }
}
```

If data were already given as variances or precisions, the corresponding lines of code should be deleted. The last line of code implements a fixed effects model by making delta fixed. It should be omitted in a random effects model, and instead the random effects code given in the generic model should be used. In addition we also need to specify the upper bound for the between-trial standard deviation. This value will depend on the specific example and the plausible values for the relative effects being considered, that is, mean differences, log odds ratios, and so on.

If trials with four or more arms were included, a further multivariate normal likelihood statement would need to be added for each trial type, and the corresponding variance–covariance and precision matrices built (Sigma2 and Omega2, say). So, for example, if ns4 four-arm trials were available, we would add the following lines of code, taking care to change all the relevant loops to go through all trials:

```
for(i in (ns2+ns3+1):(ns2+ns3+ns4)) {  # LOOP THROUGH 4-ARM STUDIES
  for (k in 1:(na[i]-1)) {              # set variance-covariance matrix
    for (j in 1:(na[i]-1)) {  Sigma2[i,j,k] <- V[i]*(1-equals(j,k)) +
    var[i,k+1]*equals(j,k)   }
  }
  Omega2[i,1:(na[i]-1),1:(na[i]-1)] <- inverse(Sigma2[i,,])
                                                   #Precision matrix
# multivariate normal likelihood for 4-arm trials
  y[i,2:na[i]] ~ dmnorm(delta[i,2:na[i]],Omega2[i,1:(na[i]-1),
                                          1:(na[i]-1)])
}
```

Note that the code to generate absolute effects and all relative treatment effects needs to account for the scale of the relative effects. Depending on the scale, the code in previous sections can be chosen, as appropriate.

4.4.1.3   Example: Parkinson's (Treatment Differences as Data)

We now assume that the data available for the Parkinson's example were not the mean off-time reduction for patients in each arm of the trial, but rather the *differences* in off-time reduction, and their standard errors, between the intervention and control arms for each trial, presented in the last two columns of Table 4.6. We need to account for the three-arm trial, which will be given in the last row of the WinBUGS data. Given values for the mean, meanA=-0.73, and precision, `precA=21`, of the effects on treatment 1, from external sources, the absolute effects for all other treatments are calculated.

The data structure is similar to that in Section 4.3.3.4, but now the y columns represent the relative effects (in this example, the mean differences) of the treatment in arm *k* compared with the treatment in arm 1, so `y[,1]` does not need to be specified. We also have to specify the number of two-arm trials `ns2` and the number of three-arm trials `ns3`. For a trial with three treatment arms, two treatment differences will be available, so two vectors of differences (the continuous outcomes) `y[,]` and their standard errors `se[,]` are needed as well as specifying `V[]` the variance of the control treatment in that trial (needed to adjust for the correlation in multi-arm trials – note that this variable only needs to have values assigned when there are multi-arm trials), with `NA` denoting a missing observation. Note also that the three-arm trial needs to appear at the end of the column format data.

```
# Data (Parkinson's example - trial-level data: treatment differences)
list(ns2=6, ns3=1, nt=5 , meanA=-0.73, precA=21)
```

t[,1]	t[,2]	t[,3]	y[,2]	y[,3]	se[,2]	se[,3]	na[]	V[]
1	3	NA	-0.31	NA	0.668089651	NA	2	NA
1	2	NA	-1.7	NA	0.382640605	NA	2	NA
3	4	NA	-0.35	NA	0.441941738	NA	2	NA
3	4	NA	0.55	NA	0.555114559	NA	2	NA
4	5	NA	-0.3	NA	0.274276316	NA	2	NA
4	5	NA	-0.3	NA	0.320087245	NA	2	NA
1	2	4	-2.3	-0.9	0.71774604	0.694988091	3	0.254736842
END								

No particular changes are required for the structure of the initial values model, apart from accounting for the number of treatments and trials included in this particular example and removing the initial values for mu, which are not required in this model.

The full code with data and initial values for the fixed and random effects models for this example are presented in the online files *Ch4_FE_NormalDiff_id.odc* and *Ch4_RE_NormalDiff_id.odc*, respectively. When all included trials have only two arms, users can wither delete the code that only applies to three-arm studies, keeping only the lines relevant to two-arm studies, or add an empty three-arm study, ensuring the covariance matrix for the relative effects is invertible. For example, the following three-arm study could be added:

t[,1]	t[,2]	t[,3]	y[,2]	y[,3]	se[,2]	se[,3]	na[]	V[]
...	[data for all two arm studies]	...						
1	1	1	NA	NA	1	1	3	0

This will not impact on results but will allow the code to run.

Table 4.8 Parkinson's example (treatment differences): posterior median and 95% CrI for both fixed and random effects models for the treatment effects of treatments 2–5 (d_{12}–d_{15}) relative to placebo, absolute effects of placebo (T_1) and treatments 2–5 (T_2–T_5), heterogeneity parameter σ and model fit statistics for trial-level data.

	Fixed effects model		Random effects model	
	Median	95% CrI	Median	95% CrI
d_{12}	−1.81	(−2.47, −1.16)	−1.84	(−2.92, −0.84)
d_{13}	−0.48	(−1.43, 0.47)	−0.49	(−1.79, 0.75)
d_{14}	−0.52	(−1.46, 0.42)	−0.53	(−1.79, 0.72)
d_{15}	−0.82	(−1.84, 0.20)	−0.83	(−2.38, 0.69)
T_1	−0.73	(−1.16, −0.30)	−0.73	(−1.16, −0.30)
T_2	−2.54	(−3.33, −1.76)	−2.57	(−3.72, −1.49)
T_3	−1.21	(−2.25, −0.17)	−1.22	(−2.59, 0.10)
T_4	−1.25	(−2.28, −0.22)	−1.26	(−2.59, 0.06)
T_5	−1.55	(−2.66, −0.44)	−1.56	(−3.17, 0.01)
σ			0.28	(0.01, 1.56)
$\bar{D}_{res}$ [a]	6.3		6.6	
p_D	4.0		5.5	
DIC	10.3		12.1	

[a] Compare to eight data points.

4.4.1.4 Results: Parkinson's (Treatment Differences as Data)

Results (based on three chains: 100,000 iterations after a burn-in of 50,000) are presented in Table 4.8 and are the same as the results obtained in Section 4.3.3.5, where the results are discussed. Model fit results differ since we now have fewer data points and parameters to estimate, but the overall conclusions are the same.

4.5 *Multinomial Likelihoods

In some applications multiple, mutually exclusive binary outcomes are reported, and the statistical dependencies between the competing outcomes need to be taken into account in the model. These dependencies are essentially within-trial negative correlations between outcomes, applying in each arm of each trial. They arise because the occurrence of events is a stochastic process, and if more patients should by chance reach one outcome, then fewer must reach the others. The network meta-analysis model is defined as before, but the likelihood depends on data generation and interpretation and must reflect the relationships between outcomes. For example, the outcomes response and discontinuation can be thought of as having a joint multinomial likelihood but can also be implemented as two separate models with binomial likelihoods where denominators are adjusted to give independent outcomes, for example, discontinuation out of all randomised patients and response in patients who have not discontinued.

In this section we present examples of multinomial data where the models used are chosen taking into account the underlying data generation process and the interpretation of parameters. Further similar examples can be found in the literature (National Collaborating Centre for Mental Health, 2014; Alfirevic et al., 2016).

Note that in the models presented in this section, the between-study correlation in outcomes is implicit in the model structure – see Chapter 11 for further discussion.

4.5.1 Ordered Categorical Data: Multinomial Likelihood and Probit Link

In some applications, the data generated by each trial may be continuous, but the outcome measure categorised, using *one or more* predefined cut-offs. Examples include the Psoriasis Area Severity Index (PASI) and the American College of Rheumatology (ACR) scales, where it is common to report the percentage of patients who have improved by more than certain benchmark relative amounts. Thus ACR-20 would represent the proportion of patients who have improved by at least 20% on the ACR scale, PASI-75 the proportion who have improved by at least 75% on the PASI scale, and so on. Trials may report ACR-20, ACR-50 and ACR-70 or only one or two of these endpoints. We can provide a coherent model and make efficient use of such data by assuming that the treatment effect is the same regardless of the cut-off. This assumption can be checked informally by examining the relative treatment effects at different cut-offs in each trial and seeing if they are approximately the same. In particular, there should not be a systematic relationship between the relative effects at different cut-off points. The residual deviance check of model fit (Chapter 3) is also a useful guide.

Assuming that there is an underlying continuous distribution for Y_{ik}, the response of patients in arm k of trial i (e.g. the percentage improvement on the ACR or PASI scale), which has been split into a number of thresholds, nc_i, in trial i, we can define $X_0, X_1, ..., X_c$ as the ordered cut-points (thresholds) for this continuous distribution, which have been reported in any of the included trials, that is, we use the greatest number of subdivisions of the original scale Y, so that all reported thresholds are represented. X_0 will represent the scale's natural lower bound (e.g. zero or minus infinity), and c will be the total number of cut-points reported in the included trials. Figure 4.9 provides an example.

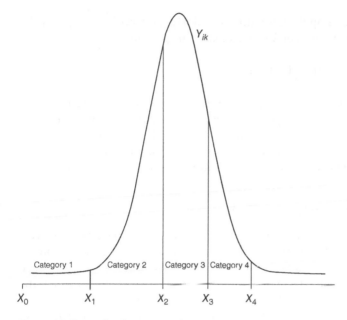

Figure 4.9 Example of categorised measurement scale as reported in arm k of trial i (assumed normal) where five categories are defined by thresholds X_0 to X_4 (the upper bound of the last category is the scale's natural upper bound).

Trials often report the number of patients in arm k of trial i achieving at least cut-point X_j. Letting π_{ikj} denote the probability that individuals in arm k of trial i achieve at least cut-point X_j, if we can assume that the underlying measure has a standard normal distribution, we may use the probit link function to map this probability onto the real line as

$$\pi_{ikj} = \Pr\left(Y_{ik} > X_j\right) = 1 - \Phi\left(\theta_{ikj}\right)$$
$$\Phi\left(\theta_{ikj}\right) = 1 - \pi_{ikj}$$

where Φ represents the cumulative normal distribution function (see Table 4.1) and $\pi_{ik0} = 1$. So the linear predictor can be written as

$$\theta_{ikj} = \Phi^{-1}\left(1 - \pi_{ikj}\right) = \mu_{ij} + \delta_{ik} \tag{4.13}$$

In this set-up, the pooled effect of taking the treatment in arm k instead of the treatment in arm 1 is to change the probit score (or Z score) of arm 1 by δ_{ik} standard deviations.

This can be translated back into probabilities of achieving at least cut-point X_j on each treatment k, $\pi_j^{(k)}$, by noting that when the pooled treatment effect

$d_{1k} > 0$, for a patient population with an underlying probability $\pi_j^{(1)}$ on the reference treatment, treatment k will decrease this probability to

$$\pi_j^{(k)} = 1 - \Phi\left(d_{1k} + \Phi^{-1}\left(1 - \pi_j^{(1)}\right)\right) \tag{4.14}$$

In general we can assume that the underlying continuous variable has been categorised by specifying different cut-offs, z_{ij} (which may be different in different trials), corresponding to the point at which an individual moves from one category to the next in trial i. Several options are available regarding the relationship between outcomes within each arm. Rewriting equation (4.13) as

$$1 - \pi_{ikj} = \Phi\left(\theta_{ikj}\right) = \Phi\left(\mu_i + z_{ij} + \delta_{ik}\right) \tag{4.15}$$

we can consider the terms z_{ij} as the cut-points on the standard normal scale defining category j in all the arms of trial i, that is, they correspond to X_j on the probit scale. One option is to assume a 'fixed effect' across trials $z_{ij} = z_j$ for each of the categories over all trials i or a 'random effect' across trials in which the trial-specific terms are drawn from a distribution, but are the same for arms within a trial. Cut-point z_1 is set to zero and prior distributions are given to the remaining z_j, taking care to ensure that they are increasing with category (i.e. are ordered). Choice of model can be made on the basis of DIC.

If we have information on the probability of belonging to the first category for patients on treatment 1, we can obtain the probabilities of being in all categories for all treatments by noting that equation (4.14) can be rewritten as

$$\pi_j^{(1)} = 1 - \Phi\left(A + z_j\right) \tag{4.16}$$

and that $z_1 = 0$ so that $A = \Phi^{-1}\left(1 - \pi_1^{(1)}\right)$ represents the probability of being in category 1 on treatment 1, on the probit scale. Given some external information on A, we can use equation (4.16) to obtain all required probabilities.

To define the likelihood we define r_{ikj} as the number of patients in arm k of trial i belonging to different mutually exclusive categories $j = 1, 2, \ldots, nc_i$, where these categories represent the different thresholds (e.g. 20, 50 or 70% improvement), on a common underlying continuous scale. In trials reporting the number of patients belonging to category j or higher (e.g. at least 20% improvement, at least 50% improvement, etc.), r_{ikj} can be obtained by subtraction of patients achieving at least threshold X_j from the number of patients achieving at least threshold X_{j-1}. The number of patients in each arm k of trial i belonging to category j will follow a multinomial distribution:

$$r_{ik,j=1,\ldots,nc_i} \sim \text{Multinomial}\left(p_{ik,j=1,\ldots,nc_i}, n_{ik}\right) \quad \text{with} \quad \sum_{j=1}^{nc_i} p_{ikj} = 1 \tag{4.17}$$

where the parameters of interest are the probabilities, p_{ikj}, that a patient in arm k of trial i belongs to category j that can be defined as

$$p_{ikj} = \Pr\left(X_{j-1} < Y_{ik} < X_j\right) = \pi_{ik(j-1)} - \pi_{ikj}$$

Because in general different trials report the number of patients achieving different cut-points (which can be transformed in the number of patients belonging to different categories) and the categories may overlap, it is useful to rewrite the multinomial likelihood in equation (4.17) as a series of conditional binomial distributions (McCullagh and Nelder, 1989). So, for a trial i reporting the number of patients r_{ikj} in category $j = 1, ..., nc_i$, we can write

$$r_{ikj} \sim \text{Binomial}\left(q_{ikj}, N_{ikj}\right) \quad j = 1, ..., nc_i - 1 \tag{4.18}$$

where q_{ikj} are the probabilities that patients in arm k of trial i belong to category j, given that they are not in the lower categories $1, ..., j - 1$, defined as

$$
\begin{aligned}
q_{ik1} &= \Pr\left(\text{individual in arm } k \text{ of trial } i \text{ is in category } 1\right) \\
&= \Pr\left(X_0 < Y_{ik} < X_1\right) \\
q_{ik2} &= \Pr\left(\text{individual in arm } k \text{ of trial } i \text{ is in category } 2 \mid \text{not in category } 1\right) \\
&= \Pr\left(X_1 < Y_{ik} < X_2 \mid Y_{ik} > X_1\right) \\
&\vdots \\
q_{ikj} &= \Pr\left(\text{individual in arm } k \text{ of trial } i \text{ is in category } j \mid \text{not in categories } 1,2,...,j-1\right) \\
&= \Pr\left(X_{j-1} < Y_{ik} < X_j \mid Y_{ik} > X_{j\,1}\right) \\
&\vdots
\end{aligned}
$$

and $N_{ikj} = n_{ik} - \sum_{u=1}^{j-1} r_{iku}$ defines the number of patients in arm k of trial i who are not in categories 1 to $j - 1$. Note that this formulation preserves the inherent correlation between the probabilities of being in each category while defining (conditionally) independent likelihoods.

It can be shown that for $j = 1, ..., nc_{i-1}$

$$q_{ikj} = 1 - \frac{\Pr\left(Y_{ik} > X_j\right)}{\Pr\left(Y_{ik} > X_{j-1}\right)} = 1 - \frac{\pi_{ikj}}{\pi_{ik(j-1)}} \tag{4.19}$$

which links the probabilities in the likelihood (equation (4.18)) to the linear predictor, which defines the treatment effects (equation (4.13)).

This model is appropriate in cases where different trials use different thresholds or when different trials report different numbers of thresholds. There is, in fact, no particular requirement for trials to even use the same underlying scale, although this could require an expansion of the number of categories.

Note that, unless the response probabilities are very extreme, the probit model suggested here (which assumes an underlying normal distribution) will be undistinguishable from the logit model (which assumes a logistic distribution) in terms of model fit or DIC. Choosing which link function to use should therefore be based on the data generating process and on the interpretability of the results.

4.5.1.1 WinBUGS Implementation

The code to fit the probit model outlined earlier is obtained by specifying the likelihood as binomial. Further changes to the generic code are required to ensure that we loop over studies, arms and the categories reported in each trial and that all deviance contributions are summed over. Thus, for a fixed effects model, the WinBUGS code is as follows.

WinBUGS code for NMA: Conditional Binomial likelihood, probit link, fixed effect model. The full code with data and initial values is presented in the online file *Ch4_FE_CondBi_probit.odc*.

```
# Conditional Binomial likelihoods, probit link (different categories)
# Fixed effect model
model{                              # *** PROGRAM STARTS
for(i in 1:ns){                     #   LOOP THROUGH STUDIES
 mu[i] ~ dnorm(0,.0001)            # vague priors for all trial
                                      baselines
  for (k in 1:na[i]) {              # LOOP THROUGH ARMS
   pi[i,k,1] <- 1                   # Pr(Yik > X0) = 1
   for (j in 1:nc[i]-1) {           # LOOP THROUGH CATEGORIES
    r[i,k,j] ~ dbin(q[i,k,j],N[i,k,j])  # binomial likelihood
    q[i,k,j] <- 1-(pi[i,k,C[i,j+1]]/pi[i,k,C[i,j]])   # conditional
                                                        probabilities
# linear predictor, assuming fixed cut-points across trials
     theta[i,k,j] <- mu[i] + d[t[i,k]] - d[t[i,1]] + z[C[i,j+1]-1]
     pi[i,k,C[i,j+1]] <- 1 - phi.adj[i,k,j+1] # link function
# adjust phi(x) for extreme values that can give numerical errors
# when x< -5, phi(x)=0, when x> 5, phi(x)=1
     phi.adj[i,k,j+1] <- step(5+theta[i,k,j])
     *( step(theta[i,k,j]-5) + step(5-theta[i,k,j])
     * (1-equals(5,theta[i,k,j])) * phi(theta[i,k,j]) )
     rhat[i,k,j] <-  q[i,k,j] * N[i,k,j]    # predicted number of events
# Deviance contribution for each category
     dv[i,k,j] <- 2 * (r[i,k,j]*(log(r[i,k,j])-log(rhat[i,k,j]))
     +(N[i,k,j]-r[i,k,j])*(log(N[i,k,j]-r[i,k,j])
     - log(N[i,k,j]-rhat[i,k,j])))
    }
   dev[i,k] <- sum(dv[i,k,1:nc[i]-1])  # deviance contribution of each arm
   }
  resdev[i] <- sum(dev[i,1:na[i]])    # deviance contribution for each
                                        trial
  }
```

```
z[1] <- 0                      # set first cut-off X1 to zero on
                                 probit scale
for (j in 2:maxc) {            # set ordered priors for z, for
                                 any number of cut points
   z.aux[j] ~ dunif(0,5)       # priors
   z[j] <- z[j-1] + z.aux[j]   # ensures z[j]~Uniform(z[j-1],
                                            z[j-1]+5)
}
totresdev <- sum(resdev[])     # Total Residual Deviance
d[1] <- 0                      # treatment effect is zero for
                                 reference treatment
for (k in 2:nt){  d[k] ~ dnorm(0,.0001) }    # vague priors for
                                               treatment effects
}                              # *** PROGRAM ENDS
```

The conditional binomial likelihoods are defined for each category reported in each trial, with the number of patients in each category and the number of patients not in lower categories given as data. The linear predictor and link function are defined as described in equation (4.15), but the link function includes an adjustment to the cumulative normal distribution function to prevent numerical errors when theta[i,k,j] is too large (>5) or too small (<−5). The conditional probabilities q[i,k,j] are defined as described in equation (4.19), where a new matrix C[i,j] is given as data. Elements of this matrix indicate the lowest threshold defining the jth category reported in trial i (since trials may report a different subset of categories). The first category in each trial is always defined by the scale's lower bound, so C[i,1] = 1 for all studies. However, the lower bound of the second category reported in trial i may be defined by different cut-points in each trial, so C[i,2] will be $l+1$ when this cut-point corresponds to X_l (note that C[i,1] = 1 corresponds to X_0). The other elements of matrix C are similarly defined. The remaining code defines the residual deviance contributions for a binomial distribution (Table 4.1), which has to be summed over category, arm and trial and sets out the appropriate prior distributions for the relative effects and cut-points on the probit scale, z[j].

Extra code to calculate the treatment differences on the probit scale or probabilities that a treatment is best can be added as before, taking care to note that positive values for d[k] will mean that treatment k is worse than the reference treatment. Absolute probabilities of achieving at least threshold j with any treatment can be obtained using equation (4.16). The implementation of the random effects model accounting for the correlations in multi-arm trials is similar (see *Ch4_RE_CondBi_probit.odc*).

4.5.1.2 Example: Psoriasis

In a health technology assessment (HTA) report to evaluate the effectiveness of treatments for moderate to severe chronic plaque psoriasis (Woolacott et al., 2006),

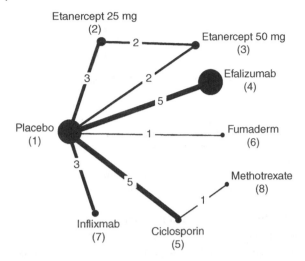

Figure 4.10 Psoriasis network: each circle represents a treatment, and connecting lines indicate pairs of treatments that have been directly compared in randomised trials. The numbers on the lines indicate the numbers of trials making that comparison, and the numbers by the treatment names are the treatment codes used in the modelling. Line thickness is proportional to the number of trials making that comparison, and the width of the circles is proportional to the number of patients randomised to that treatment.

16 trials comparing eight treatments were identified: supportive care (coded 1), etanercept 25 mg (2), etanercept 50 mg (3), efalizumab (4), ciclosporin (5), fumaderm (6), infliximab (7) and methotrexate (8). The network diagram is presented in Figure 4.10. Each trial reported the number of patients in mutually exclusive categories representing the percentage improvement in symptoms as measured by the PASI score. Different trials reported on different categories defining three cut-points, 50, 75 and 90% improvement, in addition to the scale's lower and upper bounds (0 and 100% improvement, respectively). In the following code, we define $C = 1$ representing 0% improvement (the scale's lower bound), $C = 2$ representing 50% improvement, $C = 3$ representing 75% improvement and $C = 4$ representing 90% improvement. The data are presented in Table 4.9.

Given values for the mean, 1.116, and standard deviation, 0.1007, of the probabilities of being in category 1 on placebo on the probit scale, from external sources, absolute probabilities $T[j,k]$ of having over 50, 75 or 90% improvement ($j = 1, 2, 3$, respectively) on treatment k were also calculated.

The WinBUGS code is as detailed earlier. Data that must be given are a list specifying the number of treatments nt, number of studies ns and the maximum number of cut-points maxc, with the main body of data in a vector format. Note that we need to add a column nc[] to hold number of cut-offs in that trial and matrix C[,1], ..., C[,4] that holds the cut-offs used to define the

Table 4.9 Psoriasis example: study names, treatments compared, total number of patients with different percentage improvement and total number of patients in each trial arm, where supportive care = treatment 1, etanercept 25 mg = 2, etanercept 50 mg = 3, efalizumab = 4, ciclosporin = 5, fumaderm = 6, infliximab = 7 and methotrexate = 8. Data from Woolacott et al. (2006).

Trials presenting outcomes in four categories

Trial	Arm 1 t[,1]	Arm 2 t[,2]	Arm 3 t[,3]	Arm 1: 0–50	50–75	75–90	90–100	n[,1]	Arm 2: 0–50	50–75	75–90	90–100	n[,2]	Arm 3: 0–50	50–75	75–90	90–100	n[,3]
1. Elewski 2004	1	2	3	175	12	5	1	193	70	59	46	21	196	44	54	56	40	194
2. Gottlieb 2003	1	2		49	5	1	0	55	17	23	11	6	57					
3. Lebwohl 2003	1	4		103	13	5	1	122	112	68	42	10	232					
4. Leonardi 2003	1	2	3	142	18	5	1	166	68	39	36	19	162	43	40	45	36	164

Trials presenting outcomes in three categories

Trial	Arm 1 t[,1]	Arm 2 t[,2]	Arm 3 t[,3]	Arm 1: 0–50	50–75	75–100	n[,1]	Arm 2: 0–50	50–75	75–100	n[,2]
5. Gordon 2003	1	4		161	18	8	187	153	118	98	369

Trials presenting outcomes in two categories (PASI 50)

Trial	Arm 1 t[,1]	Arm 2 t[,2]	Arm 3 t[,3]	Arm 1: 0–50	50–100	n[,1]	Arm 2: 0–50	50–100	n[,2]
6. ACD2058g	1	4		145	25	170	63	99	162
7. ACD2600g	1	4		230	33	253	216	234	450
8. Guenther 1991	1	5		10	1	11	0	12	12
9. IMP24011	1	4		226	38	264	245	284	529

(*Continued*)

Table 4.9 (Continued)

Trial	t[,1]	t[,2]	t[,3]	Trials presenting outcomes in two categories (PASI 75)								
				0–75	75–100		0–75	75–100		0–75	75–100	
10. Altmeyer 1994	1	6		50	1	51	37	12	49			
11. Chaudari 2001	1	7		9	2	11	2	9	11			
12. Ellis 1991	1	5	5	25	0	25	16	9	25	7	13	20
13. Gottlieb 2004	1	7	7	48	3	51	28	71	99	12	87	99
14. Heydendael 2003	5	8		12	30	42	17	26	43			
15. Meffert 1997	1	5		41	2	43	37	4	41			
16. Van Joost 1988	1	5		10	0	10	3	7	10			

categories reported in each trial – four columns are needed as the maximum number of cut-offs given in a trial is four:

C[,1]	C[,2]	C[,3]	C[,4]	#	Trial	Year
1	2	3	4	#	Elewski	2004
1	2	3	4	#	Gottlieb	2003
1	2	3	4	#	Lebwohl	2003
1	2	3	4	#	Leonardi	2003
1	2	3	NA	#	Gordon	2003
1	2	NA	NA	#	ACD2058g	2004
1	2	NA	NA	#	ACD2600g	2004
1	2	NA	NA	#	Guenther	1991
1	2	NA	NA	#	IMP24011	2004
1	3	NA	NA	#	Altmeyer	1994
1	3	NA	NA	#	Chaudari	2001
1	3	NA	NA	#	Ellis	1991
1	3	NA	NA	#	Gottlieb	2004
1	3	NA	NA	#	Heydendael	2003
1	3	NA	NA	#	Meffert	1997
1	3	NA	NA	#	Van Joost	1988

The number of events for the kth treatment, in category j given the number of events in categories 1 to $j-1$, r[,k,j], needs to be given as well as n[,k,1], which represents the total number of individuals in trial arm k; n[,k,2], which represents the total number of individuals in trial arm k, which were not in category 1 of that trial; and n[,k,3], which represents the total number of individuals in trial arm k, which were not in category 1 or 2 of that trial. For example, for the first trial in Table 4.9, Elewski 2004, 193 patients were included in arm 1 of the trial. These patients were split between the four categories as follows: 175 out of 193 patients had between 0 and 50% improvement, leaving 18 patients who could belong to any of the other categories, thus $r_{111} = 173$, $n_{111} = 193$; and 12 out of 18 patients had between 50 and 75% improvement, leaving six patients who could belong to any of the other categories, thus $r_{112} = 12$, $n_{112} = 193 - 175 = 18$; five out of six patients had between 75 and 90% improvement, leaving one patient (who necessarily had over 90% improvement), thus $r_{113} = 5$, $n_{112} = 18 - 12 = 6$. All other n_{ikj} were similarly calculated for the other trials. Appropriate initial values are given as before (for more details see *Ch4_FE_CondBi_probit.odc* and *Ch4_RE_CondBi_probit.odc*).

4.5.1.3 Results: Psoriasis

Fixed and random effects models were fitted. The WinBUGS DIC tool gives p_D as 24.93 and 33.11 for the fixed and random effects models, respectively, while values calculated outside WinBUGS were similar (fixed effects: 25.00 and random effects: 32.77). From the residual deviance and DIC, we conclude that the random effects model should be preferred as it is a better fit to the data (residual deviance of 63.1, compared with 74.9 for fixed effects model, for 58 data points) and has a smaller DIC (322.4 compared with 326.1, from WinBUGS). The estimated between-study heterogeneity had median 0.26 with 95% CrI (0.01, 0.86) on the probit scale, which is small relative to the estimated treatment effects (Table 4.10). The treatment effects relative to supportive care (treatment 1) are all below zero, which suggests that all treatments are better than supportive care at increasing the probability of a reduction in symptoms on the probit scale (Table 4.10). The absolute probabilities of achieving a reduction of at least 50, 75 or 90% in symptoms show that, for example, there is on average 0% probability of achieving at least a 90% reduction in symptoms with supportive care, but this probability is on average 36% with infliximab (Table 4.10).

A model that assumes the cut-points differ between trials and come from a common distribution could also be fitted and gave very similar results (see Exercise 4.4).

4.5.2 Competing Risks: Multinomial Likelihood and Log Link

Where multiple mutually exclusive endpoints have been defined, and patients leave the risk set if any one of them is reached, a competing risk analysis is appropriate. For example, in trials of treatments for schizophrenia (NICE Collaborating Centre for Mental Health, 2010), observations continued until patients either relapsed, discontinued treatment due to intolerable side effects or discontinued for other reasons. Patients who remain stable to the end of the study are censored.

Trials report r_{ikj}, the number of patients in arm k of trial i reaching each of the mutually exclusive endpoints $j = 1, 2, ..., nc_i$, at the end of follow-up in trial i, f_i. In this case the responses r_{ikj} will follow a multinomial distribution:

$$r_{ik,j=1,...,nc_i} \sim \text{Multinomial}\left(p_{ik,j=1,...,nc_i}, n_{ik}\right) \quad \text{with} \quad \sum_{u=1}^{nc_i} p_{iku} = 1$$

and the parameters of interest are the rates (hazards) at which patients move from their initial state to any of the endpoints j, λ_{ikj}. Note that the last endpoint represents the censored observations, that is, patients who do not reach any of the other endpoints before the end of follow-up.

If we assume constant hazards λ_{ikj} acting over the period of observation f_i in years, weeks or days, the probability that outcome j has occurred by the end of the observation period for arm k in trial i is

$$p_{ikj}(f_i) = \frac{\lambda_{ikj}}{\sum\limits_{u=1}^{nc_i-1} \lambda_{iku}} \left[1 - \exp\left(-f_i \sum_{u=1}^{nc_i-1} \lambda_{iku}\right)\right], \quad j = 1, 2, ..., nc_i - 1 \qquad (4.20)$$

Table 4.10 Psoriasis example: posterior median and 95% CrI from the random effects model on the probit scale for the relative effects of all treatments compared with supportive care and absolute probabilities of achieving at least 50, 70 or 90% relief in symptoms for each treatment (PASI-50, 75, 90).

Treatment	Relative effects		PASI-50		PASI-75		PASI-90	
	Median	95% CrI	Median	95% CrI	Median	95% CrI	Median	95% CrI
Supportive care	Reference		0.13	(0.09, 0.18)	0.03	(0.02, 0.05)	0.00	(0.00, 0.01)
Etanercept 25 mg	−1.52	(−2.03, −1.04)	0.66	(0.46, 0.83)	0.36	(0.19, 0.58)	0.12	(0.05, 0.27)
Etanercept 50 mg	−1.93	(−2.50, −1.37)	0.79	(0.59, 0.92)	0.52	(0.30, 0.74)	0.23	(0.09, 0.44)
Efalizumab	−1.19	(−1.56, −0.82)	0.53	(0.37, 0.69)	0.25	(0.14, 0.39)	0.07	(0.03, 0.14)
Ciclosporin	−1.99	(−2.98, −1.30)	0.81	(0.56, 0.97)	0.55	(0.27, 0.87)	0.25	(0.08, 0.63)
Fumaderm	−1.45	(−2.76, −0.34)	0.63	(0.21, 0.95)	0.34	(0.06, 0.82)	0.11	(0.01, 0.54)
Infliximab	−2.31	(−3.06, −1.57)	0.88	(0.67, 0.98)	0.67	(0.37, 0.89)	0.36	(0.13, 0.66)
Methotrexate	−1.69	(−3.10, −0.57)	0.72	(0.29, 0.98)	0.43	(0.09, 0.89)	0.16	(0.02, 0.67)

The probability of remaining in the initial state, that is, the probability of being censored, is simply 1 minus the sum of the probabilities of arriving at any of the absorbing states, that is,

$$p_{ik,nc_i}(f_i) = 1 - \sum_{u=1}^{nc_i-1} p_{iku}(f_i)$$

The hazards are modelled on the log scale $\theta_{ikj} = \log(\lambda_{ikj}) = \mu_{ij} + \delta_{ikj}$, and the trial-specific treatment effects of the treatment in arm k relative to the treatment in arm 1 of that trial for outcome j are assumed to follow a normal distribution:

$$\delta_{ikj} \sim \text{Normal}\left(d_{t_{i1}t_{ik},j}, \sigma_j^2\right)$$

The between-trial variance of the random effects distribution, σ_j^2, is specific to each outcome j. Three models for the variance can be considered: a fixed effects model, where $\sigma_j^2 = 0$; a random effects single variance model where the between-trial variance $\sigma_j^2 = \sigma^2$, reflecting the assumption that the between-trial variation is the same for each outcome; and a random effects different variances model where σ_j^2 denotes a different between-trial variation for each outcome j. Prior distributions for σ_j^2 should be chosen accounting for the scale being used.

These competing risk models share the same assumptions as the cloglog models presented in Section 4.3.2 to which they are closely related: constant hazards over time, implying proportional hazards, for each outcome. A further assumption is that the ratios of the risks attaching to each outcome must also remain constant over time (proportional competing risks). Further extensions where the assumptions are relaxed are available (Ades et al., 2010).

4.5.2.1 WinBUGS Implementation

The code to fit this model is obtained by specifying the likelihood as multinomial and the link as log, `log(lamda[i,k,j]) <- mu[i,j] + d[t[i,k],j] - d[t[i,1],j]`, ensuring that we loop over all arms and outcomes reported in each trial, and that we link the probabilities of events on each outcome with the rate of interest, through equation (4.20). Given values for the mean for each outcome in a vector of length $nc_i - 1$, `meanA` with corresponding standard deviation vector, `sdA`, of the hazards for each endpoint on treatment 1, from external sources, absolute effects and absolute probabilities of the competing outcomes occurring within a given time period, `timeA`, say, a 1-month $(1/12 = 0.083$ years$)$ interval, can be obtained for all treatments. For more details see *Ch4_FE_Multi_log.odc* and *Ch4_RE_Multi_log.odc*.

4.5.2.2   Example: Schizophrenia

In a network meta-analysis of trials of antipsychotic medication for the prevention of relapse in people with schizophrenia, 17 trials comparing eight treatments and placebo were included (Ades et al., 2010). The data available from each trial are the number of patients in each of three outcome states at the end of follow-up. The outcome states are relapse ($j = 1$), discontinuation of treatment due to intolerable side effects ($j = 2$) and discontinuation for other reasons ($j = 3$), which might include inefficacy of treatment that did not fulfil all criteria for relapse, or loss to follow-up. Patients not reaching any of these endpoints at the end of follow-up were considered as censored observations and still in remission ($j = 4$). Figure 4.11 displays the competing risks and rates of interest. The data are presented in Table 4.11 and the network diagram in Figure 4.12.

The data structure consists of a list specifying the number of treatments $nt = 9$, number of studies $ns = 15$ and the number of outcomes $nc = 4$. The main body of data is in a vector format where f [] represents the follow-up time in that trial. Readers experimenting with this example need to be aware of difficulties with initial values. We have found one set of initial values that converges, but to a different posterior distribution. Examples like this remind us of the importance of careful attention to the technical aspects of fitting models by Bayesian MCMC and the need to look at the results obtained with different starting values. This is also an example where inverse gamma prior distributions on the between-trial variance lead to faster convergence and avoid spikes in the posterior distributions (Ades et al., 2010). Full details on the data and initial values used are given in *Ch4_FE_Multi_log.odc* and *Ch4_RE_Multi_log.odc*.

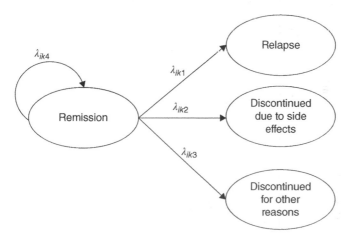

Figure 4.11 Schizophrenia example: outcomes and parameters of interest in the model.

Table 4.11 Schizophrenia example (data from Ades et al., 2010): study names, follow-up time in weeks, treatments compared, total number of events for each of the four states and total number of patients in each trial arm, where placebo = treatment 1, olanzapine = 2, amisulpride = 3, zotepine = 4, aripiprazole = 5, ziprasidone = 5, paliperidone = 6, haloperidol = 7, haloperidol = 8 and risperidone = 9.

Study	Follow-up (weeks)	Treatment		Relapse		Discontinuation due to intolerable side effects		Discontinuation for other reasons		Patient still in remission		Total no. of patients	
	f[]	Arm 1 t[1]	Arm 2 t[2]	Arm 1 r[1,1]	Arm 2 r[2,1]	Arm 1 r[1,2]	Arm 2 r[2,2]	Arm 1 r[1,3]	Arm 2 r[2,3]	Arm 1 r[1,4]	Arm 2 r[2,4]	Arm 1 n[1]	Arm 2 n[2]
1	42	1	2	28	9	12	2	15	19	47	194	102	224
2	46	1	2	7	10	0	2	4	16	2	17	13	45
3	46	1	2	5	6	2	10	5	15	2	17	14	48
4	26	1	3	5	4	5	1	39	26	23	38	72	69
5	26	1	4	21	4	4	16	24	21	9	20	58	61
6	26	1	5	85	50	13	16	12	18	45	71	155	155
7	52	1	6	43	71	11	19	7	28	10	88	71	206
8	47	1	7	52	23	1	3	7	17	41	61	101	104
9	28	2	6	11	8	6	5	44	33	10	9	71	55
10	52	2	8	87	34	54	20	170	50	316	76	627	180
11	52	2	8	28	29	9	14	26	25	78	66	141	134
12	28	2	9	20	53	17	17	36	18	99	79	172	167
13	52	3	8	5	9	3	5	2	2	19	15	29	31
14	52	8	9	65	41	29	22	80	60	14	54	188	177
15	104	8	9	8	4	0	3	4	4	18	22	30	33

Reproduced with permission of Elsevier.

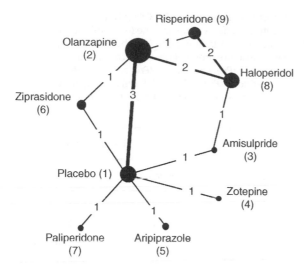

Figure 4.12 Schizophrenia network: each circle represents a treatment, and connecting lines indicate pairs of treatments that have been directly compared in randomised trials. The numbers on the lines indicate the numbers of trials making that comparison, and the numbers by the treatment names are the treatment codes used in the modelling. Line thickness is proportional to the number of trials making that comparison, and the width of the circles is proportional to the number of patients randomised to that treatment.

Given values for the mean for each outcome, meanA = c(-0.078, -1.723,-0.7185), and precision, precA = c(1.6, 1.05, 0.61), of the hazards for each endpoint on treatment 1, from external sources, absolute effects for all treatments and absolute probabilities of the competing outcomes occurring within a given time period, timeA, say, a 1-month (1/12 = 0.083 years) interval, were also calculated.

4.5.2.3 Results: Schizophrenia

A random effects model with different between-trial variations for each outcome and a fixed effects model were fitted. Results (based on three chains: 100,000 iterations after a burn-in of 50,000 for the fixed effects model and 100,000 iterations after a burn-in of 10,000 for the random effects model) are presented in Table 4.12 where the values of p_D are those calculate outside WinBUGS (the DIC tool gave very similar values: fixed effects 68.5 and random effects 80.2). Note that the follow-up time data are entered in weeks, while the analysis delivers annual rates. The model fit statistics suggest that the random effects model is a better fit to the data and the DIC shows that this model should be preferred.

The log hazard rates for each of the competing events and the absolute probabilities of the competing outcomes occurring within 1 month (assuming a log hazard for the reference treatment as detailed earlier) are given in Table 4.12.

Table 4.12 Schizophrenia example: posterior median and 95% CrI for both fixed and random effects models for the treatment effects of olanzapine (d_{12}), amisulpride (d_{13}), zotepine (d_{14}), aripiprazole (d_{15}), ziprasidone (d_{16}), paliperidone (d_{17}), haloperidol (d_{18}) and risperidone (d_{19}) relative to placebo, absolute probabilities of reaching each of the outcomes for placebo (Pr_1), olanzapine (Pr_2), amisulpride (Pr_3), zotepine (Pr_4), aripiprazole (Pr_5), ziprasidone (Pr_6), paliperidone (Pr_7), haloperidol (Pr_8) and risperidone (Pr_9); heterogeneity parameter τ for each of the three outcomes; and model fit statistics for the fixed and random effects models.

	Fixed effects model		Random effects model	
	Median	95% CrI	Median	95% CrI
Relapse				
d_{12}	−1.57	(−2.01, −1.13)	−1.47	(−2.39, −0.50)
d_{13}	−1.14	(−2.10, −0.22)	−0.97	(−2.53, 0.63)
d_{14}	−0.52	(−0.96, −0.06)	−2.08	(−4.11, −0.19)
d_{15}	−0.72	(−1.10, −0.36)	−0.73	(−2.37, 0.92)
d_{16}	−1.12	(−1.50, −0.73)	−1.23	(−2.56, 0.05)
d_{17}	−1.01	(−1.53, −0.53)	−1.02	(−2.70, 0.66)
d_{18}	−0.90	(−1.41, −0.38)	−0.71	(−1.93, 0.63)
d_{19}	−1.27	(−1.82, −0.71)	−1.13	(−2.59, 0.37)
Pr_1	0.07	(0.02, 0.29)	0.07	(0.02, 0.29)
Pr_2	0.02	(0.00, 0.08)	0.02	(0.00, 0.10)
Pr_3	0.02	(0.00, 0.14)	0.03	(0.00, 0.23)
Pr_4	0.01	(0.00, 0.06)	0.01	(0.00, 0.10)
Pr_5	0.03	(0.01, 0.16)	0.03	(0.00, 0.27)
Pr_6	0.02	(0.00, 0.11)	0.02	(0.00, 0.15)
Pr_7	0.02	(0.00, 0.12)	0.02	(0.00, 0.20)
Pr_8	0.03	(0.01, 0.14)	0.04	(0.00, 0.24)
Pr_9	0.02	(0.00, 0.10)	0.02	(0.00, 0.19)
σ			0.66	(0.30, 1.53)
Discontinuation due to side effects				
d_{12}	−1.33	(−2.01, −0.67)	−1.14	(−2.67, 0.97)
d_{13}	−1.64	(−3.09, −0.41)	−1.72	(−4.63, 0.99)
d_{14}	1.12	(0.06, 2.41)	1.12	(−2.05, 4.34)
d_{15}	0.02	(−0.72, 0.78)	0.02	(−3.10, 3.12)
d_{16}	−1.06	(−1.71, −0.35)	−1.02	(−3.28, 1.52)
d_{17}	1.16	(−1.10, 4.58)	1.24	(−2.48, 5.76)
d_{18}	−0.90	(−1.67, −0.15)	−0.86	(−3.16, 1.56)

Table 4.12 (Continued)

	Fixed effects model		Random effects model	
	Median	95% CrI	Median	95% CrI
d_{19}	−1.33	(−2.17, −0.52)	−0.89	(−3.08, 2.63)
Pr_1	0.01	(0.00, 0.09)	0.01	(0.00, 0.09)
Pr_2	0.00	(0.00, 0.03)	0.00	(0.00, 0.07)
Pr_3	0.00	(0.00, 0.03)	0.00	(0.00, 0.06)
Pr_4	0.04	(0.00, 0.35)	0.04	(0.00, 0.82)
Pr_5	0.01	(0.00, 0.10)	0.01	(0.00, 0.38)
Pr_6	0.00	(0.00, 0.04)	0.01	(0.00, 0.11)
Pr_7	0.04	(0.00, 0.84)	0.05	(0.00, 0.97)
Pr_8	0.01	(0.00, 0.04)	0.01	(0.00, 0.11)
Pr_9	0.00	(0.00, 0.03)	0.01	(0.00, 0.26)
σ			1.03	(0.14, 3.31)
Discontinuation due to other reasons				
d_{12}	−0.52	(−0.96, −0.06)	−0.52	(−0.99, 0.03)
d_{13}	−0.60	(−1.11, −0.11)	−0.60	(−1.23, 0.07)
d_{14}	−0.48	(−1.10, 0.15)	−0.49	(1.25, 0.30)
d_{15}	0.23	(−0.50, 1.00)	0.24	(−0.64, 1.11)
d_{16}	−0.44	(−0.96, 0.10)	−0.45	(−1.06, 0.24)
d_{17}	0.73	(−0.13, 1.70)	0.76	(−0.26, 1.79)
d_{18}	−0.44	(−0.94, 0.08)	−0.43	(−1.04, 0.22)
d_{19}	−1.07	(−1.64, −0.51)	−1.10	(−1.76, −0.33)
Pr_1	0.04	(0.00, 0.37)	0.04	(0.00, 0.37)
Pr_2	0.02	(0.00, 0.26)	0.02	(0.00, 0.27)
Pr_3	0.02	(0.00, 0.24)	0.02	(0.00, 0.25)
Pr_4	0.02	(0.00, 0.27)	0.02	(0.00, 0.27)
Pr_5	0.05	(0.00, 0.49)	0.05	(0.00, 0.49)
Pr_6	0.03	(0.00, 0.28)	0.02	(0.00, 0.29)
Pr_7	0.07	(0.00, 0.66)	0.07	(0.00, 0.66)
Pr_8	0.03	(0.00, 0.28)	0.02	(0.00, 0.29)
Pr_9	0.01	(0.00, 0.16)	0.01	(0.00, 0.16)
σ			0.13	(0.00, 0.59)
$\bar{D}_{res}$[a]	120.0		89.1	
p_D	70.2		80.9	
DIC	190.2		170.0	

[a] Compare to 90 data points.

For a graphical representation of which treatment is best for each of the competing outcomes and further comments, see Ades et al. (2010).

4.6   *Shared Parameter Models

Shared parameter models allow the user to generate a single coherent synthesis when trials report results in different formats. For example, some trials may report binomial data for each arm, while others report only the estimated log odds ratios and their variances; or some may report numbers of events and time at risk, while others give binomial data at given follow-up times. In either case the trial-specific relative effects δ_{ik} represent the shared parameters, which are generated from a common distribution regardless of which format trial i is reported in.

So if, in a meta-analysis of M trials, M_1 trials report the mean of a continuous outcome for each arm of the trial, and the remaining trials report only the difference in the means of each experimental arm relative to control, a shared parameter model to obtain a single pooled estimate, can be written as a combination of the models presented in Section 4.3.3 with the likelihood given in equation (4.8) where

$$\theta_{ik} = \begin{cases} \mu_i + \delta_{ik} & \text{for } i = 1, \ldots, M_1; \ k = 1, 2, \ldots, a_i \\ \delta_{ik} & \text{for } i = M_1 + 1, \ldots, M; \ k = 2, \ldots, a_i \end{cases}$$

and a_i represents the number of arms in trial i ($a_i = 2, 3, \ldots$). The trial-specific treatment effects δ_{ik} come from a common random effects distribution (equation (4.2)).

Separate likelihood statements could also be defined, so, for example, in a meta-analysis with a binomial outcome, the M_1 trials reporting the binomial counts in each trial arm could be combined with the trials reporting only the log odds ratio of each experimental treatment relative to control and its variance. In this case the binomial data would be modelled as in Chapter 2, and the continuous log odds ratio data could be modelled as in Section 4.4.1, with the shared parameter being the trial-specific treatment effects δ_{ik} as before. For a fixed effects model, δ_{ik} can be replaced by $d_{t_{i1}t_{ik}} = d_{1,t_{ik}} - d_{1,t_{i1}}$ in the model specification.

These models can be easily coded in WinBUGS by having different loops for each of the data types, taking care to index the trial-specific treatment effects appropriately.

Examples of shared parameter models will primarily include cases where some trials report results for each arm, whether proportions, rates or continuous outcomes, and other trials report only the between-arm differences but can also include models for aggregate and individual patient data meta-analysis (Saramago et al., 2014). A common model for log rates could be shared between trials with Poisson outcomes and time at risk and trials with binomial data with a cloglog link (National Clinical Guideline Centre, 2015); log rate ratios with identity link and normal likelihood could form a third type of data for a shared log rate model. Note that for this type of model, if a study reports both the events and person-years at risk (Poisson data) and the number of events out of the number of randomised individuals (binomial data), the former should be preferred as these will account for censoring and, if relevant, multiple events per individual (although note the assumption that the rate stays constant regardless of previous events).

Shared parameter models can also be used to combine studies reporting outcomes as mean differences or as binomial data (Dominici et al., 1999) and to combine data on survival endpoints that have been summarised either by using a hazard ratio or as number of events out of the total number of patients (Woods et al., 2010; Oba et al., 2016). Another possibility would be to combine trials reporting test results at one or more cut-points using a probit link with binomial or multinomial likelihoods, with data on continuous outcomes transformed to a standard normal deviate scale.

To combine trials that report continuous outcome measures on different scales with trials reporting binary outcomes created by dichotomising the underlying continuous scale, authors have suggested converting the odds ratios calculated from the dichotomous response into an SMD (Chinn, 2000; Higgins and Green, 2008) or converting both the binary and continuous measures into log odds ratios for pooling (Whitehead et al., 1999). These methods could be used within a shared parameter model.

4.6.1 Example: Parkinson's (Mixed Treatment Difference and Arm-Level Data)

To illustrate a meta-analysis with a shared parameter model (Section 4.6), we will assume that the data available for the Parkinson's example were the mean off-time reduction for patients in each arm of the trial for the first three trials, but only the differences between the intervention and control arms (and their standard errors) were available for the remaining trials (Table 4.6).

Fixed and random effects models were fitted. The code used consists of a combination of the code used for Section 4.3.3.4 (arm-level likelihood) and the code used for Section 4.4.1.3 (the trial-level likelihood). The data structure consists of three parts. First is a list giving the number of studies with arm-level information, ns.a; the number of studies with trial-level information, ns.t; and the number of treatments, nt. Two sections of column format data follow: one with the arm-level data, with the structure described in Section 4.3.3.4, and another with the trial-level data, with the structure described in Section 4.4.1.3.

All three data components need to be loaded into WinBUGS for the program to run. Note that because two separate sets of data are being read into WinBUGS, different variable names need to be given. In the WinBUGS code files (*Ch4_FE_NormalShared_id.odc* and *Ch4_RE_NormalShared_id.odc*), the variable names referring only to the arm-level data have the added suffix . a to distinguish them from the trial-level data.

4.6.2 Results: Parkinson's (Mixed Treatment Difference and Arm-Level Data)

Results (based on three chains: 100,000 iterations after a burn-in of 20,000 and 50,000 for the fixed effects and random effects models, respectively) are presented in Table 4.13 and are the same as the results obtained in Section 4.3.3.5. As noted in Section 4.4.1.4, model fit results differ although the overall conclusion is the same.

Table 4.13 Parkinson's example (shared parameter model): posterior median and 95% CrI for both fixed and random effects models for the treatment effects of treatments 2–5 (d_{12}–d_{15}) relative to placebo, absolute effects of placebo (T_1) and treatments 2–5 (T_2–T_5), heterogeneity parameter τ and model fit statistics for different data types.

	Fixed effects model		Random effects model	
	Median	95% CrI	Median	95% CrI
d_{12}	−1.81	(−2.46, −1.16)	−1.83	(−2.91, −0.86)
d_{13}	−0.48	(−1.43, 0.48)	−0.50	(−1.79, 0.75)
d_{14}	−0.52	(−1.47, 0.41)	−0.54	(−1.78, 0.70)
d_{15}	−0.82	(−1.85, 0.20)	−0.84	(−2.35, 0.69)
T_1	−0.73	(−1.16, −0.30)	−0.73	(−1.16, −0.30)
T_2	−2.54	(−3.32, −1.77)	−2.57	(−3.71, −1.49)
T_3	−1.21	(−2.25, −0.17)	−1.23	(−2.58, 0.09)
T_4	−1.25	(−2.29, −0.23)	−1.27	(−2.58, 0.05)
T_5	−1.56	(−2.67, −0.45)	−1.57	(−3.14, 0.02)
σ			0.28	(0.01, 1.53)
$\bar{D}_{res}$ [a]	9.3		9.6	
p_D	7.0		8.5	
DIC	16.4		18.1	

[a] Compare to 11 data points.

4.7   Choice of Prior Distributions

In all the models presented in this chapter, we assume unrelated, non-informative normal prior distributions for the trial-specific baselines and relative treatment effects, as described in equations (2.7) and (2.10). However, care needs to be taken when deciding on a suitable prior distribution for the between-study heterogeneity in random effects models, which depends on the choice of scale. For rate models, whether with log or cloglog link functions (Sections 4.3.1 and 4.3.2), uniform prior distributions on σ may be used as in Chapter 2, but investigators need to be aware of the scale: a prior distribution that is vague for a rate per month may not be so vague for a rate per year. For continuous outcomes, close attention to the scale of measurement is essential. For example, for trials reporting weight changes in kilograms, a Uniform(0,30) may be considered vague, whereas for trials reporting blood pressure outcomes it might be too restrictive, and it may be better to consider a Uniform(0,100) prior distribution. Informative prior distributions based on expert opinion or meta-epidemiological studies should be considered, particularly when there are few studies per comparison and few evidence loops in the network. Rhodes et al. (2015) proposed empirically based prior distributions for the heterogeneity variance that could be used when outcomes are analysed on the SMD scale (see Exercise 4.2 and Chapter 2).

Whatever prior distribution is used for the between-study heterogeneity, the WinBUGS results should be checked to ensure that the prior distribution was not overly informative (i.e. that it overly constrained the posterior distribution) and that it has been updated in light of the data (i.e. that the posterior distribution is not equal to the prior distribution), as recommended in Chapter 2.

4.8   Zero Cells

Because binomial and Poisson likelihoods with zero cells are allowed, special precautions do not usually need to be taken in the case of the occasional trial with a zero cell count, and this is a major strength of the Bayesian MCMC approach with arm-level data. However, in extreme cases where several trials have zero cells and many of the trials are small, the models we have recommended can be numerically unstable, either failing to converge or converging to a posterior distribution with very high standard deviation on some of the treatment effects. This is unlikely to happen with fixed effects models, and it can often be remedied in random effects models by using a (more) informative prior distribution on the variance parameter. A last resort, recognising the assumptions being made, is to put a random effects model on the trial-specific baselines μ_i as well as the relative treatment effects d_{1k}. Note also that the residual deviance will usually be higher for zero cell studies (Chapter 3) and this should be taken into account when interpreting model fit results. See Chapter 6 for a detailed discussion of these and other issues related to sparse data and zero cells.

4.9   Summary of Key Points and Further Reading

We have shown that it is relatively simple to extend the core network meta-analyses models in Chapter 2 to all the commonly used data types. In our framework, extension of the exchangeability and consistency assumption required for pairwise meta-analysis automatically deliver the assumptions required for multiple-treatment meta-analysis, and vice versa, so that the core synthesis model remains the same regardless of the type of data or scale being used.

We have presented models for different outcome types based on the GLM framework, which have the particularly advantage of allowing arm-based (exact) likelihoods to be used (Chapter 2 and Sections 4.3 and 4.5). Implementation of these models in the Bayesian framework is straightforward and is a major strength of the approach. A similar approach with arm-based likelihoods has also been proposed in a frequentist setting using generalised linear mixed models (Tu, 2014).

We have also highlighted the need to jointly analyse multiple outcomes to produce a coherent set of estimates for all outcomes (Section 4.5). The models suggested automatically account for any within-study and between-study correlation across the outcomes through the multinomial likelihood link functions and model specification. Similar models that estimate a between-study correlation in relative effects of different outcomes across trials have been proposed in the pairwise and network meta-analysis contexts (Trikalinos and Olkin, 2008; Schmid et al., 2014).

Shared parameter models are proposed that take full advantage of the modular nature of the WinBUGS code and the fact that the core synthesis model is unchanged. They provide a powerful tool to analyse data reported in different formats and can be extended further to include data presented in multiple ways, as long as a common parameter is being informed.

Regardless of the type of outcome being analysed, model fit and consistency checks should be performed, as described in Chapters 3 and 7, using the adaptations to the model and code described here.

It was mentioned earlier (Section 4.2 and Section 2.3.3) that the appropriate scale of measurement, and thus the appropriate link function, was the one in which effects were linear. To perform an arm-based analysis using the risk difference or log relative risk requires special care, because unlike the 'canonical' (McCullagh and Nelder, 1989) logit models, there is otherwise nothing to prevent the fitted probabilities in a risk difference or log risk model from being outside the zero-to-one range. Section 2.3.3 describes a possible solution based on Warn et al. (2002). However, our experience with these models is that they can sometimes be less stable and issues of convergence and starting values need especially close attention (see Chapter 2). One can readily avoid their use, of course, by taking the estimates of the log risk ratio or risk difference as data and using the models described in Section 4.4.1, but this approach is cumbersome when multi-arm trials are included and can lead to biased estimates if the event is rare (see Chapter 6). Alternatively, an analysis on the odds ratio scale could be undertaken, with the final results transformed onto the risk difference or risk ratio scales, given a probability of event (risk) in the reference treatment, as detailed in Section 2.2.3.1.

As previously discussed (Chapter 2), care should be taken when specifying prior distributions, particularly for the heterogeneity parameter. Sensitivity to the chosen prior distributions should be assessed, and it should be checked that these were not too restrictive (see Exercises 4.1 and 4.2).

In Chapter 6 issues that arise when the event of interest is rare or when evidence is sparse (i.e. there are few studies per comparison and few evidence loops) are discussed. Chapter 10 extends the core model to survival data, and Chapter 11 discusses extension of the ideas presented here to more complex syntheses involving multiple related outcomes.

4.10   Exercises

4.1   A systematic review of RCTs extracted data on studies comparing three different exercise modalities to improve weight loss: aerobic exercise training, coded 1; resistance training, coded 2; and combined (aerobic and resistance) training, coded 3 (Schwingshackl et al., 2013). Fifteen studies reported weight at follow-up or the change in weight from baseline (i.e. the point of entry to the trial) and their standard deviations, along with the number of participant assessed in each arm of each trial. The data are given in *Ch4_Weight.txt*.

a)   Draw the treatment network and decide which of the likelihood/link models described in this chapter should be used.

b)   If a random effects model is fitted, what should be the upper bound of the uniform prior distribution for the between-study heterogeneity?

c)   Fit the appropriate random effects model to these data. Choose the upper bound of the uniform prior distribution for the between-study heterogeneity to be 2. Inspect the posterior distribution of the heterogeneity parameter. What do you conclude?

d)   Increase the upper bound of the uniform prior distribution for the between-study heterogeneity to be 5 and then 10. Compare the results.

e) *If the aim was to obtain a predictive interval for the relative effect of treatment 2 compared with 1, how would you use the posterior summaries of the mean effect and between-study heterogeneity to obtain an approximate value for the credible interval for this relative effect?

f) *Add code to calculate the predictive distribution for the relative effect of treatment 2 compared with 1 (see Chapter 3) and obtain the summaries from WinBUGS. Compare to the approximate values calculated earlier. [Note: An appropriate prior distribution for the between-study heterogeneity should be used.]

4.2 In a systematic review of interventions to reduce caries in children, data are available on the mean number of caries in each treatment arm and their standard errors (Salanti et al., 2009; Dias et al., 2010c). We wish to model these data assuming the mean number of caries in each arm has a normal likelihood but want results to be pooled on the SMD scale (though see Chapter 9 for an alternative model). Data along with a standardising constant for each study (Pooled.sd[]) are given in *Ch4_FluorNormal.txt*. The network plot and full details of the treatment names and codes are given in Section 9.3.1.

a) Adapt the random effects code for arm-based continuous data to pool SMDs, as suggested in Section 4.3.3.2, and run the model. Take care to specify an appropriate upper bound for the between-study standard deviation.

b) Interpret the results, including the between-study heterogeneity, noting that they are on the SMD scale.

c) *Change the prior distribution for the between-study heterogeneity to the predictive distribution suggested for 'a future meta-analysis in a generic setting' by Rhodes et al. (2015): $\log(\sigma^2) \sim t(-3.44, 2.59^2, 5)$, that is, the log of the between-study variance has a t-distribution with location parameter -3.44 and scale parameter 2.59. [Hint: Check how WinBUGS parameterises the t-distribution in the User Manual.] Compare the results to those in (b).

4.3 An HTA report identified 18 RCTs comparing four treatments for non-small cell lung cancer (Brown et al., 2013). The treatments compared were GEM + PLAT, coded 1; VNB + PLAT, coded 2; PAX + PLAT, coded 3; and DOC + PLAT, coded 4. For further details see the original HTA report (Brown et al., 2013). Data on overall survival were extracted from tables presented in the report as log hazard ratios with standard errors and are presented in *Ch4_lnHR.txt*. Note that there were 3 three-arm studies in the dataset, which report all pairwise relative effects and their standard errors.

a) Decide which of the likelihood/link models described in this chapter should be used and ensure the data are in the correct format for WinBUGS. Note: Extra calculations may be required.
b) Draw the treatment network.
c) Fit both the fixed and random effects models and compare them. Choose the preferred model and interpret the results.

4.4 Suppose a manufacturer of a treatment for rheumatoid arthritis presents the following results for the predicted ACR probabilities of response on their product and their main competitor, which they claim are based on a network meta-analysis:

% Response	ACR-20	ACR-50	ACR-70
Manufacturer product	77	47	29
Competitor	62	68	46

Look at the pattern of results. Comment on what conclusions can be drawn. Is it possible to make a coherent recommendation of the 'best' treatment (considering efficacy alone) based on these results?

4.5 *In Section 4.5.1.2 a multinomial/probit model was used for the PASI improvement outcome in the psoriasis example. The model fitted assumed the cut-points from different trials were the same.
a) Adapt the code to fit a model that assumes the cut-points differ between trials but come from a common distribution.
b) Compare the results, including model fit statistics, to Section 4.5.1.3 and note any differences.

5

Network Meta-Analysis Within Cost-Effectiveness Analysis

5.1 Introduction

This chapter looks at the questions that arise when a network meta-analysis is used to inform a cost-effectiveness analysis (CEA). This is a topic that has both technical and conceptual aspects.

Throughout the entire tradition of meta-analysis, going right back to the earliest papers, it has been seen as a way of pooling the *relative* treatment effects in trials. The focus on relative effects, it should be said, follows the huge body of classic work in epidemiological biostatistics on pooling measures of association in 2×2 tables (Mantel and Haenszel, 1959; Zelen, 1971; Robbins et al., 1986). The network meta-analysis models presented throughout this book belong firmly within this line of thought. Decision makers, however, have to attend more to the absolute differences between arms in events or outcomes of interest rather than to differences expressed with ratio measures like the odds ratio or relative risk. It is the absolute difference between treatments in the number of positive outcomes achieved, or negative outcomes avoided, that will determine the value of the treatment to patients and to society. This remains true regardless if the decision is based on an economic analysis or, for example, on numbers needed to treat.

We therefore distinguish between the model for relative treatment effects, as set out in Chapters 2 and 4, and what we refer to as the *baseline model*. Technically, this is a model for the effect of the reference treatment in the target population. The first issue addressed in this chapter is how this baseline model is informed and estimated and how the model for the relative treatment effects is integrated with the baseline model to generate absolute effects on every treatment. Sections 5.2 and 5.3 will look at alternative proposals about how information on absolute effects can be identified and estimated, because some methods may impact on the relative effect estimates.

The evaluation of treatments usually relies on randomised controlled trials, often with quite short-term outcomes. As a result, the baseline model must

Network Meta-Analysis for Decision-Making, First Edition. Sofia Dias, A. E. Ades, Nicky J. Welton, Jeroen P. Jansen and Alexander J. Sutton.
© 2018 John Wiley & Sons Ltd. Published 2018 by John Wiley & Sons Ltd.
Companion website: www.wiley.com/go/dias/networkmeta-analysis

almost always be extended to represent longer-term outcomes, often 'extrapolating' trial outcomes to longer-term lifetime and health benefits. The embedding of the network meta-analysis model within a wider natural history model is therefore covered in Sections 5.4 and 5.5. Our focus here is on appropriate data sources for this extended baseline model and how the relative effect model generated by the network meta-analysis is combined with the extended baseline model to generate long-term outcome predictions on all treatments. For wider issues on CEA modelling methods, readers are referred to recent textbooks (Hunink et al., 2001; Briggs et al., 2006) and good-practice statements (Weinstein et al., 2003; Briggs et al., 2012; Eddy et al., 2012).

In Section 5.6 we illustrate the embedding of a network meta-analysis within a highly stylised CEA. This allows us to show how cost-effectiveness acceptability curves (CEACs) can be generated by adding simple WinBUGS code. It also presents us with an opportunity to discuss which outputs from a random effects model should be used in a CEA.

Finally, again, based on the assumption that investigators will wish to adopt probabilistic decision analytic methods, as recommended by the National Institute for Health and Care Excellence (NICE, 2013a) and leading texts (Briggs et al., 2006), Section 5.7 outlines alternative methods of statistical estimation that are suitable for embedding network meta-analysis in decision analytic models.

5.2 Sources of Evidence for Relative Treatment Effects and the Baseline Model

Most CEAs consist of two separate components: the baseline model that represents the absolute effect of the reference treatment, often placebo or standard care, and a model for relative treatment effects. Note that the baseline model referred to here is *not* a model on the control arms of each trial, even though we refer to these a 'trial-specific baselines'. It is a model for the reference treatment in the meta-analysis. Both models are on the scale of the chosen linear predictor (see Chapter 4). The baseline model may be based on trial or cohort evidence, while the relative effect model is generally based on RCT data. The absolute effect under each treatment is then obtained by adding the relative treatment effects from the network meta-analysis to the absolute effect of the reference treatment from the baseline model. This addition takes place on the linear predictor scale, so the results must finally be converted back to the appropriate scale by inverting the link function (Table 4.1).

For example, if the baseline model is that the probability of an undesirable event under standard care is 0.25 and the odds ratio for a given treatment compared to standard care is 0.8 (favouring the treatment), then, ignoring the uncertainty in these quantities, the probability, p, of an event with the treatment can be obtained from

$$\text{logit}(p) = \text{logit}(0.25) + \ln(0.8) \tag{5.1}$$

which readers will recognise as the model statement for a standard evidence synthesis based on binomial data from Chapter 2, except that the parameter μ_i for the absolute effect on the 'control' arm of trial i has been replaced by the effect given by the baseline model. This gives $p = 0.21$.

We begin by noting that the two questions 'which is the most effective treatment?' and 'what happens under standard care?' have to be considered separately. There are two reasons for this view. As we emphasised in the introduction of this chapter, the entire basis for meta-analysis is that it pools *relative* treatment effects, not absolute effects. This in turn is based on the assumption that, while the absolute efficacy of a treatment may vary with the trial population, the relative effect remains relatively stable. Note that the concept of a relative treatment effect implies further assumptions about the scale. For example, when we claim that the relative treatment effect is more stable, we need to state whether it is to be measured as a risk difference, a risk ratio, an odds ratio or a hazard ratio. A relative effect that is stable if measured as a relative risk may be highly variable if measured as a risk difference (Deeks, 2002; Caldwell et al., 2012) and the consistency assumptions made on one scale will not transfer to another (Norton et al., 2012; van Valkenhoef and Ades, 2013).

The second reason has to do with identifying appropriate sources of evidence. We know that the appropriate source of evidence for relative treatment effects is the ensemble of trials including comparisons of two or more of the treatments in the comparator set (Chapter 1). But the trial evidence may not constitute the best evidence about absolute outcomes on placebo or standard care or on any treatment. Investigators should identify evidence sources to inform the baseline model based on a protocol-driven systematic search (Sculpher et al., 2000; Weinstein et al., 2003; Petrou and Gray, 2011). In particular, the baseline model should be as specific as possible to the target population of interest (Briggs et al., 2006; National Institute for Health and Clinical Excellence, 2008b). As such, it may be that the best evidence is from a carefully selected subset of the trials or from relevant cohort studies, register studies (Golder et al., 2005) or expert opinion (Petrou and Gray, 2011).

If trials are to be used to inform the baseline model, investigators are not obliged to use the same set of trials that inform the relative treatment effects or even a subset of these trials. Nor are they restricted to studies with data on the treatment chosen as the reference treatment in the network meta-analysis. This is because, given a set of relative treatment effect estimates, a baseline model for the reference treatment can be constructed from a model on the absolute effects of any of the treatments in the network. This is illustrated in Section 5.3.

5.3   The Baseline Model

The baseline model for the reference treatment used to define the absolute efficacy will generally be the same GLM as the model for the treatment effect. Given a set of studies that represent the absolute effect on a particular treatment J, which is among those in the network informing the relative effects, a typical approach is to put a hierarchical model on the absolute effects of treatments. If the GLM for treatment effects is $g(\gamma_{ik}) = \theta_{ik} = \mu_i + \delta_{ik}$, for $k > 1$, as defined in Chapter 4, one might model the absolute effects for treatment J using the same link function, as follows:

$$g\left(\gamma_{ik}\right) = \theta_{iJ} = \mu_{iJ}$$
$$\mu_{iJ} \sim \text{Normal}\left(m_J, \sigma_m^2\right)$$

(5.2)

Absolute effects for any other treatment k, T_k, can then be reconstructed as

$$g\left(T_k\right) = m_J + d_{1k} - d_{1J}$$

(5.3)

as in the relationship in equation (5.1).

This formulation calculates the absolute efficacy of treatment J at the mean of the distribution of effects in the selected data. But there is no reason why the effect seen in the future should be equated to the mean of what has been seen in the past. We recommend instead that investigators use the predictive distribution (see Chapter 3) where a new baseline effect, $\mu_{J.new}$, is sampled from the posterior distribution:

$$\mu_{J.new} \sim \text{Normal}\left(m_J, \sigma_m^2\right)$$
$$g\left(T_k\right) = \mu_{J.new} + d_{1k} - d_{1J}$$

(5.4)

We emphasise again that there is nothing in equation (5.4) to suggest that the absolute effects should be informed by the same set of trials as the relative effects. It would, in fact, be a strange coincidence if the sources of evidence on the relative treatment effects were also the most appropriate to inform the absolute effects in the target population.

5.3.1   Estimating the Baseline Model in WinBUGS

We argue against using the same studies for relative effects and the baseline model on epidemiological grounds. However, in cases where it can be justified – perhaps in syntheses of new classes of treatments, whose position in the treatment pathway has yet to be decided – the precise estimation method requires little care. Use of a hierarchical baseline model as in equations (5.2) and (5.4) could introduce bias in the estimated relative treatment effects if they are estimated jointly, unless the baseline model is correct. Suppose, for example, that the condition of patients included in trials with mortality as the outcome was increasingly less severe over the time period of the data. The effect of equation (5.2) would be to shrink the absolute effect in recent trials towards higher mortality, thereby magnifying relative treatment effects. In earlier trials, the opposite effect would occur, and relative treatment effects would be underestimated.

It is precisely to avoid any such systematic bias that traditional meta-analysis has always conditioned out the effect on the control treatment by operating on treatment contrasts in the form of estimated log-odds ratios, (standardised) mean differences or other relative treatment effect measures. In our network meta-analysis models (Chapters 2 and 4) with arm-based likelihoods, we put vague and unrelated priors $\mu_i \sim \text{Normal}(0, 100^2)$ on the trial-specific control arm effects, for precisely the same reason.

As implied by equations (5.2) and (5.4), the absolute outcomes may be informed by several studies, requiring their own synthesis on the scale of the linear predictor. In order to have the benefits of vague and unrelated priors on the study-specific effects on the control arms when estimating the relative treatment effect model, while still constructing a hierarchical baseline model, we run them as separate analyses. This can be done by running a single WinBUGS program that incorporates the network meta-analysis model *and* a baseline model. This program must, of course, include both a dataset for estimating relative effects *and* a dataset for estimating the baseline model, so that in some cases the same trial arms may occur twice in the data. Note that different variable names must be used in the two sections. An example of this separate estimation model is given in *Ch5_Smoking_separatebaselineNMA.odc*, which uses the Smoking Cessation data (Hasselblad, 1998).

The Smoking Cessation data consists of 24 trials on four treatments: no contact, self-help, individual counselling and group counselling (Table 5.1). The network plot is presented in Figure 5.1. No contact is taken as the reference treatment, and it is included in 19 trials. The code for the separate baseline model therefore loops through the ns=24 trials for the standard random effects model and then loops through the nsb=19 (number of studies with treatment 1 arms) from a separate dataset.

Although we do not recommend joint estimation of baseline and network meta-analysis parameters except in specific circumstances (see Chapter 6 for some examples), Table 5.2 compares the results from joint estimation and the recommended approach of separate estimation of baseline model and treatment effects. The joint estimation model (code previously published in Dias et al. (2011d, 2013c)) has a better fit, lower between-trial variation in both relative and absolute treatment effects and lower posterior standard deviation. Nevertheless, for the aforementioned reasons, we

Table 5.1 Smoking Cessation Data (Lu and Ades 2006): events, r, are the number of individuals with successful smoking cessation at 6–12 months out of the total individuals randomised to each trial arm, n.

| | Number of arms | Treatment | | | Arm 1 | | Arm 2 | | Arm 3 | |
| | | Arm 1 | Arm 2 | Arm 3 | | | | | | |
Study ID	na[]	t[,1]	t[,2]	t[,3]	r[,1]	n[,1]	r[,2]	n[,2]	r[,3]	n[,3]
1	3	1	3	4	9	140	23	140	10	138
2	3	2	3	4	11	78	12	85	29	170
3	2	1	3	NA	75	731	363	714	NA	1
4	2	1	3	NA	2	106	9	205	NA	1
5	2	1	3	NA	58	549	237	1561	NA	1
6	2	1	3	NA	0	33	9	48	NA	1
7	2	1	3	NA	3	100	31	98	NA	1
8	2	1	3	NA	1	31	26	95	NA	1
9	2	1	3	NA	6	39	17	77	NA	1
10	2	1	2	NA	79	702	77	694	NA	1
11	2	1	2	NA	18	671	21	535	NA	1
12	2	1	3	NA	64	642	107	761	NA	1
13	2	1	3	NA	5	62	8	90	NA	1
14	2	1	3	NA	20	234	34	237	NA	1
15	2	1	4	NA	0	20	9	20	NA	1
16	2	1	2	NA	8	116	19	149	NA	1
17	2	1	3	NA	95	1107	143	1031	NA	1
18	2	1	3	NA	15	187	36	504	NA	1
19	2	1	3	NA	78	584	73	675	NA	1
20	2	1	3	NA	69	1177	54	888	NA	1
21	2	2	3	NA	20	49	16	43	NA	1
22	2	2	4	NA	7	66	32	127	NA	1
23	2	3	4	NA	12	76	20	74	NA	1
24	2	3	4	NA	9	55	3	26	NA	1

Reproduced with permission of Taylor & Francis.
Treatment codes are defined in Figure 5.1.

would contend that the simultaneous estimation method is subject to bias. In this dataset, the differences in the estimated treatment effects, in the order of 5–15%, are noticeable, but not large, which seems to be a fairly typical finding.

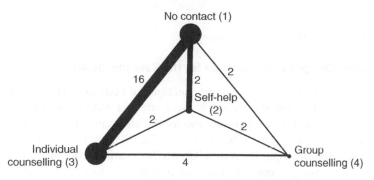

Figure 5.1 Smoking Cessation network: there are 22 two-arm and 2 three-arm trials. Each circle represents a treatment; connecting lines indicate pairs of treatments, which have been directly compared in randomised trials. The numbers on the lines indicate the numbers of trials making that comparison, and the numbers by the treatment names are the treatment codes used in the modelling. Line thickness is proportional to the number of trials making that comparison, and the width of the circles is proportional to the number of patients randomised to that treatment.

Table 5.2 Comparison of separate estimation of absolute and relative effects, the preferred approach, with joint estimation of absolute and relative effect.

	Separate models			Simultaneous modelling		
	Mean/ median	sd	95% CrI	Mean/ median	sd	95% CrI
Baseline model parameters 'no contact'						
m	−2.59	0.16	(−2.94, −2.30)	−2.49	0.13	(−2.75, −2.25)
σ_m	0.54	0.16	(0.32, 0.93)	0.45	0.11	(0.29, 0.71)
μ_{new}	−2.59	0.60	(−3.82, −1.41)	−2.49	0.49	(−3.48, −1.52)
Relative treatment effects compared to 'no contact'						
Self-help	0.49	0.40	(−0.29, 1.31)	0.53	0.33	(−0.11, 1.18)
Individual counselling	0.84	0.24	(0.39, 1.34)	0.78	0.19	(0.41, 1.17)
Group counselling	1.10	0.44	(0.26, 2.01)	1.05	0.34	(0.39, 1.72)
σ	0.82	0.19	(0.55, 1.27)	0.71	0.13	(0.51, 1.02)
Absolute probabilities of response based on the posterior distribution of the baseline probability						
No contact	0.07	0.01	(0.05, 0.09)	0.08	0.01	(0.06, 0.10)
Self-help	0.12	0.05	(0.05, 0.23)	0.13	0.04	(0.07, 0.21)
Individual counselling	0.15	0.04	(0.09, 0.24)	0.15	0.03	(0.11, 0.21)
Group counselling	0.19	0.07	(0.08, 0.37)	0.20	0.05	(0.11, 0.31)
Model fit and DIC						
Residual deviance	54.1			47.4		
p_D	45.0			40.1		
DIC	99.1			87.5		

Posterior summaries of the parameters of the baseline model and the relative treatment effects model.

5.3.2   Alternative Computation Methods for the Baseline Model

The previous section proposes two separate sections of code and data in the same WinBUGS run, one for the network meta-analysis model for relative effects and another for a separate baseline model. An alternative is to run the two models as separate programs (Dias et al., 2011d, 2013c). There are two options: one is to bring the separate WinBUGS MCMC outputs together later, for example, in a CEA programmed in R, at which point the absolute effects T_k for each treatment k are composed from its constituents as in equation (5.4). This can be particularly effective if different baseline models are to be used, for example, in sensitivity analyses. A second, perhaps even simpler, alternative is to take the posterior summary mean and standard deviation of the predictive effects of the reference treatment from the WinBUGS output and use it to identify the parameters of a normal distribution, with, say, mean M_J for reference treatment J and variance S^2. This can then be plugged into the WinBUGS code for the network meta-analysis to generate the absolute outcomes for treatment k, T_k, as described in Chapters 2 and 4:

$$\mu_{J.new} \sim \text{Normal}\left(M_J, S^2\right)$$
$$g\left(T_k\right) = \mu_{J.new} + d_{1k} - d_{1J}$$

This involves a degree of approximation, as the posterior of $\mu_{J.new}$ may not be exactly normally distributed, but the approximation is likely to be acceptable except in extreme cases. This is the approach we have adopted in the exercises at the end of this chapter. Using the Smoking Cessation dataset as an example, the program *Ch5_Smoking_nocontactbaseline.odc* estimates the predictive distribution of the log-odds of the 'no contact' treatment. The predictive mean is −2.589, corresponding to a 7.1% chance of smoking cessation, with a standard deviation of 0.6021. These figures can then be plugged into the programs for CEA (see Exercises 5.2 and 5.3).

Of course, investigators are not necessarily obliged to carry out any formal synthesis to inform the baseline model. Expert opinion, perhaps based on evidence from clinical studies, can be used to characterise a baseline event rate, using a beta distribution for probabilities or a gamma for hazard rates. These should then be put on the appropriate scale so that equation (5.3) can be used to obtain absolute effects for all treatments.

5.3.3   *Arm-Based Meta-Analytic Models

The arm-based meta-analytic models that have been proposed in a series of recent papers (Hedges, 1981; Carlin and Hong, 2014; Zhang et al., 2014; Hong et al., 2016) are a particular form of baseline model. These models, which must of course be distinguished from the arm-based *likelihood* approaches that form the backbone of this book (Chapters 2 and 4), provide a seemingly attractive solution that generates both relative and absolute treatment effect estimates from a single model. For binomial data, with $\Phi()$ representing the standard normal cumulative distribution function, the joint model is represented as

$$\Phi^{-1}(p_{ik}) = \mu_k + \sigma_k \nu_{ik}$$
$$\mathbf{v_i} \sim MVN(\mathbf{0}, \mathbf{R}) \tag{5.5}$$

Here the probits, $\Phi^{-1}()$, of the observed proportion responding on trial i and treatment k, ν_{ik} have a multivariate normal distribution, with treatment-specific means μ_k and with random effects ν_{ik} having a between-trial standard deviation σ_k and a between-trial correlation matrix $\mathbf{R}$. The population mean absolute response probability on treatment k, π_k, can be recovered from these parameters as

$$\pi_k = \Phi\left(\frac{\mu_k}{\sqrt{1+\sigma_k^2}}\right)$$

From here the 'treatment effects' of, say, Y compared to X can then be estimated, as a risk difference $\pi_Y - \pi_X$, as an odds ratio $\pi_Y(1-\pi_X)/(\pi_X(1-\pi_Y))$ or as a relative risk π_Y/π_X. The probit model in equation (5.5) could be replaced with a logit link without changing the fundamental properties of the model, but there is no commitment to a scale of measurement, or equivalently to a link function and linear predictor on which relative treatment effects are additive, and comparatively stable across different trial populations.

The arm-based models, therefore, represent an intriguing challenge to the accepted wisdom that meta-analysis should be pooling relative effects and that these effects should be calculated by comparison to a group of patients selected at the same time according to the same protocol and differing only in the treatment they were randomised to (Senn et al., 2013). The proposed models discard the concept of 'treatment effect' and with it, apparently, the randomisation structure in the evidence. Covariates are readily incorporated, but the distinction between modifiers of relative effects and covariates acting on the baseline model is gone. The extensive literature on heterogeneity of treatment effects in meta-analysis or indeed heterogeneity in 2×2 tables, which has played such a prominent role in epidemiology, is apparently rendered redundant.

Suffice it to say that, from a traditional meta-analysis viewpoint, arm-based models risk generating biased estimates of treatment effects if the model for the absolute effects is wrong or if trials have unequal allocation between arms (Senn, 2010; Senn et al., 2013). A second undesirable consequence is that the posterior variance of the relative effect estimates is greatly increased because they are exposed to the high between-arm variances (Hong et al., 2016). This is precisely what meta-analysis on the relative effects was designed to avoid. Against this, it must be conceded that traditional meta-analysis will itself generate biased estimates if the wrong scale is chosen for the linear predictor.

Whether or not arm-based models are considered attractive from a statistical or epidemiological perspective, they are not well suited to the current practice in decision making, because they would oblige investigators to use the same trial data to inform both the baseline model and the relative treatment effects. This is quite contrary to the guidance on how natural history models should be populated and to the way decision models are conceptualised: the idea of a relative treatment effect that can be estimated in one context and applied in another is quite fundamental to all formal decision-making methods we are aware of. To take one example, the long-standing public health debate on statin use has revolved around the baseline risk at which they should be recommended, not on their relative efficacy compared to placebo (Stone et al., 2013; National Clinical Guideline Centre, 2014). This debate is premised, of course, on having separable relative and absolute effect models. Further discussion of the advantages and disadvantages on arm-based models can be found in Dias and Ades (2016) and Hong et al. (2016).

5.3.4 Baseline Models with Covariates

5.3.4.1 Using Aggregate Data
Covariates may be included in the baseline model by including terms in the linear predictor. For a covariate C, which could either be a continuous covariate or a dummy covariate, we would have, for arm k of trial i

$$\theta_{ik} = \mu_i + \beta C_i + \delta_{ik}$$

where $\delta_{i1} = 0$. But, again, we would want the baseline model with its covariate to be estimated separately from the relative effects. As before, the estimate of the covariate effect β, like the estimate of μ, could be obtained from the trial data or externally. Govan et al. (2010) give an example where the covariate on the baseline is estimated from aggregate trial data with the purpose of reducing aggregation bias (Rothman et al., 2012). This is a phenomenon in which the presence of a strong covariate, even if balanced across arms and not a relative effect modifier, causes a bias in the estimation of the relative treatment effects, towards the null. Some degree of aggregation bias inevitably occurs whenever treatment effects on the probability scale are not linear in the scale of the GLM linear predictor – in other words, whenever the link function is not the identity link.

Covariate effects are seldom reported in every study. Govan et al. (2010) analyse an example where some studies report a breakdown by severity, others by age and others by both. They illustrate a method for dealing with missing data on covariates, although, like other methods for missing covariates (Dominici et al., 1997; Dominici, 2000), it can only be applied when there is at least one study that provides results at the finest breakdown of the covariates (see also Section 8.2.1).

5.3.4.2 Risk Equations for the Baseline Model Based on Individual Patient Data

A far more reliable approach to informing a baseline model, which expresses difference in baseline progression due to covariates such as age, sex and disease severity at the onset of treatment, is to use individual patient data. This is considered superior to aggregate data as the coefficients can be estimated more precisely and with less risk of ecological bias. The results are often presented as 'risk equations' based on multiple regressions from large trial databases, registers or cohort studies. Natural histories for each treatment are then generated by simply adding the treatment effects based on trial data to the risk equations as if they were another risk factor. Examples are the Framingham risk equations (Anon, 2015), which are used to predict coronary heart disease risk, and the UK Prospective Diabetes Study (UKPDS Office), which is used to inform absolute event rates in patients with diabetes. Both these models are used routinely in clinical guidelines. The main difficulty facing the cost-effectiveness analyst here is in justifying the choice of data source and its relevance to the target population. At NICE, organisations making submissions to the Health Technology Appraisals process are obliged to set out a detailed justification for data sources used to inform baseline risk (National Institute for Health and Clinical Excellence, 2012).

5.4   The Natural History Model

Generally speaking, the source of evidence used for each natural history parameter should be determined by a protocol-driven review (Sculpher et al., 2000; National Institute for Health and Clinical Excellence, 2008b; Petrou and Gray, 2011). Previous CEAs are an important source of information on the data sources that can inform natural history. In practice, the full systematic review is generally reserved for data to inform relative treatment effects (Kaltenthaler et al., 2011). For other parameters in the natural history model, a series of strategies may be adopted, which aim as much for efficiency as for sensitivity (Bates, 1989; Pirolli and Card, 1999; Cooper et al., 2007).

Exactly how a natural history model is constructed and how the assembled information is used within it is beyond the scope of this book. However, we provide some comments on one particular aspect of these models: how treatment differences based generally on trial evidence that is typically somewhat short-term are propagated through the extrapolation model.

In Section 5.3, we described how the relative treatment effects are put together with the baseline model to generate absolute effects for all treatments. The role of the natural history model is to extrapolate these absolute effects on trial outcomes to the full range of lifetime health benefits and clinical care costs. The simplest strategy is to assume that there are no differences between treatments in the 'downstream' outcomes, conditional on the shorter-term trial outcomes. We can call this the 'single-mapping hypothesis' as the implication is that, given the information on the short-term differences, longer-term differences can be obtained by a single mapping applicable to all treatments.

The single-mapping hypothesis is effectively the assumption of surrogacy. For example, that the effect of cholesterol-lowering or blood pressure-lowering drugs on cardiovascular outcomes is entirely predictable from their effect on cholesterol levels and blood pressure. Or, to take a more complex example, in a model to assess cost-effectiveness of various antiviral drugs for the treatment of influenza, the base-case analysis assumed that the use of antivirals only affected short-term outcomes and had no additional impact on longer-term complication and hospitalisation rates (Burch et al., 2010). Models with this property are attractive; although they make strong assumptions that the short-term outcomes are 'perfect surrogates', there is no difference between longer-term outcomes that cannot be predicted on the basis of short-term differences.

The use of 'surrogate endpoint' arguments in HTA extends far beyond the outcomes classically understood as 'surrogates' in the clinical and statistical literature (Taylor and Elston, 2009). HTA literature makes frequent use of 'mapping' from short-term to longer-term outcomes, as this allows modellers to base the modelled treatment differences on short-term evidence. Needless to say, the strong assumptions required need to be carefully reviewed and justified, particularly if the treatments under consideration are based on different physiological mechanisms.

If the assumption that all downstream differences between treatments outcomes are due exclusively to differences in shorter-term trial outcomes is not supported by the evidence, then the first option is to use available randomised evidence to drive longer-term outcomes. This necessarily implies different 'mappings' for each treatment, but it does at least base this on randomised evidence. The second and least preferred option is the use of non-randomised evidence. However, as with short-term outcomes, it is essential that any use of non-randomised data that directly impacts on differential treatment effects within the model is carefully justified and that the increased uncertainty and the possibility of bias are recognised and addressed (National Institute for Health and Clinical Excellence, 2008b).

In Chapter 11 we set out a series of extensions to network meta-analysis to encompass multiple outcomes as well as multiple treatments. These models are designed to capture and exploit the clinical and logical relationships between trial outcomes. Where the outcomes occur at different time points in the natural history, the synthesis models for the relative treatment effects can no longer be kept entirely separate from the estimation of the natural history model. There is a potential advantage in having models express the logical relationships between outcomes, as well as basing treatment effects of all outcomes on randomised evidence. But, equally, it is essential that such relationships are supported by expert clinical advice.

Against this, there will be a need to extrapolate anyway, with or without additional information on longer-term treatment effects, and it is clearly preferable that data on longer-term outcomes, if available, is allowed to contribute to predictions on longer-term treatment effects. Unless integrated models for relative treatment effects at different time points are adopted, investigators will be obliged to ignore the most informative trial evidence – those with the longer follow-up – and fall back on short-term studies to make inferences about long-term relative effects.

5.5 Model Validation and Calibration Through Multi-Parameter Synthesis

Natural history models should be validated against independent data wherever possible. For example, in CEAs comparing a new cancer treatment to a standard comparator, the survival predicted in the standard arm could be compared with the published survival, perhaps after suitable adjustment on age or other covariates. With other conditions, given an initial estimate of incidence or prevalence, together with statistics on the size of the population, the natural history model may deliver predictions on absolute numbers of those admitted to hospitals with certain sequelæ, complications or mortality. Once again these predictions could be checked against independent data to provide a form of validation.

A more sophisticated approach is use this external data to 'calibrate' the natural history model. This entails changing the 'progression rate' parameters within the model so that the model accurately predicts the independent calibrating data. Calibration, in a Bayesian framework particularly, can also be seen as a form of evidence synthesis (Welton and Ades, 2005; Ades and Sutton, 2006). In this case the calibrating data is characterised as providing an estimate of a complex function of model parameters. This approach offers a remarkably simple form of calibration because, in principle, all that are required are that the investigator specifies the function of model parameters that the calibrating data estimates and that a term for the likelihood for the additional data is added to the model. The information then propagates 'backwards' through the model to inform the basic parameters. There are many advantages of this method over standard methods of calibration, which have recently been reviewed (Vanni et al., 2011):

1) It gives an appropriate weight to the calibrating data, taking account of sampling error.
2) It avoids the 'tweaking' of model parameters until they 'fit' the calibrating data, a procedure that fails to capture the uncertainty in the data.
3) It can simultaneously accommodate data informing more than one function of parameters that could be used for calibration.
4) It avoids forcing the investigator to decide *which* of several natural history parameters should be changed.
5) Assessment of whether the validating data conflicts with the rest of the model and the data supporting it can proceed using standard model diagnostics, such as residual deviance, DIC or cross-validation (Ades, 2003; Dias et al., 2011a, 2011c) (see also Chapter 3).

Examples of Bayesian calibration approaches have appeared in descriptive epidemiology (Goubar et al., 2008; Presanis et al., 2008; Sweeting et al., 2008; Presanis et al., 2012), particularly in screening applications. In a model of early-onset neonatal group B streptococcus (EOGBS) disease, a chain of decision-tree parameters on maternal infection, neonatal infection and neonatal disease have been calibrated to British Isles surveillance data on the frequency of EOBGS disease (Colbourn et al., 2007a). A particularly important area of application is in cancer models: cancer registry data has been used to recalibrate parameters informed by colorectal cancer screening trials (Whyte et al., 2011). The effect of this kind of calibration is to put rather weak constraints on the individual progression parameters and to place quite strong constraints on complex functions of progression parameters.

Another example is the use of external cancer registry data on conditional survival, which, for many cancers, stabilises 5 or 6 years after diagnosis (Merrill and Hunter, 2010; Yu et al., 2012), to impose constraints on spline-based (Royston and Parmar, 2002) extrapolation of survival curves in cancer treatment trials (Guyot, 2014). Such calibration is necessary to extrapolate short-term trials to obtain life expectancy estimates and is distinctly superior to extrapolation by fitting standard parametric survival curves where results are notoriously sensitive to the choice of curve (Latimer, 2013).

Bayesian calibration is a very powerful application of multi-parameter synthesis. It could be applied in a number of clinical areas to harmonise model predictions with observed data on disease incidence, which is essential for coherent decision-making.

5.6 Generating the Outputs Required for Cost-Effectiveness Analysis

In this section we illustrate how a network meta-analysis can be embedded in a CEA and how the relevant outputs can be generated. We then take the opportunity to discuss which outputs correctly capture the sources of uncertainty and variation, particularly when random effects models are used for the relative treatment effects.

5.6.1 Generating a CEA

Here we build a highly simplistic incremental CEA around the Smoking Cessation network meta-analysis introduced earlier in this chapter (Table 5.2). In addition to the network meta-analysis itself, we define the absolute log-odds of Smoking Cessation using the predictive distribution, as in equation (5.4), plugging in a mean and precision estimated in a separate WinBUGS run. This delivers a vector of absolute probabilities T_k of giving up smoking on intervention k.

The next step is to elaborate the costs and benefits attaching to each treatment. The expected costs of the four treatments are as follows: no contact = 0, self-help = £200, individual counselling = £6000 and group therapy = £600. We can already anticipate that group counselling will be the most cost-effective treatment as it is the most effective, but not the most costly. We assume that the benefits of Smoking Cessation are an additional 15 quality adjusted life years gained, with a standard error of 4 years. A decision tree (Figure 5.2) can be drawn up showing exactly how the costs and benefits will be quantified for each treatment. Notice that the benefit accrues only to the proportion who give up smoking, while the costs on each strategy apply whether the treatment is successful or not. We may now define our objective function, which will be the net benefit, monetised health gain less cost (Claxton and Posnett, 1996; Stinnett and Mullahy, 1998):

$$NB(k, w) = T_k w - C_k$$

The parameter w represents the 'willingness to pay' of the decision maker, the amount of money a decision maker is prepared to spend to obtain a unit gain in health outcome, measured here in quality adjusted life years (QALYs). A typical value, used for example by NICE, is £20,000 per additional QALY. A CEAC is generated by plotting out the probability that each treatment will be cost-effective, against w, varied, for example, from 0 to £50,000. These probabilities are generated by counting, at each value of w, the proportion of iterations in which $NB(k, w)$ for the k-th treatment is higher than all the others.

To do this we can monitor the node p.ce[k,w] in the following WinBUGS code (*Ch5_Smoking_CEA.odc*):

```
for (k in 1:nt)  {
 ly[k] <- T[k] * lyg
 for (will in 1:50){
  nb[k,will] <- ly[k]*will*1000 - C[k]
  p.ce[k,will] <- equals(rank(nb[,will],k),nt)
 }
}
```

The required probabilities of being cost-effective can then be obtained from the posterior means of the matrix p.ce[,] (Welton et al., 2012). These can be copied to an external package, to be plotted against *w* to form CEACs (Figure 5.3). At any willingness to pay, group counselling is most likely to be the cost-effective strategy.

5.6.2   Heterogeneity in the Context of Decision-Making

In Chapter 3 we introduced the predictive distribution of the relative treatment effect and interpreted it as the (true) effect we would expect in a 'new trial' whose relative treatment effect was drawn from the same random effects distribution estimated from the existing trial evidence. Its role there was to assist

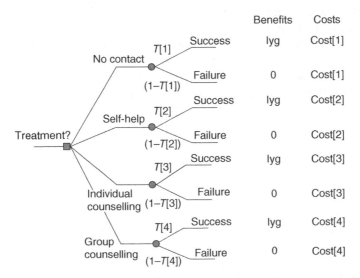

Figure 5.2 Smoking Cessation: decision tree for cost-effectiveness analysis of the four strategies (Welton et al. 2012). Reproduced with permission of John Wiley & sons.

in predictive model checking. In this section we consider this, and other interpretations of 'random treatment effects', from a decision maker's point of view.

In Section 5.3 we suggested that, if data from several studies was used to inform the absolute effect on the reference treatment 1, then the predictive distribution should be used, rather than the distribution of the mean. This is in order to capture more appropriately the actual uncertainty about what the response to the reference treatment might be in a future scenario. To understand the intuition behind this, consider a situation where there have been hundreds of trials, so that the uncertainty regarding the mean absolute response on the reference treatment will be negligible. But, in the presence of heterogeneity, our uncertainty about what the baseline effect will be in a future scenario is not really changed: we know this is highly variable from the present collection of trials. This intuition is correctly captured if we use the predictive distribution, which is dominated by the variance of the random effects distribution, however much (or little) certainty there is about the mean.

For the relative treatment effect, decision modellers have almost invariably used the random effects mean. However, some authorities have suggested that the predictive effect might be more appropriate in a decision-making context (Spiegelhalter et al., 2004). In fact there are several outputs from a random effects model that could be used as inputs to a CEA, each reflecting a particular interpretation and origin for the random effect (Ades et al., 2005; Welton et al., 2007; Welton and Ades, 2012):

a) *Random effects mean.* If the true relative treatment effect is fixed but is observed under some form of noise, then the random effects mean is exactly appropriate in a decision problem. For example, the observed variation between trials might be due to variation in the way an outcome is defined or to differences in the way a test instrument is scored.

b) *Shrunken study-specific estimate.* Here we suppose that among the M trials, there are one or more in which the circumstances (setting, clinical population) are a very close match to the target population for the decision problem. The random effects model expresses the belief that the different circumstances impact on the treatment effect in an unpredictable way, but that they are drawn from a common distribution. The specific estimate for the trial (or trials) whose circumstances match the target will be influenced by being in the model and will be 'shrunk' towards the overall mean (see Section 2.2).

c) *Predictive distribution.* In this scenario the target population/setting bears no specific relation to any of the previous trials. The target treatment effect could be considered as another sample from the random effects distribution. It should be noted that this trial-specific effect is, however, to be considered as a 'fixed' effect, in the sense that, if the target population/setting was to be reproduced in 20 'new' trials, these trials would all estimate the same 'fixed' effect. There are, in effect, two models: a random effects model for the previous data and a fixed effects model for an imaginary ensemble of trials on the new target population. For example, if the predictive distribution was used to inform a prior distribution for the treatment effect (Spiegelhalter et al., 2004), it could only be updated by the likelihood of the data in the new trial if it was believed that both represented the same 'fixed' treatment effect.

In a CEA, the use of the predictive distribution in a decision model, illustrated in *Ch5_Smoking_CEApred.odc*, rather than the distribution of the mean, will produce markedly different results (Figure 5.3).

d) *Whole distribution.* This final option is subtly different from the predictive distribution. The key difference is that the future scenario does not consist

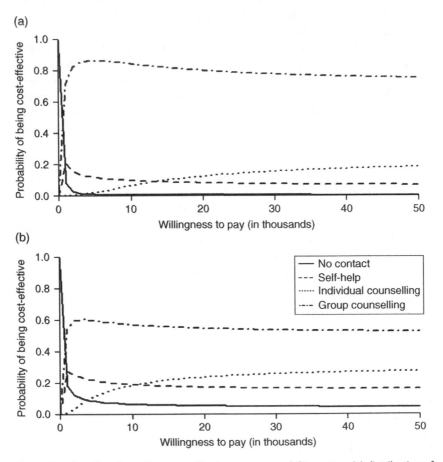

Figure 5.3 Smoking Cessation: cost-effectiveness acceptability curves, (a) distribution of mean treatment effects and (b) distribution of predictive treatment effects.

of a single sample from the random effects distribution, but the entire distribution itself. A good model for this would be a multicentre trial of a psychotherapy protocol in which therapists in different centres came 'off protocol' to different random extents. This would generate a distribution of effects that we might expect to be duplicated in any future roll-out of the therapy. Possibly random differences between the patient populations in different centres could generate a similar result.

This fourth way of using a random effects distribution requires a further level of integration within each MCMC cycle, to account for what is, in effect, patient or centre heterogeneity in the treatment effect. This has been illustrated in a stylized example (Ades et al., 2005; Welton et al., 2007), but it has yet to be applied to relative treatment effects in applications. There are examples, however, where it has been applied to account for heterogeneity in the baseline model (Welton et al., 2008a).

In practice, of course, investigators faced with a random effects model on treatment effects will have an estimate of between-studies variation that may include both random error in observation, true heterogeneity between patient populations, random variation from internal biases and random variation from centre or care differences. We have no specific advice to offer, other than to encourage investigators to give some thought to the sources of unexplained variation in treatment efficacy and how this is captured in the CEA.

5.7 Strategies to Implement Cost-Effectiveness Analyses

In this and previous chapters, we have proposed Bayesian posterior simulation as the 'engine' with which to estimate both the relative treatment effects in pairwise and network syntheses, the baseline model and any further parameters required to extrapolate treatment effects in a cost-effectiveness model. The decision maker must then choose an intervention strategy under uncertainty, selecting the one that delivers the highest expected net benefit. Put formally, there is a net benefit function $NB(S, \theta)$ with uncertain parameters θ. The optimal decision S^* is the one that delivers the highest expected net benefit

$$S^* = \underset{S}{Max}\, E_\theta \left[NB(S, \theta) \right] \tag{5.6}$$

The expectation over the uncertain parameters is therefore an expectation over a joint posterior distribution that is likely to comprise complex correlations, because parameters have been estimated simultaneously from the same input data. In addition, the net benefit function is likely to be non-linear in its parameters. This will necessarily be true whenever log or logistic link functions feature in the synthesis model, or when the natural history model includes a Markov model. We note in passing that the concept of 'optimal decision' is entirely conditional on the model.

If there are correlations between parameters in the net benefit function or if it is non-linear in any of its parameters, the expectation of the net benefit is not the same as the net benefit at the expected value of the parameters. It is for this general reason that decision-making under parameter uncertainty requires integration over net benefit. Posterior simulation is a simple and popular method for carrying out the evaluation of those integrals. The posterior correlations are also very important in the incremental analysis, as they bear on the differences between the net benefits of alternative strategies (see Exercise 5.3).

There are several ways in which the results of the evidence synthesis can be incorporated into the probabilistic CEA.

5.7.1 Bayesian Posterior Simulation: One-Stage Approach

When estimation of the synthesis parameters is via sampling from a Bayesian posterior distribution of the relevant parameters, this can be integrated with the CEA as a single process within a single programming package. Bayesian MCMC simulation (Gilks et al., 1996) using WinBUGS (Lunn et al., 2000, 2013), OpenBUGS (Lunn et al., 2009) or other MCMC packages provides the obvious example. The advantage of this approach is that it not only estimates a Bayesian posterior distribution but also is simulation-based, so that its outputs are perfectly compatible with the MC sampling approach that has become the standard modelling method in so many areas of science. Samples from the joint posterior distribution can be put directly through the decision analysis, so that net benefit and other outputs can be evaluated for each set of parameter samples, without requirements for assumptions about its distributional form. Distributions of additional parameters and costs can be readily incorporated.

Development of MCMC algorithms and sampling schemes is a specialised area of research. For completeness it is worth mentioning that a broad range of *non*-MCMC simulation-based Bayesian updating schemes have also been proposed, including the sample importance resampling algorithm (Rubin, 1988), Bayesian Melding (Raftery et al., 1995; Poole and Raftery, 2000) and Bayesian Monte Carlo (MC) (Brand and Small, 1995). All these have the same properties as Bayesian MCMC in that they all feature both Bayesian estimation and sampling from joint posterior distributions. The latter two were specifically designed for evidence synthesis. We describe some of them in Section 5.7.2.

5.7.2 Bayesian Posterior Simulation: Two-Stage Approach

If investigators have a preferred software for CEA, another option is to take the posterior samples from the Bayesian MCMC, or other posterior sampling scheme, and use them as input to the CEA package. This has the same technical properties as the Bayesian one-stage approach since the full posterior distribution is preserved. From WinBUGS, the convergence diagnostics and output analysis (CODA) output, which lists all values generated from the full posterior distribution, can be exported into a spreadsheet-based program such as EXCEL, using BUGS Utility for Spreadsheets (BUS) (Hahn, 2001). The CODA output can also be converted to the freely available statistical software R (R Development Core Team, 2010) for convergence diagnostics, further analysis and plotting using add-on packages such as Bayesian Output Analysis Program (BOA) (Smith, 2005) or CODA (Plummer et al., 2006). When using the CODA output for a CEA, it is important that the correlations in the parameter estimates are preserved. This is done by ensuring that *all parameter values are sampled from the same MCMC iteration.* If the CODA output is stored as separate columns for each parameter with iteration values along the rows, this would correspond to sampling all the parameter values in one row, each time.

The two-stage approach is particularly useful when there is substantial autocorrelation between successive MCMC samples. This can arise in many situations but usually depends on the statistical model, the way it is parameterised and sparseness of the data. The effect of high levels of autocorrelation is to increase the degree of MC error, with the result that it may require hundreds of thousands of simulations, rather than tens of thousands, before stable estimates are obtained. A common practice in decision modelling has been to 'thin' the posterior sampling. For example, rather than store every posterior value from the MCMC process, one might store every 10th or every 20th. This will usually be enough to reduce autocorrelation substantially, so that the decision model can be run with, say, 25,000 samples from a thinned chain, rather than with the 500,000 original samples. This is particularly relevant for computationally expensive models, although users should ensure that MC error is still appropriately small when using a reduced number of samples. MC error at less than 5% of the posterior standard deviation is the recommended target (see Chapter 2).

Finally, another approach to embedding an evidence synthesis in a CEA is to characterise the posterior distributions in an algebraic form, most often as a multivariate normal distribution, and then simulate from this distribution to implement the CEA. Care must be taken, of course, to ensure that the assumption of multivariate normality is justified (see Exercise 5.3).

5.7.3 Multiple Software Platforms and Automation of Network Meta-Analysis

Interfacing between software packages leads to greater flexibility and facilitates a multidisciplinary approach to CEA. For example, statisticians may wish to use general statistical software, whereas decision modellers may wish to use those designed specifically for decision modelling. Or, while WinBUGS may be chosen to conduct the network meta-analysis, packages with more advanced graphical capability such as R may be needed to display the results. In this section we briefly review the potential to communicate between software platforms, integrated use of different platforms and also on progress towards automating network meta-analysis analysis.

Transfer of data between packages can be effected in a variety of ways. If the analysis is to be carried out in WinBUGS, data columns can be copied directly from spreadsheet software into WinBUGS and pasted by selecting **Paste Special** from the WinBUGS **Edit** menu and choosing the **Plain text** option. Alternatively, *XL2BUGS* (Misra, 2011) is an EXCEL add-in that converts EXCEL data into WinBUGS vector format, while *BAUW* (Zhang and Wang, 2006) converts data in text format into WinBUGS vector or matrix format. If data are stored in R, *R2WinBUGS* (Sturtz et al., 2005) can be used to convert R objects into WinBUGS list data using the *bugs.data* function. Details on software capable of communicating with WinBUGS are available in the WinBUGS web pages (MRC Biostatistics Unit, 2015a).

Integrated platforms reduce the need to copy data and intermediate results from one screen/system to another and thereby reduce the risk of transcription errors. They also facilitate rerunning analyses on updated datasets, conducting sensitivity analyses and, more generally, promoting transparency. It is possible to integrate input, analysis and the display of results using multiple packages, into a single step, adding an interface that facilitates access by clinicians. To this end, a transparent interactive decision interrogator (Bujkiewicz et al., 2011), which integrated syntheses conducted in WinBUGS with graphical displays and the decision model conducted in R and a 'point and click' interface in EXCEL was developed for use by the NICE Appraisals Committee, allowing members to rerun the analyses with different parameters and different synthesis models in real time in the committee meetings (Thompson et al., 2006).

Several freely available code routines have been developed for commonly used packages in HTA, which allow them to communicate with other packages, and these can be utilised in the creation of integrated analyses (Thompson et al., 2006; Heiberger and Neuwirth, 2009; Yan and Prates, 2013).

Finally, procedures have been written that take appropriately formatted datasets as their input, and which automatically generate initial values and computer code, using Bayesian MCMC as the basic computational engine (van Valkenhoef et al., 2012a, 2012b). Further similar software to carry out node-splitting checks for inconsistency (see Chapter 7) are also available (van Valkenhoef et al., 2012a, 2012b; van Valkenhoef and Kuiper, 2016).

5.8   Summary and Further Reading

In this chapter we have proposed that the choice of evidence to inform the baseline model should be considered entirely separately from the choice of data to choose the relative treatment effect, although this can be less convenient from a modelling viewpoint. We have briefly discussed other aspects of the natural history model and have proposed that, wherever possible, the natural history of the disease beyond the trial outcomes should be identical for all treatments conditional on the estimates of the trial outcomes. The 'single-mapping hypothesis' is essentially an assumption of perfect surrogacy.

Given the properties of the joint posterior distribution of treatment effects, we have proposed that the simplest way to ensure that parameter uncertainty is correctly propagated through a decision model is to use the MCMC outputs directly. Frequentist network meta-analysis is an alternative that many investigators may find more convenient, but generation of a joint distribution of parameter values would then be an extra step if inputs for a CEA were required.

This is a good opportunity to briefly summarise why we believe that Bayesian MCMC is the statistical method of choice for network meta-analysis. Among the purely statistical reasons we would cite the following:

- The generation of joint parameter distributions given the data, rather than a distribution of the data at the maximum likelihood parameter values
- Better performance with sparse data and zero cells (see Chapter 6)
- Flexibility to introduce informative priors for variance parameters (see Chapters 2 and 4)
- Modular structure with relatively simple extensions to shared parameter models, meta-regression, inconsistency analysis and multiple outcomes reported irregularly (see Chapters 4, 7, 8, 10 and 11)

These reasons would hold even if the network meta-analysis was not going to be embedded in further decision analytic procedures. If there is to be such embedding, the further advantages are as follows:

- Flexibility to combine treatment effects with natural history parameters, which might also have required complex synthesis.
- Posterior simulation is compatible with probabilistic decision analytic methods.
- Correct uncertainty propagation through to any objective function, in one step.

Two topics are recommended for further reading. Firstly, equation (5.6) defines the 'optimal decision based on current evidence'. It is important to appreciate that this is not necessarily the 'correct decision' that would be taken under conditions where there was no uncertainty. The value to the decision maker in reducing uncertainty is the subject of value of information theory (Raiffa, 1961; Raiffa and Schlaiffer, 1967; Pratt et al., 1995). This fascinating theory has attracted increasing interest in the health-care field (Thompson and Evans, 1997; Felli and Hazen, 1998; Claxton, 1999; Claxton et al., 2000, 2005a). The majority of the recent work has been devoted to solving the considerable computational problems in its implementation (Brennan and Kharroubi, 2007; Welton et al., 2008a, 2014; Oakley et al., 2010; Madan et al., 2014a; Strong et al., 2015).

Secondly, we have discussed the embedding of a network meta-analysis within a CEA at some length, but net benefit (Stinnett and Mullahy, 1998) is only one example of an objective function that can be used to decide between treatments. These same principles can be applied to any objective function. Perhaps the best known decision-making method that does not consider costs is MCDA. Here outcomes can be weighted in various ways, and the objective function is the sum of a linear combination of weights and treatment effects on absolute scales. There are a wide range of variants including fixed weights, weight distributions or imprecise weights that can be considered similar to preferences (Lahdelma and Salminen, 2001; Tervonen and Lahdelma, 2007). Examples where network meta-analysis has been embedded in MCDAs are beginning to be published (van Valkenhoef et al., 2012c; Naci et al., 2014b; Tervonen et al., 2015).

5.9 Exercises

5.1 For the Smoking Cessation example, using the WinBUGS files *Ch5_Smoking_CEA.odc* and *Ch5_Smoking_CEApred.odc*, examine the posterior correlations between the T[k]: (a) when the T[k] are based on the mean treatment effect and (b) when they are based on the predictive distribution of the relative treatment effects. Explain why one set of correlations is higher than the other.

NB: monitor d[] and record their posterior correlations for use in Exercise 5.3.

5.2 For the Smoking Cessation example, enter the posterior summaries of the log-odds ratios d[k] based on the mean treatment effects as 'data' into a cost-effectiveness analysis, rather than embedding the network meta-analysis within the CEA. Import a distribution for the baseline model for T[1] as suggested in Section 5.3.2. Note that this means ignoring the correlations. Construct the CEACs and comment on how and why they differ from the CEACs (a) in Figure 5.3.

5.3 Repeat Exercise 5.2, but now incorporate the posterior correlations of the absolute treatment effects as data. Compare to the results from *Ch5_Smoking_CEA.odc* and Figure 5.3.

6

Adverse Events and Other Sparse Outcome Data

6.1   Introduction

This chapter focuses on the specific challenges for pairwise and network meta-analysis of sparse outcomes (i.e., low numbers of events) and few data (i.e., small numbers of studies, especially a lack of multiple estimates for specific comparisons). Sparse outcomes are predominantly, but not exclusively, encountered with adverse event outcomes. In addition to evaluating efficacy, network meta-analysis is increasingly being used to synthesise information on adverse events relating to the interventions of interest. The aim of such an analysis could be to explore concerns regarding the safety of a drug or to quantify the risks and problems associated with an intervention. This could potentially inform either a clinical decision model weighing up the benefits and risks associated with an intervention strategy (Sutton et al., 2005; Braithwaite et al., 2008) or an economic decision model of the form discussed in Chapter 5. We consider the synthesis of RCT adverse event data and the unique technical challenges a quantitative synthesis of such data presents, although we acknowledge that RCT data alone may not always be the optimal choice of data to use (Loke et al., 2011). The predominant issues considered relate to the sparseness of such data and specifically to where one or more arms of trials observe zero events (and similarly for very common events where everyone in a trial arm has an event). For a more general coverage of the evaluation of adverse events, including the challenges of non-standardised reporting practices observed for adverse events in RCTs and the use of different sources of data, the interested reader is referred elsewhere (Zorzela et al., 2014 and references 9–28 cited within it).

Network Meta-Analysis for Decision-Making, First Edition. Sofia Dias, A. E. Ades, Nicky J. Welton, Jeroen P. Jansen and Alexander J. Sutton.
© 2018 John Wiley & Sons Ltd. Published 2018 by John Wiley & Sons Ltd.
Companion website: www.wiley.com/go/dias/networkmeta-analysis

6.2 Challenges Regarding the Analysis of Sparse Data in Pairwise and Network Meta-Analysis

Consider a single two-arm RCT that randomised 100 patients to each arm, in which three patients were observed to have the adverse event of interest in the intervention group and 0 in the control group. In both the calculation of the (ln) OR or RR, the denominator in the control group is 0, and thus the outcome measure is undefined, as is the associated variance (Sutton et al., 2000). If we replace the zero by increasingly small (positive) numbers for the number of events in the control group into either outcome measure calculation, it can be seen that such measures of relative effect tend to infinity as the number of events in one arm tends to 0. It is precisely this issue that causes analysis problems when including such trials in a meta-analysis context.

Historically, a routine approach, when dealing with zero events in meta-analysis, was to apply a 'continuity correction' to each trial in which 0 events were observed in one arm (of two-arm trials) to enable relative outcomes of interest such as the OR or RR to be calculable and combined using standard frequentist (inverse variance weighting) methods (Higgins and Green, 2008). The 'correction' was typically to add 0.5 to the numbers of patients who do and do not experience an event in each arm of the associated trials. This approach has also been applied to network meta-analysis, where the correction is added to all arms in a study. However, as discussed in the Cochrane Handbook for Systematic Reviews of Interventions (Higgins et al., 2008a, section 16.9) and informed by simulation studies (Sweeting et al., 2004, 2006; Bradburn et al., 2007), such methods are to be avoided in favour of methods that do not need continuity corrections including 'exact' likelihood methods that model the data directly (see Chapters 2 and 4) because adding continuity correction factors can introduce bias into the analysis, particularly when the number of patients in each trial arm differs considerably from each other (i.e., if using an allocation ratio other than $1:1$, e.g., $2:1$ to randomise patients or when losses to follow-up are greater in one of the arms). This bias can be quite extensive if there are many trials with zero events in the same intervention arm. Such 'exact' likelihood methods also remove the reliance on the assumptions that the effect estimates are normally distributed or that the standard errors of the effect sizes are known when they are in fact estimated (which is done using standard inverse variance methods) (Simmonds and Higgins, 2014).

The Bayesian MCMC models used throughout this book (Chapters 2 and 4) use exact likelihood specifications, where data availability allows, and hence provide an appealing approach to synthesising sparse data that can be used without modification (Sweeting et al., 2004, 2006). However, as already mentioned in Chapter 2, care needs to be taken when specifying prior distributions, which are intended to be vague, particularly for the heterogeneity variance, as these can be more informative than intended when data are few and/or sparse (Lambert et al., 2005). For example, often data will be available in binary form, indicating the number of patients in each treatment group and the number who experience the adverse event of interest. In such situations a model with a binomial likelihood is appropriate (Section 2.2.1). However if information on follow-up time is available, this can be incorporated into the analysis using a Poisson likelihood, as described in Chapter 4 and by Ohlssen et al. (2014). Recall the binomial model assumes constant (relative) proportional odds of experiencing the adverse event of interest over time, whereas the Poisson likelihood assumes a constant hazard in each arm (and also allows for multiple (assumed independent) events per person). Using MCMC simulation it is even possible to specify exact binomial likelihoods for the data when conducting the meta-analysis on the relative risk or risk difference scale (see Section 2.3.3 for further details of this approach). Note that the risk of experiencing some adverse events may be time dependent, for example, with higher risks occurring at the start of treatment until the body adjusts to the effects of a given drug. Estimating time-dependent effects will often be challenging using summary data alone but may be possible if enough data are reported (see Chapters 10 and 11) and greatly facilitated by the availability of individual participant data (IPD) as illustrated elsewhere (Askling et al., 2011).

A little discussed issue is that it is sometimes not possible to get anything other than zero events on some treatments for particular adverse event outcomes. For example, febrile neutropenia is a well-documented adverse event following chemotherapy, but trial arms in the network in which patients do not receive chemotherapy should always have zero events (unless patients have switched-over treatment groups and given other interventions at some point during follow-up) since such an event is impossible to experience for such patients. In these treatment-specific adverse event examples, relative treatment effects comparing treatments on which the event cannot be experienced really are not meaningful or defined. We suggest such inappropriate treatment comparisons should be removed from the network for that outcome. If a study then only includes a single relevant treatment arm, that study is removed from the network entirely as it is no longer comparative. (In a corresponding decision model the event rate is simply specified as zero, with no uncertainty, for treatment regimens on which the event is impossible.) In outcomes where adverse events are not treatment-specific and can be experienced on all treatment regimens (e.g., stroke), all trials should be included.

A further issue relates to how to deal with 'all-zero' studies, that is, studies with no events of interest in any trial arm. Collective wisdom is that such trials do not contain information regarding the relative effects when using the standard fixed and random effects models for meta-analysis (Chapter 2) and thus can be excluded from the analysis. Broadly speaking, we would make the same recommendation. However, there has been a recent challenge to this position (Kuss, 2015) that appears to be based on meta-analytic models which are quite different from the standard models for pooling relative effects used throughout this book and elsewhere (Higgins and Green, 2008). These models assume that the probabilities of events in trial control groups are exchangeable, and the treatment effect is fixed (see Section 6.3.3 for consideration of a related model). The hierarchical model placed on the event rates in the control groups allows studies with 0 events in the control group to 'borrow strength' via the information available in the distribution of event rates across the non-zero control groups and thus contribute to the analysis. In a similar vein, we note that in a Bayesian context, if a strong prior distribution is put on the event rate in the control arm, an all-zero study will provide information on the event rate in the treatment arm.

6.2.1 Network Structure and Connectivity

A further consideration for network meta-analysis is the *distribution* of the some-zero (i.e., zero observations in one or more arm of the trial, but not in all arms) and all-zero (if not excluded) studies throughout the network as these can impact on network connectivity (Chapter 1). Care needs to be taken to ensure not only that intervention nodes are connected to the network, which should be done routinely in all contexts but also that these connections are not created only by study comparisons containing zeros, as this is highly likely to lead to numerical instability and potentially lack of convergence of the MCMC sampler. Our suggestion is to initially construct a network diagram excluding all pairwise comparisons involving single- and double-zeros. (Note that, for a three-arm trial comparing A, B and C, in which there were zero events observed in the A arm, the B vs C comparison can be included in the network diagram, but not the A vs B or A vs C comparisons). Then, identify any disconnected treatments in the network that are only connected when adding in all the single- and double-zero pairwise comparisons. Essentially, these treatments should not be considered to be connected, except via prior distributions. If the prior distributions are vague, then numerical instability problems will almost certainly occur that can lead to a lack of convergence and/or extremely large or small treatment effects estimated with extremely large variances.

6.2.2 Assessing Convergence and Model Fit

Even for networks that are not very sparse, problems with estimation can potentially exist, especially for models that include random effects, and care needs to be exercised to ensure 'sensible' answers from converged estimation algorithms are obtained. To this end, we stress the importance of examining the MCMC parameter chain histories, including those for the between study variance parameter (if using a random effects model), the trial-specific event rates on the reference treatment and trial-specific treatment effects, to make sure they have all been identified and converged and that none are 'drifting' in an unconstrained manner towards +/- infinity or have very large variances. Section 6.3 considers several particular approaches that may be applied to network meta-analysis of sparse and few data, which may be particularly valuable in instances where estimation and numerical stability/convergence are a concern.

In addition to these challenges regarding robust parameter estimation, it should be appreciated that the diagnostic approaches to assessing model fit described in Chapter 3 can be problematic and themselves unstable in the presence of sparse data. The way the residual deviance has been coded in the examples in this book ensures that the correct values of the residual deviance are calculated and that no numerical errors occur when there are zero events (see Section 3.2.3). However, the residual deviance will always appear large (i.e., >1) for individual data points where there are zero cells because none of the models presented can actually predict a zero cell since probabilities at zero or one are ruled out. Thus we would expect model fit statistics to be larger in general for models with sparse data, and these need to be interpreted accordingly. Also no leverage can be calculated for these points (see Section 3.3.2.1).

6.3 Strategies to Improve the Robustness of Estimation of Effects from Sparse Data in Network Meta-Analysis

A recurring theme in the overview of approaches that follows is the desire to make estimation more robust and reduce uncertainty by 'borrowing strength' across information sources; this often comes at the cost of having to make stronger modelling assumptions than the standard network meta-analysis models covered in Chapters 2 and 4. Note that, if interest lies in a particular intervention, network meta-analysis itself can be seen as an approach to including more information in an analysis (Ohlssen et al., 2014) compared to limiting the analysis to just direct evidence, and this may improve robustness of the estimation.

6.3.1 Specifying Informative Prior Distributions for Response in Trial Reference Groups

An informative prior distribution could be specified for each unconstrained reference group (e.g., placebo group) response. This could be in the form of information external to the network meta-analysis (Ohlssen et al., 2014), perhaps from observational studies (Soares et al., 2014), or elicited expert clinical opinion. An application utilising observational data in this way is presented by

Soares et al. (2014), although careful assessments of model fit identified inconsistency between the trial and observational data, motivating the authors to exclude the observational data in the final analysis.

6.3.2 Specifying an Informative Prior Distribution for the Between Study Variance Parameters

Another alternative strategy for making estimation more robust, in a random effects modelling context, is to specify an informative prior distribution for the between study variance (heterogeneity) parameter (see Section 2.3.2). Work has been done, in a pairwise context, on producing and using estimates from previous similar meta-analyses as the empirical basis for a prior distribution in a new meta-analysis, particularly when estimation is poor due to few studies in the meta-analysis in question (Turner et al., 2012, 2015b; Rhodes et al., 2015). Turner et al. (2015b) present prior distributions for the between-study variance parameters in 80 settings, based on predictive distributions derived from their modelling of previous meta-analyses of binary outcomes, including some in adverse event contexts. For example, for an adverse event outcome in 'pharmacological versus pharmacological' trials, the proposed prior distribution for the between-study variance is log-Normal(-2.10, 1.58^2) (where -2.10 is the mean and 1.58 is the standard deviation on the log scale). In principle, such an approach could be used for network meta-analysis of sparse (adverse) event data although to use the approach in an unmodified form all intervention comparisons would have to be from the same category, so that 'pharmacological versus pharmacological' is a category but 'pharmacological versus placebo/control' is a different category and would need to have a different prior distribution or a decision would need to be made as to which prior distribution to consider (see Exercise 2.4).

6.3.3 Specifying Reference Group Responses as Exchangeable with Random Effects

The models presented in Chapters 2 and 4 assumed that the reference group response rates in the trials were not related to one another and hence an unrelated model parameter was specified for each one; that is to say their estimation was unconstrained. An alternative is to express a model for reference group response, which requires fewer model parameters but makes stronger assumptions. For example, it has been suggested that reference group response could be assumed to be exchangeable across studies using a random effects model on the reference group arms (see Chapter 5), in order to assist parameter estimation in sparse data problems (Dias et al., 2013c; Ohlssen et al., 2014). For clarity, this is a model that was *not* generally recommended in Chapter 5, for integration of network meta-analysis and cost-effectiveness analysis. This is because the estimation of the reference group response used in the decision modelling should ideally be kept independent of treatment effects. In addition, in a network meta-analysis, care needs to be taken when specifying the model as the treatment in the reference arm (arm 1) can differ across trials (Dias et al., 2011d; Achana et al., 2013; Dias et al., 2013c).

6.3.4 Situational Modelling Extensions

In addition to the aforementioned general strategies for improving estimation in networks of sparse data, several context-specific modelling ideas have been presented in the literature, and these are summarised in the following text. Ohlssen et al. (2014) consider a simultaneous analysis of multiple (binary) safety outcomes, in a multivariate network meta-analysis model in order to borrow strength across (potentially) both outcomes and treatments in an effort to reduce uncertainty (see Chapter 11 for a full account of multiple outcome methods). Warren et al. (2014) describe alternative approaches to modelling safety data relevant when one or more classes of treatments are of interest. They consider alternative modelling assumptions regarding how drugs relate to one another and incorporate varying dose levels. Rather than considering each drug and dose level within a class of drugs as independent, random effects are specified to place hierarchical exchangeable assumptions across drug effects within the same class and dose levels, again to borrow strength to reduce uncertainty (but make stronger modelling assumptions in the process) as well as placing constraints on the effect of dose level (e.g., the risk of an adverse event cannot be higher for a lower dose of the same drug). Fu et al. (2013) consider variable doses of drugs and incorporate specific parametric dose–response models into their network meta-analysis model, and Mawdsley et al. (2016) provide a framework for incorporating dose–response models into a network meta-analysis. In Section 8.6 treatment level covariates are considered, which can reduce the number of treatment nodes in the model. This includes modelling dose effects parametrically and assuming the effects of combinations of intervention components are additive. Again, at the expense of making stronger assumptions, these approaches could be used to reduce the number of treatment effect parameters and thus increase their identifiability from sparse data.

Soares et al. (2014) also explore the impact of assuming an exchangeable treatment class effect across different types of wound dressing, but not across all interventions included in the network, in order to link treatments into a network with zero observed events. They acknowledge that sensitivity analysis is essential with sparse data as modelling assumptions are difficult to verify. However, as an alternative to verification, they provide an illustration of how the consequences of the assumptions can be tested (via internal validity checks) and how external sources (such as observational studies and formally elicited expert opinion) can be used to verify these assumptions in the context of a leg ulcer healing application. Soares et al. (2011) and the cited references, which consider different components of the work, are recommended reading.

6.3.5 Specification of Informative Prior Distributions Versus Use of Continuity Corrections

Although continuity corrections could be used in network meta-analysis to assist convergence and help stabilise the MCMC sampler, we do not recommend automatic uncritical use of them over the alternative approaches suggested earlier, including the use of carefully considered prior distributions. However, they do provide an option for 'connecting' network nodes that would otherwise not be connected (see Exercise 6.2). While some may take issue with informative prior distributions, in an informal sense, the use of continuity corrections is akin to the use of informative prior distributions in the sense that external 'information' (in the form of 0.5 events or whatever number is used) is being 'combined' with the data in a similar way to which the prior distribution and likelihood are combined. We would argue that the uncritical use of continuity corrections is more arbitrary and less explicit than using an informative prior distribution, particularly if 0.5 is used and there is imbalance in the sizes of the study arms the corrections are applied to. In fact, simulation has shown advantages of applying corrections proportional to the size of the study arm over a constant such as 0.5 leading to less bias (Sweeting et al., 2004).

6.4 Summary and Further Reading

Zero events in study arms can potentially cause problems in network meta-analysis of (adverse) event data, and researchers are encouraged to explore and understand the distribution of such arms across the network before any analysis. In addition, careful examination of the MCMC chain output is recommended to check stability of the estimates produced. A number of approaches to improving model estimation stability, while making further modelling assumptions, are available, including the use of informative prior distributions. See Chapter 8 for more on dose models, additive models and class effects models.

A guideline paper on Bayesian network meta-analysis for drug safety (Ohlssen et al., 2014) is recommended reading, providing (among other things) a helpful reporting checklist. In this guideline much emphasis is placed on the need for sensitivity analysis and checking robustness across modelling assumptions, a sentiment which we fully endorse.

6.5 Exercises

6.1 The data in the following table come from the Cochrane database of systematic reviews (Smaill, 1996, analysis 1.2). The data come from four studies comparing antibiotics with placebo for preventing early-onset group B strep disease in newborns (#EOGBS) out of those for which there is maternal colonisation (#M.Col). When conducting a (pairwise) meta-analysis of these data, care needs to be taken because only one EOGBS event is observed in the antibiotic group across all four trials.

Study (first author)	Antibiotics (#EOGBS/#M.Col)	Placebo (#EOGBS/#M.Col)
Boyer	0/85	5/79
Matorras	0/60	3/65
Morales	0/135	3/128
Tuppurainen	1/88	10/111

a) The original publication reported an odds ratio from a meta-analysis using the Peto fixed effects approach. You may want to reanalyse the data using the same approach to confirm that the estimate for the pooled odds ratio obtained is 0.17 with 95% confidence interval (0.07, 0.39).

b) If frequentist software with meta-analysis capabilities (e.g., Review Manager, Stata or R) is available to you, conduct the meta-analysis using the Mantel–Haenszel fixed effects approach and the inverse variance fixed effects approach to produce meta-analysed odds ratios and 95% confidence intervals. For the inverse variance method, add a continuity correction of 0.5 to each 2×2 cell for each study with an arm with zero events, and similarly estimate the weighting in the Mantel–Haenszel method using the same correction (note that some software may do this for you automatically). Compare the results obtained to those in (a).

c) Conduct the analysis in WinBUGS using the network meta-analysis fixed effects code for a binomial outcome presented in Chapter 2. Confirm that the MCMC sampler appears to converge without problem. Contrast the results obtained in WinBUGS with those in (a) and (b). Check the difference observed is not due to the influence of the prior distributions by increasing the variance of all prior distributions by 100-fold and reanalysing the data. What do you conclude?

d) Would it make sense to conduct a random effects analysis for this data? Discuss and justify your answer.

6.2 The table below presents a fictitious dataset with five treatments (A, B, C, D, E) in which a binary outcome has sparse numbers of events.

Study number	Treatment		Number with event		Number of patients		Treatment code	
	Arm 1	Arm 2	Arm 1	Arm 2	Arm 1	Arm 2	Arm 1	Arm 2
1	A	B	3	1	100	101	1	2
2	A	B	2	1	75	74	1	2
3	A	C	1	0	48	50	1	3
4	A	C	0	1	22	24	1	3
5	A	C	2	3	151	154	1	3
6	B	C	1	2	120	120	2	3
7	C	D	0	2	60	63	3	4
8	D	E	1	2	150	153	4	5
9	D	E	2	3	226	228	4	5

a) Sketch out a network diagram (as presented in Chapters 1 and 2) for this dataset excluding any trial comparisons, which have 0 events in one study arm. Is the network connected?

b) Augment the network sketched in (a) to include all the trials in the dataset. Is the network now connected? Identify any connections that are made only by trials with 0 events in one arm.

c) Conduct an analysis in WinBUGS using the network meta-analysis fixed effects code for a binomial outcome presented in Chapter 2 (using all the data). Carefully assess whether the model converges or not and whether reliable parameter estimates are obtained.

d) If you are concerned about the results obtained in (c) discuss ways in which the analysis could be modified to improve it.

7

Checking for Inconsistency

7.1 Introduction

Chapters 1, 2 and 4 have shown how, given a connected network of comparisons, network meta-analysis produces an internally coherent set of estimates of the efficacy of any treatment in the network relative to any other. A key assumption of network meta-analysis is that of evidence *consistency*. The requirement, in effect, is that in every trial i in the network, regardless of the actual treatments that were compared, the true effect of treatment Y relative to treatment X is the same in a fixed effects model or exchangeable between-trials in a random effects model (Chapter 2). From this exchangeability assumption, the 'consistency equations' (equation (2.9)) can be deduced.

Where doubts have been expressed about network meta-analysis, these have focused on the consistency equations because, unlike the exchangeability assumptions from which they are derived, which are notoriously difficult to verify, the consistency equations offer a clear prediction about relationships in the data that can be statistically tested. However, it is important to note that failure to detect inconsistency does not imply consistency since tests for inconsistency are inherently underpowered (Veroniki et al., 2014) (see also Chapter 12). As with other interaction effects, the evidence required to confidently rule out any but the most glaring inconsistency is seldom available, and in many cases, such as networks without any loops, there is no way of testing the consistency assumptions at all, even though we still rely on them to perform indirect comparisons. It is therefore important to clarify the measures that can be taken to minimise the risk of drawing incorrect conclusions from indirect comparisons and network meta-analysis, and we suggest some empirical indicators that might help assess what that risk might be. A more thorough discussion of these issues is presented in Chapter 12. In this chapter we will focus mostly on the technical aspects of checking for consistency.

Network Meta-Analysis for Decision-Making, First Edition. Sofia Dias, A. E. Ades, Nicky J. Welton, Jeroen P. Jansen and Alexander J. Sutton.
© 2018 John Wiley & Sons Ltd. Published 2018 by John Wiley & Sons Ltd.
Companion website: www.wiley.com/go/dias/networkmeta-analysis

Inconsistency and heterogeneity can be conceptualised as manifesting in different ways. Inconsistency is defined as conflict between estimates of the same contrast based on direct or indirect evidence, and heterogeneity is variability in estimates within the same contrast. Thus inconsistency is a form of heterogeneity that is linked to the network structure. However, since both concepts are derived directly from the same exchangeability assumption (Chapter 2), there is a close relationship between between-trial heterogeneity and inconsistency between 'direct' and 'indirect' evidence. Like heterogeneity, inconsistency is caused by effect modifiers and specifically by an imbalance in the distribution of effect modifiers in the direct and indirect evidence (see also Chapter 12). Checking for inconsistency therefore logically comes alongside a consideration of the extent of heterogeneity and its sources and the possibility of adjustment by meta-regression (Chapter 8) or bias adjustment (Chapter 9). On some occasions, heterogeneity and/or inconsistency can be eliminated by meta-regression or bias adjustment, and these techniques should be used whenever known effect modifiers, or risk of bias components, are present.

We define inconsistency as a property of 'loops' of evidence (Chapter 1) in accordance with the common sense notion that has been at the heart of previous methodological (Lumley, 2002) and empirical work (Song et al., 2003, 2011; Veroniki et al., 2013) that evaluated whether the relative effects obtained using only direct evidence (i.e., studies directly comparing the treatments of interest) conflicted with the relative effects estimated from indirect evidence. We set out different approaches that can be applied depending on the network structure, favouring simpler methods where they can be justified.

The generalised linear modelling framework for different likelihoods and link functions set out in Chapters 2 and 4 carries over entirely to the models developed for inconsistency. Thus, all consistency checking methods proposed in this chapter can be applied to different types of data. We will also make use of the measures of model fit and comparison described in Chapter 3. It is however important to note that, although we will propose models that relax the consistency assumptions for the purposes of detecting inconsistency and exploring its sources, we are not recommending that the outputs (e.g., relative treatment effects) from these models be used to inform decision-making, whether there is evidence of inconsistency or not. Decisions should be based on relative effect estimates estimated from consistent evidence, and we consider ways of addressing inconsistency in Section 7.5. Chapter 12 provides a thorough discussion of the consistency assumption and when it should be expected to hold.

7.2 Network Structure

The first step in checking for inconsistency should be to examine the network diagrams carefully because inconsistency can only be detected when there is both direct and indirect evidence contributing to a relative effect estimate (i.e., when there are loops in the network). For example, the networks in Figure 1.1a–d do not have any loops, and therefore only 'direct' evidence is available on a particular contrast. In these networks there is no potential to detect inconsistency, although depending on the number and nature of trials and outcomes, there is still potential to identify between-trial (within-comparisons) heterogeneity. The networks in Figure 1.1e and f have loops, and therefore there is potential for direct and indirect evidence on particular contrasts to disagree.

Another reason is that it can reveal particular features that may assist in the choice of analysis method such as how many loops there are and whether they are independent. Care should also be taken to identify any multi-arm trials and their position in the network. In a network plot, multi-arm trials will define loops since they make comparisons between all included treatments. As such, when multi-arm trials are present, it may appear that there may be potential inconsistency, when in fact a multi-arm trial cannot be inconsistent with itself (see also Section 7.2.2).

To help with examining the network structure, it is particularly useful if care is taken when drawing the network plots to ensure that evidence loops and comparisons with no indirect evidence are clearly distinguishable. Of course this may not always be possible in large networks or those with complex structures, but should be attempted whenever possible (Rücker and Schwarzer, 2016).

7.2.1 Inconsistency Degrees of Freedom

The inconsistency degrees of freedom (ICDF) can be a useful guide as to how many independent inconsistent loops there may be in a given network (Lu and Ades, 2006).

Consider a triangular network that consists only of two-arm trials, such as that presented in Figure 1.1e, where each edge represents the direct evidence comparing the treatments it connects. If we take A as our reference treatment, the standard network meta-analysis consistency model defined in Chapter 2 has two basic parameters, d_{AB} and d_{AC}, but we have data on three contrasts AB, AC and BC. The latter, however, does not estimate an independent parameter, since d_{BC} is wholly determined by the other two parameters through the consistency equations. Setting aside the question of the number of trials informing each pairwise contrast, we can see that there are two independent parameters to estimate and three sources of data. This generates one degree of freedom with which to detect inconsistency, that is, conflict between the evidence sources. More generally, we can define the ICDF as the number of pairwise contrasts on which there are data, N, minus the number of basic parameters, the latter being one less than the number of treatments, S (Lu and Ades, 2006). Thus if all trials are two-arm trials, the ICDF can be calculated as

$$ICDF = N - (S - 1)$$

Every additional *independent* loop in a network of two-arm trials represents one additional ICDF, and one further way in which potential inconsistency can be realised. For example, the networks presented in Figure 1.1a–d have no loops, so that no comparison is informed by both direct and indirect evidence and therefore $ICDF = 0$, for all of them. In Figure 1.1f we can count two triangle loops and a quadrilateral, although the latter is wholly determined by the two triangles, so we can have two possible sources of inconsistency and $ICDF = 2$.

Ignoring for the moment the presence of multi-arm trials, in the Parkinson's network (Figure 4.6) there are two independent loops, and $ICDF = 6 - 4 = 2$ although we can count two triangle loops and one quadrilateral. In the Schizophrenia network (Figure 4.12), there are three loops (two triangles and one quadrilateral) and $ICDF = 11 - 8 = 3$.

As networks get more complex, it is harder to identify loops but the ICDF can still be calculated. For example, for the diabetes network (Figure 4.3), $ICDF = 12 - 5 = 7$. Note that the ICDF is equal to the number of *independent* loops so, for example, in the diabetes network although we are able to count more than seven triangles, they are not all independent as once all edges in seven of the triangles are known, all edges of the remaining triangles are also known. Similarly, in the smoking cessation network (Figure 5.1), one can count a total of seven loops: 4 three-treatment loops and 3 four-treatment loops. However, there are only three independent loops, since once the edges of three of the triangles are known, all others are also fully determined and $ICDF = 6 - 3 = 3$. When loops have edges in common, it is not possible to specify *which* loops are independent, only how many independent loops there are.

7.2.2 Defining Inconsistency in the Presence of Multi-Arm Trials

When multi-arm trials are included in the network, the definition of inconsistency and ICDF becomes more complex. For example, a three-arm trial provides evidence on all three edges of an ABC triangle, and yet it cannot be inconsistent with itself. In other words, although trial i estimates three parameters, $\delta_{i,AB}$, $\delta_{i,AC}$ and $\delta_{i,BC}$, only two are independent because $\delta_{i,BC} = \delta_{i,AC} - \delta_{i,AB}$. There can therefore be no inconsistency within a multi-arm trial. Similarly, if *all* the evidence was from three-arm trials on the same three treatments, there could be no inconsistency in the network, only between-trial heterogeneity. A thorough discussion of these issues is available in van Valkenhoef et al. (2016).

The difficulty in defining inconsistency comes when we have both two-arm and multi-arm trial evidence, for example AB, AC, BC and ABC trials. This raises two questions. The first question is 'do we wish to consider evidence from an ABC trial as potentially inconsistent with evidence from an AB trial?' Although one of the very first treatments of these data structures *did* regard this as a form of inconsistency (Gleser and Olkin, 1994) and recent work has explored this as 'design inconsistency' (Higgins et al., 2012; Jackson et al., 2014), it appears that in practice AB evidence from AB, ABC and ABD studies has been synthesised without any special consideration being given to the presence of the further arms. Of course, there could be some instances where a multi-arm study of ABC might be different from a two-arm AB study. For example, dose finding studies are more likely to be multi-arm, or studies including certain interventions (e.g., surgery) are more likely to have specific issues such as the inability to blind. However, in systematic reviews and meta-analyses of AB evidence, when multi-arm trials are available, only the AB arms are included and any further arms discarded. In all the published meta-analyses and systematic reviews, the issue of whether the presence of multiple arms might be associated with greater heterogeneity, that is, inconsistency, has not, to our knowledge, been raised. Should data from multi-arm trials have particular features that give cause for such concerns, these should be addressed through the use of covariate or bias adjustment methods such as those described in Chapters 8 and 9.

Thus, in everything that follows, inconsistency will be used only to refer to conflict within evidence loops. Loops of evidence that are potentially inconsistent can only arise from structures in which there are at least three independent sources of evidence from distinct trials or sets of trials.

Another issue is parameterisation. When AB, AC and BC evidence arises from three separate sources, it can be inconsistent. But suppose there are *also* ABC trials. We know that these can contribute *independent* evidence on only two treatment effects, but it is not clear *which* two to choose. One way to see what the implications are is to consider between-trial heterogeneity. If we are interested in the heterogeneity of the AB, AC and BC effects, we might begin by looking at each set of two-arm trials separately. But how should the three-arm trials be used? They contribute further information on between-trial heterogeneity, but strictly speaking, they can only provide independent information on *two* of the three contrasts, and these are correlated (see Chapters 2 and 4). Clearly, the choice of contrasts to use will have an impact on both estimates of between-trial heterogeneity *and* the detection of inconsistency.

For example, looking carefully at the Parkinson's network (Figure 4.6), we see that $ICDF = 2$ but the loop connecting treatments 1, 2 and 4 is formed by a three-arm trial comparing these three treatments, which also contains a two-arm trial comparing treatments 1 and 2 only. One could conceptualise that there is potential for inconsistency between the 'indirect' evidence on d_{12} calculated from the three-arm trial and the 'direct' evidence on d_{12} calculated from the two-arm trial. However, this 'indirect' evidence is not really indirect at all, since it comes from a three-arm trial that studies it directly, so any conflict between these estimates of d_{12} is captured as heterogeneity and not inconsistency. On the other hand, we might consider that the 'direct' evidence on d_{24}, which comes from the three-arm trial, may be in conflict with the indirect evidence provided through the four-way path (2, 1, 3, 4), which involves other trials. We might also think that the direct evidence on d_{12} from the two-arm study comparing it could conflict with the indirect evidence from the three-arm study comparing treatments 1, 2 and 4. However, as discussed in Section 7.1, in the models proposed in this chapter, such conflict would be captured as heterogeneity in d_{12}, and not as inconsistency (see Sections 7.3 and 7.4).

The Psoriasis network (Figure 4.10) has a similar situation. There is a single loop connecting placebo and etanercept at 25 and 50 mg and $ICDF = 1$. However, the loop is formed only by two three-arm trials that compared these three treatments and a single trial comparing placebo with etanercept 25 mg. Following the same reasoning, there is no potential for inconsistency in this network, only heterogeneity. The distinction between these two sources of conflict is discussed further in Section 7.6.

Unfortunately, where there are mixtures of two-arm and multi-arm trials, our definition of inconsistency as arising in loops creates inherent technical difficulties that, as far as it is known, cannot be avoided (Lu and Ades, 2006; Higgins et al., 2012; van Valkenhoef et al., 2016). On the other hand, the inclusion of multi-arm trials strengthens the estimates from network meta-analyses precisely because they cannot suffer from inconsistency. Thus, in the presence of multi-arm trials, the ICDF is not well defined and any loop formed by a multi-arm trial alone is not counted as an independent loop and must be discounted from the total ICDF (Lu and Ades, 2006). When loops are formed both by multi-arm and two-arm studies, inconsistency is possible. However, defining the sources of direct and indirect evidence for a particular contrast is complicated since multi-arm trials contribute to both, and their estimates are not independent (van Valkenhoef et al., 2016).

Whatever the circumstances, the ICDF should still be calculated as it indicates the maximum number of possibly inconsistent loops in a network. The networks should then be carefully examined to identify any loops containing multi-arm trials and to determine whether or not there is potential for inconsistency in them. We return to this issue as we explain approaches to detecting inconsistency in Sections 7.3.1.4 and 7.4. The solutions we suggest are simple and practical, and, while still not entirely satisfactory, they are predicated on the assumption that the majority of trials are two-arm trials and there is unlikely to be any material impact on detection of inconsistency from trials with more arms. Conversely, if the proportion of multi-arm trials becomes higher, the distinction between inconsistency, conceptualised as systematic differences between 'direct' and 'indirect' evidence, and heterogeneity become harder to draw and less relevant. In this situation, other approaches such as careful consideration of model fit and the amount of between-study heterogeneity become more relevant.

7.3   Loop Specific Tests for Inconsistency

A key consideration in consistency assessment in networks of evidence is whether independent tests for inconsistency can be constructed. This can usually be determined by careful inspection of the network structure. In Section 7.3.1 we describe how to construct independent tests for inconsistency and the circumstances when this is possible. When independent tests cannot be constructed, Section 7.3.2 sets out methods that can be applied to any network. However, we emphasise that the simpler methods should be used wherever possible as they provide the simplest, most complete and easiest to interpret analysis of inconsistency possible.

7.3.1   Networks with Independent Tests for Inconsistency

Calculation of the ICDF and a careful inspection of the evidence structure should always be done, before attempting to check for inconsistency. In simple cases, networks may contain loops that are independent, in the sense that separate 'indirect' estimates can be constructed for each loop without using the same evidence. The simplest example is the triangle network in Figure 1.1c. However, the network in Figure 7.1 is also an example where there are two completely independent loops of evidence: the assessment of inconsistency in loop A, B, C is completely unrelated (has no data in common) with the assessment of inconsistency in loop D, E, F. We can therefore perform two independent tests for inconsistency in the two loops in this network.

7.3.1.1   Bucher Method for Single Loops of Evidence
The first and simplest method for testing consistency of evidence in a loop is due to Bucher et al. (1997b). It is essentially a 'two-stage' method. The first stage is to synthesise the evidence in each pairwise contrast without the consistency assumption (e.g., from separate pairwise meta-analyses; see also Section 7.4); the second stage is to form the indirect estimate using equation (1.1) and test

Figure 7.1 Hypothetical network with two independent loops.

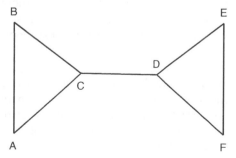

whether direct and indirect evidence are in conflict. An estimate of the inconsistency, ω, and its variance can then be calculated by simply subtracting the direct and indirect estimates and adding their variances, since they arise from independent data sources:

$$\hat{\omega}_{BC} = \hat{d}_{BC}^{Dir} - \hat{d}_{BC}^{Ind}$$

$$Var(\hat{\omega}_{BC}) = Var(\hat{d}_{BC}^{Dir}) + Var(\hat{d}_{BC}^{Ind}) = Var(\hat{d}_{BC}^{Dir}) + Var(\hat{d}_{AB}^{Dir}) + Var(\hat{d}_{AC}^{Dir})$$

(7.1)

Under the null hypothesis of consistency, we would expect this estimate to be zero, allowing for its standard error. An approximate test of this hypothesis can be obtained by assuming the inconsistency estimate is approximately normally distributed and referring

$$z_{BC} = \frac{\hat{\omega}_{BC}}{\sqrt{Var(\hat{\omega}_{BC})}}$$

(7.2)

to the standard normal distribution.

It is easy to confirm that it makes no difference whether we compare the direct BC evidence with the indirect evidence formed through AB and AC or compare the direct AB evidence with the indirect AC and BC, or the AC with the AB and BC. The absolute values of the inconsistency estimates will be identical, and they will all have the same variance, thus the z-statistic will always yield the same conclusion (see Table 7.6 and Exercise 7.2). This agrees with the intuition that, in a single loop, there can only be one inconsistency. However, this method can only be applied to three *independent* sources of data. When pairwise estimates from three-arm trials are included, because they are internally consistent, they reduce the chances of detecting inconsistency between the remaining sources of evidence.

This method generalises naturally to quadrilaterals and higher order loops that, like the triangle and any other simple 'circuit' structure, have $ICDF = 1$. An indirect estimate of any edge can be formed from the remaining edges, and the variance of the inconsistency term is the sum of the variances of all the comparisons. Consequently, as the number of edges in the loop increases, it becomes less likely that a real inconsistency will be detected due to the higher variance calculated for the inconsistency estimate.

7.3.1.2 Example: HIV

We illustrate the Bucher methods with network meta-analysis of three treatments for virologic suppression in patients with human immunodeficiency virus (HIV) (Chou et al., 2006) (see also Exercise 1.1). Direct estimates for all three comparisons are given in Table 7.1, and this corresponds to a triangle network (Figure 1.1e).

Table 7.1 Direct estimates for virologic suppression in patients with HIV (data from Chou et al., 2006).

	Log hazard ratio	95% CI	Standard error
$\hat{d}_{BC}$	0.47	(0.27, 0.67)	0.10
$\hat{d}_{AB}$	2.79	(1.69, 3.89)	0.56
$\hat{d}_{AC}$	1.42	(0.76, 2.08)	0.34

We calculate the indirect estimate for the relative effect of treatment C compared to B as

$$\hat{d}_{BC}^{Ind} = 1.42 - 2.79 = -1.37 \text{ with } Var\left(\hat{d}_{BC}^{Ind}\right) = 0.56^2 + 0.34^2 = 0.429$$

Subtracting this from the direct estimate in Table 7.1, we obtain $\hat{\omega}_{BC} = 0.47 - (-1.37) = 1.84$ with $Var(\hat{\omega}_{BC}) = 0.10^2 + 0.429$ and $z_{BC} = 1.84/\sqrt{0.439} = 2.78$ indicating strong evidence of inconsistency (p-value < 0.01). For further comments on this example, see Section 7.6.

7.3.1.3 Extension of Bucher Method to Networks with Multiple Loops: Enuresis Example

Figure 7.2 shows an example (Russell and Kiddoo, 2006; Caldwell et al., 2010) where data are available on 10 contrasts involving eight treatments for childhood enuresis, and therefore $ICDF = 3$. The special feature of this network is that all the inconsistencies can be seen as concerning estimates of the alarm versus no treatment effect. In particular, there are four independent estimates of this parameter: one direct estimate and three indirect estimates via psychological therapy, imipramine and dry bed training, respectively. Since the three indirect estimates are independent, the Bucher method (Section 7.3.1.1) can be extended to perform an approximate chi-square (χ^2) test of inconsistency (Caldwell et al., 2010). The evidence on each of the relevant edges obtained from separate fixed effect pairwise meta-analyses on the log relative risk scale is presented in Table 7.2.

To analyse the potential inconsistencies in this network, we need to compare the direct estimate of the relative effect of alarm versus no treatment (Table 7.2) with the three possible indirect estimates, obtained from repeated application of equation (7.1), presented in Table 7.3.

Given four independent estimates $\hat{\alpha}_1, \hat{\alpha}_2, \hat{\alpha}_3, \hat{\alpha}_4$ of the relative treatment effect of alarm versus no treatment, and their variances V_1, V_2, V_3 and V_4, an average treatment effect $\tilde{\alpha}$ is estimated by inverse variance weighting as $\tilde{\alpha} = \sum_{i=1}^{4} w_i \hat{\alpha}_i / \sum_{i=1}^{4} w_i$, where $w_i = 1/V_i$, $i = 1, 2, 3, 4$. In this case $\tilde{\alpha} = -0.776$ and

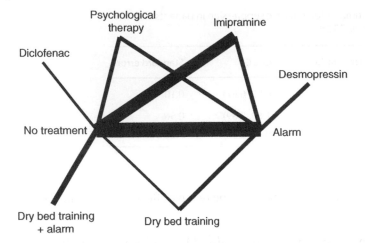

Figure 7.2 Enuresis treatment network (Caldwell et al., 2010): lines connecting two treatments indicate that a comparison between these treatments has been made and the thickness of the lines is proportional to the number or RCTs making that comparison. Reproduced with permission of Elsevier.

Table 7.2 Enuresis example: all possible direct estimates presented as log relative risks, ln(RR), obtained from separate meta-analysis using a fixed effects model. Data from Caldwell et al., 2010.

			Direct estimates	
Control	Treatment	Number of RCTs	ln(RR)	Variance of ln(RR)
No treatment	Alarm	14	−0.968	0.006
No treatment	Psychological therapy	3	−0.371	0.012
Alarm	Psychological therapy	3	0.386	0.038
No treatment	Imipramine	11	−0.261	0.001
Alarm	Imipramine	3	0.315	0.009
No treatment	Dry bed training	2	−0.198	0.012
Alarm	Dry bed training	3	−0.285	0.071

Table 7.3 Enuresis example: indirect estimates of alarm versus no treatment effect presented as log relative risks, ln(RR).

	ln(RR)	Variance of ln(RR)
Via Psychological therapy	−0.757	0.050
Via Imipramine	−0.576	0.010
Via Dry bed training	0.087	0.083

an approximate χ^2 statistic is given by $Q = \sum_{i=1}^{4} w_i (\hat{\alpha}_i - \tilde{\alpha})^2 = 18.8$. Referring Q to a χ_3^2 distribution (the degrees of freedom are given by the number of independent estimates – 1, which is also the ICDF) suggests that there is evidence of inconsistency (p-value < 0.01).

This method can be extended to any number of indirect estimates, as long as they are independent, that is, use different studies.

7.3.1.4 Obtaining the 'Direct' Estimates of Inconsistency

The Bucher method for triangle structures, and its extension to larger loops and to χ^2 tests, are all based on two-arm trials. Inclusion of multi-arm trials will lower the power of these tests to detect inconsistency. We have suggested that the direct estimates for each contrast could be obtained from separate pairwise meta-analyses, although accounting for multi-arm studies is not straightforward. One suggestion is that when a test on a loop ABC is being constructed, evidence from ABC trials is excluded. However, ABC evidence on AB *should* be included when testing, for example, the ABD loop. Similarly ABCD trials would be excluded from tests on the ABCD loop but included in studies of the ABCE, BCDE loops and so on. This substantially complicates the implementation of this method in large, complex networks. In addition it will no longer result in a completely fair test of the inconsistency since it will not use the same set of studies as the network meta-analysis. However, performing the separate pairwise meta-analyses using the treatment effects from multi-arm trials multiple times is also undesirable as this will result in 'double-counting' the evidence. One way to handle the problem of correlated treatment effect estimates resulting from multi-arm studies would be to split each arm of the multi-arm study into two parts and entering each pair or comparisons as if they were separate (and independent) two-arm studies (Higgins et al., 2008a).

A better option to mitigate the problem of multi-arm trials while avoiding the issue of 'double-counting' participants from multi-arm trials and hence inflating the precision of estimates has recently been proposed in a frequentist framework (Rücker, 2012; Rücker and Schwarzer, 2014). The idea is to adjust the weight multi-arm trials are assigned in the meta-analysis, by adjusting the variance of all their relative effect estimates (Rücker, 2012; Rücker and Schwarzer, 2014). This works well in a Bayesian context when fixed effect models are used and can be implemented in WinBUGS by including multi-arm trials in the form of treatment differences with adjusted standard errors. If the remaining data are modelled using arm-level likelihoods, a shared parameter model will be required (see Chapter 4, Section 7.4.2 and Exercise 7.2). However, when random effects models are used in a Bayesian framework, this approach will underestimate the uncertainty arising from multi-arm trials as, for a random effects model, the adjusted variances are a function of the between-study heterogeneity, which in a frequentist analysis is assumed fixed and known

(Rücker, 2012; Rücker and Schwarzer, 2014), but in Bayesian modelling is assumed to have a distribution with uncertainty that needs to be propagated to the relative effect estimates.

Having decided how to handle multi-arm trials, the pairwise meta-analyses can be conducted using the models and codes presented in Chapters 2 and 4, depending on the outcome type, taking care to use only the studies making each comparison. When fixed effect models are used to obtain estimates for all contrasts, equation (1.1) provides a consistent estimate of the relative effect based on indirect evidence, which is fair to compare to the direct estimate. However, if random effects models are used for some, or all, of the comparisons, different between-study variances will be estimated, and it is likely that, for comparisons informed by one or only a few trials, we will be forced to fit a fixed effects model, thus setting the between-study heterogeneity to zero. Reliably estimating the between-studies variance in random effects meta-analyses is recognised to be a difficult problem in both frequentist and Bayesian statistics (see Chapters 2 and 4). Estimation of variances in networks is particularly complex, because under the null hypothesis of consistency that we set out to test, the *true* variances have to conform to special relationships known as 'triangle inequalities' (Lu and Ades, 2009). These state that, when three comparisons are made, corresponding to a triangle loop, it is not possible to have zero variances for two of the comparisons, but a non-zero variance for the third. Similarly, if one comparison has zero variance, then that implies that the variances of the other two must be equal. Therefore, there is always a risk of breaking triangle inequalities if each contrast is analysed independently of the others, especially when there is a zero variance estimate. A strictly fair test of the consistency hypothesis cannot be constructed unless the triangle inequalities are followed. One possible suggestion would be to obtain a posterior distribution for the between-study heterogeneity, perhaps estimated from a network meta-analysis of the full dataset (Chapters 2 and 4) or from meta-epidemiological studies (Rhodes et al., 2015; Turner et al., 2015b) and impose that as the heterogeneity distribution in each of the pairwise meta-analyses, taking care not to allow any posterior updating. Alternatively a model that allows for separate synthesis of effects without assuming consistency, while sharing the between-study variance, would automatically provide pairwise estimates with a common between-study heterogeneity, which can then be used to obtain indirect and inconsistency estimates. Section 7.4 suggests a model of this type and highlights some of its advantages.

7.3.2 Methods for General Networks

We have so far considered methods for relatively simple networks with independent loops. However, networks can be far more complex than that and more general methods are required. The Smoking Cessation network in

Figure 5.1 is a four treatment network in which there are data on every contrast and three possible inconsistencies (Section 7.2.1). The difference between this network and the network in Figure 7.2 is that, in the Smoking Cessation network, there are four three-treatment loops and three four-treatment loops, but the loops are not statistically independent. This makes it impossible to construct a set of independent tests to examine the three inconsistencies.

7.3.2.1   Repeat Application of the Bucher Method

One possible approach is to apply the Bucher method to each of the loops in turn, even if they are not independent. This has the advantage of being simple, if tedious, to implement (see Exercise 7.1). When this is done, the number of loops, and hence the number of tests, will far exceed the number of inconsistencies that the network can actually have.

Each application of the Bucher test will be a valid test at its stated significance level. *If no inconsistencies are found*, at a significance level of, say, 0.05, when applying the test to all loops in the network, one can correctly claim to have failed to reject the null hypothesis of no inconsistency at this level, *or higher*. This is not necessarily very reassuring because of the inherently high variance of indirect evidence, especially in multi-sided loops (the situation is analogous to concluding 'no difference' from a small, under-powered trial). Difficulties in the interpretation of statistical tests arise if any of the loops show significant inconsistency, at the chosen significance level. One cannot immediately reject the null hypothesis at this level because a certain degree of multiple testing has taken place, and adjustment of significance levels would need to be considered. However, because the tests are not independent, calculating the correct level of adjustment becomes a complex task.

In a network where $N = 14$, $S = 6$ and $ICDF = 9$, the Bucher method was applied to every three-way, four-way and five-way loop, leading to 20 tests of inconsistency, none of which suggested that inconsistency might be present (Salanti et al., 2009). In another example (Cipriani et al., 2009), $N = 42$, $S = 12$ and $ICDF = 31$ but repeated use of the Bucher method on each of the three-way loops in the network gave 70 estimates of inconsistency for one of the outcomes of interest (response) and 63 estimates of inconsistency for the other outcome of interest (acceptability). Thus, 133 separate tests for inconsistency were performed. In total six loops showed statistically significant inconsistency and the authors concluded that this was compatible with chance. However, this conclusion could be questioned on the grounds that the 133 tests were not independent. Assuming the response and acceptability outcomes are unrelated, there could not be more than 62 independent tests (31 for each outcome).

The excessive number of tests carried out is a problem common to most methods for detecting inconsistency, not just the Bucher approach. However, a further drawback in networks with many treatments is that the total number of triangular, quadrilateral and higher order loops may be extremely large. In addition it can be argued that, in a complex network, such as the Smoking Cessation example, this method does not use all of the available indirect evidence to obtain the 'indirect' estimate. For example, looking at Figure 5.1, if we wanted to check consistency of the direct estimate of the relative effect of treatment 3 compared with 1, with the indirect estimate formed from the relative effect estimates of 2 versus 1 and 3 versus 2, we would be ignoring the remaining network that also contributes to the estimates of 2 versus 1 and 3 versus 2. Thus we could argue that this comparison does not truly capture all the 'indirect' evidence in the network and is thus not a fair assessment of consistency.

7.3.2.2 A Back-Calculation Method

We can extend the Bucher approach so that it accounts for the whole network when calculating the indirect estimate. This method can then be applied to networks of any complexity. Suppose we have an estimate $\hat{d}_{XY}^{Dir}$ based only on direct evidence on the contrast (X,Y), with variance V_{XY}^{Dir}, and the network meta-analysis estimate, based on the entire network of evidence, $\hat{d}_{XY}^{NMA}$, with posterior variance V_{XY}^{NMA}. Since direct and indirect evidence come from separate data sources and are therefore independent (provided no 'indirect' evidence from multi-arm trials is used), we can consider $\hat{d}_{XY}^{NMA}$ to be an inverse-variance weighted average of the direct estimate $\hat{d}_{XY}^{Dir}$ and an indirect estimate $\hat{d}_{XY}^{Ind}$ with variance V_{XY}^{Ind}. Manipulating the standard formulae for weighted averages, we can see that (Dias et al., 2010b):

$$\frac{1}{V_{XY}^{Ind}} = \frac{1}{V_{XY}^{NMA}} - \frac{1}{V_{XY}^{Dir}} \text{ and } \hat{d}_{XY}^{Ind} = \left(\frac{\hat{d}_{XY}^{NMA}}{V_{XY}^{NMA}} - \frac{\hat{d}_{XY}^{Dir}}{V_{XY}^{Dir}} \right) \times V_{XY}^{Ind}$$

These can be solved to find $\hat{d}_{XY}^{Ind}$ and its variance, which can be interpreted as the implied *additional* information that the network meta-analysis 'adds' to the direct information and its variance. The difference between the direct and indirect estimates for each loop are estimates of inconsistency, which can be used to construct the test statistic in equation (7.2).

Although simple and intuitive, it should be noted that this approach performs quite poorly when multi-arm trials are present and is hard to generalise to random effects models due to the different estimates of heterogeneity in the network and pairwise meta-analysis models. It is therefore not generally recommended. The node-split method (see Section 7.3.2.4) is a better way of accounting for the network structure and has been recommended over back-calculation by Dias et al. (2010b).

7.3.2.3 *Variance Measures of Inconsistency

In this method, the idea is to fit a model that relaxes the consistency assumption by introducing inconsistency factors, ω, which allow comparisons informed by both direct and indirect evidence to deviate from the consistency equations, thus allowing some inconsistency (Lu and Ades, 2006). For example, in the four-treatment Smoking Cessation network (Figure 5.1), we would estimate six parameters: three basic parameters d_{12}, d_{13} and d_{14} and three inconsistency factors (Lu and Ades, 2006):

$$
\begin{cases}
\omega_{23} = d_{23} - \left(d_{13} - d_{12}\right) \\
\omega_{24} = d_{24} - \left(d_{14} - d_{12}\right) \\
\omega_{34} = d_{34} - \left(d_{14} - d_{13}\right)
\end{cases}
$$

The $\omega_{23}, \omega_{24}, \omega_{34}$ parameters reflect the 'inconsistencies' between the direct and indirect evidence on these three edges. This model can be used as a global assessment of inconsistency by inspecting the between-study heterogeneity alongside goodness-of-fit statistics and comparing them with the network meta-analysis model under consistency (see Section 7.4 for another version of this model).

Rather than considering the three inconsistency parameters as unrelated, we might assume that they all come from a random distribution (Lu and Ades, 2006), for example, $\omega_{XY} \sim \text{Normal}(0, \sigma_\omega^2)$. The additional variance term in this model has been called 'incoherence variance' (Lumley, 2002) and 'inconsistency variance' (Lu and Ades, 2006), and it has been suggested that this additional between-contrast variance can serve as a measure of inconsistency. We do not recommend this however, because measures of variance will have very wide credible intervals unless the ICDF is extremely high, and, even then, large numbers of large trials on each contrast would be required to obtain a meaningful estimate. Furthermore, where there is a single loop (*ICDF* = 1), it should be impossible to obtain any estimate of σ_ω^2. However, a number of published applications of the using the model as suggested by Lumley (2002) have reported estimates of inconsistency variance in networks consisting of only a single loop (Elliott and Meyer, 2007; Trikalinos and Olkin, 2008), and it seems likely that the model has not always been implemented in a way that takes account of the number of inconsistencies. The reader is referred to the literature for further comments on this issue (Salanti et al., 2008a; Jackson et al., 2014).

7.3.2.4 *Node-Splitting

The dependency structure in Bayesian hierarchical models can be represented graphically using a directed acyclic graph (DAG), where parameters to estimate correspond to nodes (Lauritzen, 2003). O'Hagan (2003) suggested node-based methods to detect conflict between different sources of information on a given parameter, in a model criticism context. He suggested that splitting different sources of information about a node in a DAG and comparing the resulting posterior distributions can be useful to examine conflict between these different sources of information, and this is where the term 'node-splitting' originates

(Marshall and Spiegelhalter, 2007; Presanis et al., 2013). The basic idea is that, for the node of interest, two posterior distributions are generated from independent sources. Measures of conflict between these sources are the measures of compatibility between these two posterior distributions (O'Hagan, 2003). Other authors have used the terms edge-splitting or side-splitting (White, 2015) when implementing the same ideas in a frequentist context.

In the network meta-analysis context, node-splitting can be used to split the information contributing to estimates of a relative effect parameter (node), d_{XY}, into two distinct components: the 'direct' based on the XY data only (which may come from XY, XYZ, WXY trials) and the 'indirect' based on *all* the remaining evidence that is pooled using a network meta-analysis model (Dias et al., 2010b). Two posterior distributions are obtained for the mean treatment effect d_{XY}: one based on studies comparing treatments X and Y directly, with mean d_{XY}^{Dir}, and another with mean d_{XY}^{Ind}, from a network meta-analysis of all the remaining studies. The process can be applied to any contrast in the network (node in a DAG) where there is both direct and indirect evidence, and in networks of any complexity. A shared variance term solves the difficulties created in a random effects model when some contrasts are supported by only one or two trials. This method ensures that the 'indirect' estimates take the whole of the network structure into account, rather than being simply based on the indirect evidence from a specific loop, which may be attached to or nested within other loops of evidence.

More formally, the underlying network meta-analysis model in equation (4.1) remains the same, but now for any trial i comparing treatments X and Y, the trial-specific treatment effects are drawn from a normal distribution with variance σ^2 and mean d_{XY}^{Dir} when node (X,Y) is being split. At the same time, the evidence from the remaining studies is used in a standard network meta-analysis model that estimates the indirect treatment effect d_{XY}^{Ind}, using the consistency equations. We consider whether the hypothesis that the posterior distributions generated from direct and indirect evidence are in agreement can be reasonably supported by the data by examining the posterior distribution of the inconsistency parameter $\omega_{XY} = d_{XY}^{Dir} - d_{XY}^{Ind}$. We obtain a measure of conflict, P, defined as the probability that the ω_{XY} is greater than 0 (Marshall and Spiegelhalter, 2007; Presanis et al., 2013). For symmetric unimodal distributions, this is implemented using MCMC by counting the proportion of times d_{XY}^{Dir} exceeds d_{XY}^{Ind} (or $\omega_{XY} > 0$) in the sample, *prob*, and taking

$$P = 2 \times \min\left(prob, 1 - prob\right) \tag{7.3}$$

representing a two-sided Bayesian p-value for the probability of inconsistency (Marshall and Spiegelhalter, 2007; Presanis et al., 2013). These computations need to be done separately for each pair of treatments (X,Y) for which consistency is being assessed. When the node to split is included in a multi-arm trial, the arms being split are used to form the direct estimate and the remaining arms are included in the indirect comparisons (Dias et al., 2010b; van Valkenhoef et al., 2016).

The posterior mean of the residual deviance and the DIC are used to compare the full network meta-analysis model with the model where a particular node has been split. A reduction in residual deviance or DIC indicates that there may be inconsistency between the different sources of evidence on that node. Plots of each point's contribution to the DIC can also identify which points in the data contribute to the poor model fit and how their contribution changes when different nodes are split. For random effects models, a substantial reduction in heterogeneity when a particular node is split can also indicate inconsistency: the heterogeneity estimated by the network meta-analysis model could have been high to allow for inconsistent data points. Node-splitting can also generate intuitive graphics showing the posterior distributions obtained from the direct and indirect evidence, which can be displayed together with the posterior distributions from the network meta-analysis model with full consistency (Dias et al., 2010b).

However, node-split models are not easy to parameterise, especially when multi-arm trials are present. As with other inconsistency models, there may be more than one possible parameterisation and care should be taken to ensure that the split nodes refer to contrasts involved in generating potential inconsistencies. The presence of multi-arm trials complicates both the coding of the model and the decision of which nodes to split (Dias et al., 2010b; van Valkenhoef et al., 2016). The R package *gemtc* (van Valkenhoef et al., 2012a; van Valkenhoef and Kuiper, 2016) has been developed to automate and simplify the decision of which nodes to split and to implement the algorithm for every split node for the univariate arm-based and contrast-based likelihood models presented in Chapter 4. Inconsistency is defined as occurring only when there are at least three independent sources of evidence supporting the comparisons in a loop. The nodes to split and basic parameters are chosen so that there is a unique specification of models when multi-arm trials are present (van Valkenhoef and Kuiper, 2016; van Valkenhoef et al., 2016).

We recommend this method as the best way to check for and present conflict between direct and indirect evidence in complex networks. However, there will still be more *p*-values generated than there are potential inconsistencies and application of this method to large networks can be extremely time-consuming.

7.4 A Global Test for Loop Inconsistency

Another way of testing for inconsistency is to approach it from a model critique and comparison point of view (Chapter 3) rather than aiming to compare estimates based on direct and indirect evidence for every loop. To check whether the consistency assumption as assumed in the network meta-analysis models presented in Chapters 2 and 4 is reasonable for our data, we can compare that model with a model where no such consistency is assumed. We can then look at model fit statistics (Chapter 3) and assess whether eliminating the assumption of consistency for all contrasts has resulted in a better fitting model. In a random effects model, we can also compare estimates of between-study heterogeneity – a substantial reduction in the between-study heterogeneity when a model without the consistency assumption is fitted may indicate the presence of inconsistency, which was being masked by high between-study heterogeneity. We return to discuss the relationship between these two notions of conflict in Section 7.6.

7.4.1 Inconsistency Model with Unrelated Mean Relative Effects

In complex networks where independent tests cannot be constructed, or where the node-split approach would be too time-consuming, we propose that the standard consistency model (Chapters 2 and 4) be compared with an *inconsistency* model. The 'inconsistency factors' model in Section 7.3.2.3 is an example of an inconsistency model as it allows for inconsistency in the network. However, instead of introducing an extra variance parameter, in this section we suggest a simpler inconsistency model that assumes 'unrelated mean (relative) effects' (UME) (Dias et al., 2011e, 2013d).

In the network meta-analysis model with consistency, a network with S treatments defines S-1 basic parameters, which estimate the effects of all treatments relative to the reference treatment. Prior distributions are placed on these parameters and all other contrasts are derived through the consistency assumption. In the proposed UME model, each of the N contrasts for which evidence is available represents a separate, unrelated basic parameter to be estimated: no consistency is assumed. So, for the Smoking Cessation network (Figure 5.1), the consistency model introduced in Chapter 2 would estimate three relative treatment effect parameters from evidence on six contrasts with the other contrasts defined from the consistency equations. A UME inconsistency model would estimate six relative treatment effect parameters from the evidence on six contrasts, without assuming any relationship between the parameters. Note that the extra number of parameters in the UME model is exactly the ICDF in this network, which is three.

More formally, suppose we have a set of M trials comparing S treatments in any connected network. In a random effects model, the study-specific treatment effects of the treatment in arm k relative to the treatment in arm 1 of that study, δ_{ik}, are assumed to follow a normal distribution (equation (2.8)). In a consistency model, $S-1$ basic parameters are given vague prior distributions and the consistency equations define all other possible contrasts (equations (2.9) and (2.10)).

In a random effects UME model, each of the mean treatment effects in equation (2.9) is treated as a separate (independent) parameter to be estimated, sharing a common variance σ^2. These treatment effects are *all* given vague prior distributions

$$d_{ck} \sim \text{Normal}\left(0,\,100^2\right) \quad \text{with} \quad c = 1,\ldots,(nt-1);\ k = 2,\ldots,nt \quad (k > c)$$

(7.4)

and are estimated using only direct evidence – any indirect evidence is ignored. We also need to ensure that we set $d_{kk} = 0$ (the effects of treatment k compared to itself) for $k = 1, \ldots, nt$.

It can be seen that this model is a re-parameterisation of the model in Section 7.3.2.3, where for the Smoking Cessation example, instead of estimating three basic parameters and three inconsistency factors (d_{12}, d_{13}, d_{14}, ω_{23}, ω_{24}, ω_{34}), we estimate six unrelated relative effects (d_{12}, d_{13}, d_{14}, d_{23}, d_{24}, d_{34}).

The WinBUGS code to implement this model can be adapted from the code presented in Chapters 2 and 4 by removing the consistency equations. So for a given likelihood and link function, the generic code in Chapter 4 becomes

Generic WinBUGS code, UME inconsistency model: fixed effect. The likelihood, link function and residual deviance formula need to be specified in each case.

The # symbol is used for comments – text after this symbol is ignored by WinBUGS. Note that WinBUGS specifies the normal distribution in terms of its mean and precision.

```
# Fixed effect model for multi-arm trials
model{                          # *** PROGRAM STARTS
for(i in 1:ns){                 # LOOP THROUGH STUDIES
 mu[i] ~ dnorm(0,.0001)         # vague priors for all trial baselines
    for (k in 1:na[i]) {        # LOOP THROUGH ARMS
    likelihood                  # define likelihood
    link function <- mu[i] + d[t[i,1], t[i,k]]   # model for
                                                 linear
                                                 predictor
        dev[i,k] <- residual deviance formula  # Deviance
                                                 contribution
        }
    resdev[i] <- sum(dev[i,1:na[i]])   # summed residual deviance
                                         contribution for this trial
    }
totresdev <- sum(resdev[])      # Total Residual Deviance
for (k in 1:nt) { d[k,k] <- 0 } # set effects of k vs k to zero
for (c in 1:(nt-1)) {
```

```
for (k in (c+1):nt) { d[c,k] ~ dnorm(0,.0001)   }  # priors for all mean
                                                     treatment effects
   }
}                          # *** PROGRAM ENDS
```

The only difference between this fixed effect model code and the network meta-analysis code presented in Chapter 4 is that instead of defining the linear predictor in terms of the consistency equations where all relative treatment effects are written in terms of the basic parameters, d is now a matrix with d[c,k] representing the relative effect of treatment k compared with treatment c, which are all given non-informative prior distributions. Note that this code requires the data to have the treatments coded in ascending order so that $k \geq c$ in the definition of matrix d. This also means that for multi-arm trials, only the relative effects compared with the treatment in arm 1 will be used.

The same change is applied to the code for the random effects model, although this can be simplified compared with the code presented in Chapter 4 since we do not need to allow for the correlation in the random effects in multi-arm trials as we are considering them to be independent. The code therefore can be written generically as follows:

Generic WinBUGS code, UME inconsistency model: random effects. The likelihood, link function, residual deviance formula and upper bound for the uniform prior distribution need to be specified in each case.

The # symbol is used for comments – text after this symbol is ignored by WinBUGS. Note that WinBUGS specifies the normal distribution in terms of its mean and precision.

```
# Random effects model
model{                    # *** PROGRAM STARTS
for(i in 1:ns){           # LOOP THROUGH STUDIES
 delta[i,1] <- 0          # treatment effect is zero for control arm
 mu[i] ~ dnorm(0,.0001)   # vague priors for all trial baselines
     for (k in 1:na[i]) { # LOOP THROUGH ARMS
     likelihood           # define likelihood
     link function <- mu[i] + delta[i,k]        # model for linear
                                                  predictor
         dev[i,k] <- residual deviance formula  # Deviance
                                                  contribution
     }
     resdev[i] <- sum(dev[i,1:na[i]]) # summed residual deviance
contribution for this trial
# random effects distribution (independent mean effects)
     for (k in 2:na[i]) { delta[i,k] ~ dnorm(d[t[i,1],t[i,k]], tau) }
   }
```

```
totresdev <- sum(resdev[])        # Total Residual Deviance
for (k in 1:nt) { d[k,k] <- 0 }   # set effects of k vs k to zero
for (c in 1:(nt-1)) {
   for (k in (c+1):nt)   { d[c,k] ~ dnorm(0,.0001) }  # priors for
                                               all mean treatment effects
 }
sd ~ dunif(0, appropriate upper bound)  # vague prior for between-
                                          trial SD
tau <- pow(sd,-2)                 # between-trial precision =
                                    (1/between-trial variance)
}                                 # *** PROGRAM ENDS
```

Alternatively the multi-arm trial correction for the random effects distribution could be included (Chapter 2). In the generic code presented here, for networks where not all treatment pairs have been compared in trials, some of the relative effects will not have any information and will therefore not be estimable. Their posterior distribution will then be equal to the prior distribution in equation (7.4). Adjustments could be made to the code to ensure that only contrasts with evidence are included in the code. However, this would need to be done on a case-by-case basis with the code having to be adjusted for each network and would not have any meaningful impact on the estimates, model fit statistics or between-study heterogeneity, which are of primary interest. In such cases all the user needs to do is to ignore the estimates produced for contrasts for which no direct evidence exists, which can be identified as those having very large standard deviations (and wide credible intervals).

In essence the UME inconsistency model is equivalent to performing separate pairwise meta-analyses on each contrast, but in random effects models we allow for a shared variance parameter to be estimated. It can also be seen as a node-splitting model in which all nodes are split at the same time. The shared variance term solves the difficulties created in a random effects model when some contrasts are supported by only one or two trials. The between-study variance will be informed by all the studies. This avoids problems with the consistency of indirect estimates from fixed and random effects models or from random effects models with different variances, which may not follow the triangle inequalities (Section 7.3.1.4). In a fixed effects model, the inconsistency UME model is equivalent to performing completely separate pairwise meta-analysis of the data each using a fixed effects model. However, fitting a single UME model to all the data has the advantage of requiring only one run of the code, as well as providing global measures of model fit, for comparison to the network meta-analysis consistency model and allowing direct calculation of probabilities of conflict for simple networks (see Exercise 7.2).

We therefore recommend that, for simple networks, the outputs from a UME inconsistency model be used as the 'direct' estimates used to calculate the measures of inconsistency described in Section 7.3.1. In fact, in simple networks, calculation of indirect estimates, inconsistency estimates (ω) and *p*-values for inconsistency can be coded into WinBUGS so that the required tests for inconsistency in addition to the model comparison statistics can be produced (see Exercise 7.2).

When multi-arm trials are included in the evidence, the UME inconsistency model can have different parameterisations depending on which of the multiple contrasts defined by a multi-arm trial are chosen (see Section 7.2.2). Choice of parameterisation will affect parameter estimates, and the tests of inconsistency, although in most cases it is unlikely to affect the overall conclusion. For example, a three-arm trial ABC can inform the AB and AC independent effects, or it can be chosen to inform the AB and BC effects (if B was the reference treatment), or the AC and BC effects (with C as reference). The model and code implemented here arbitrarily choose the contrasts relative to the 'first' treatment in the trial. Thus, ABC trials inform the AB and AC contrasts, BCD trials inform BC and BD and so on. This avoids multiple use of data from multi-arm trials but is a limitation since if the ABC trial is the only trial comparing treatments B and C, no data will be used to inform this relative effect. If a fixed effects model is being considered, the re-weighting approach to adjust standard errors proposed by Rücker and Schwarzer (2014) can be used to incorporate all comparisons from multi-arm trials (see Section 7.3.1.4).

7.4.2 Example: Full Thrombolytic Treatments Network

The full network of thrombolytic treatments was introduced in Figure 3.1. Not all treatment contrasts have been compared in a trial, as there are treatment pairs that are not connected. There are nine treatments in total and information on 16 pairwise comparisons, which would suggest an ICDF of eight. However, there is one loop, SK, Acc t-PA, SK + t-PA that is only informed by a three-arm and therefore cannot contribute to the number of possible inconsistencies. Discounting this loop gives $ICDF = 7$.

A fixed effects network meta-analysis consistency model with a binomial likelihood and logit link was fitted to the data, taking SK as the reference treatment, that is, the eight treatment effects relative to SK are the basic parameters and have been estimated, while the remaining relative effects were obtained from the consistency assumption (see Chapters 2 and 3). A fixed effects inconsistency UME model was also fitted to the same data, which estimated 15 independent mean treatment effects. The full code and results are given in *Ch7_FE_Bi_logit_incon.odc*.

Results for the 15 contrasts on which there is information for both models are presented in Table 7.4, along with measures of model fit. These are based on 50,000 iterations on two chains after a burn-in period of 50,000 for the

Table 7.4 Thrombolytics example: posterior summaries (mean and 95% credible interval) on the log odds ratio scale for treatments Y compared with X for all contrasts that are informed by direct and indirect evidence; and posterior mean of the residual deviance (resdev), p_D and DIC, from the fixed effects consistency and UME inconsistency models.

Treatments		Network meta-analysis (consistency)		Inconsistency model (UME, original data)		Inconsistency model (UME, adjusted data)	
X	Y	Mean	95% CrI	Mean	95% CrI	Mean	95% CrI
SK	t-PA	0.002	(−0.06, 0.06)	−0.004	(−0.06, 0.06)	0.001	(−0.06, 0.06)
SK	Acc t-PA	−0.177	(−0.26, −0.09)	−0.158	(−0.25, −0.06)	−0.148	(−0.25, −0.04)
SK	SK + t-PA	−0.049	(−0.14, 0.04)	−0.044	(−0.14, 0.05)	−0.039	(−0.14, 0.06)
SK	r-PA	−0.124	(−0.24, −0.01)	−0.06	(−0.23, 0.11)	−0.060	(−0.23, 0.11)
SK	PTCA	−0.173	(−0.32, −0.02)	−0.665	(−1.03, −0.31)	−0.664	(−1.03, −0.31)
SK	UK	−0.476	(−0.67, −0.28)	−0.369	(−1.41, 0.63)	−0.363	(−1.41, 0.63)
SK	ASPAC	−0.203	(−0.64, 0.23)	0.005	(−0.07, 0.08)	−0.005	(−0.09, 0.08)
t-PA	PTCA	0.016	(−0.06, 0.09)	−0.544	(−1.38, 0.25)	−0.537	(−1.38, 0.25)
t-PA	UK	−0.18	(−0.28, −0.08)	−0.294	(−0.99, 0.37)	−0.291	(−0.99, 0.37)
t-PA	ASPAC	−0.052	(−0.16, 0.06)	−0.29	(−1.01, 0.41)	0.014	(−0.07, 0.10)
Acc t-PA	SK + t-PA	0.128	(0.02, 0.23)	NA		0.106	(−0.04, 0.25)
Acc t-PA	r-PA	−0.126	(−0.26, 0.01)	0.019	(−0.11, 0.15)	0.019	(−0.11, 0.15)
Acc t-PA	TNK	−0.175	(−0.34, −0.01)	0.006	(−0.12, 0.13)	0.006	(−0.12, 0.13)

(Continued)

Table 7.4 (Continued)

Treatments		Network meta-analysis (consistency)		Inconsistency model (UME, original data)		Inconsistency model (UME, adjusted data)	
X	Y	Mean	95% CrI	Mean	95% CrI	Mean	95% CrI
Acc t-PA	PTCA	−0.478	(−0.68, −0.27)	−0.216	(−0.45, 0.02)	−0.215	(−0.45, 0.02)
Acc t-PA	UK	−0.206	(−0.64, 0.23)	0.146	(−0.54, 0.86)	0.134	(−0.55, 0.85)
Acc t-PA	ASPAC	0.013	(−0.06, 0.09)	1.405	(0.63, 2.27)	1.391	(0.64, 2.26)
$resdev^a$		105.9		99.7		99.2	
p_D		58		65		64	
DIC		163.9		164.7		163.2	

UME models were fitted to the original arm-level data and to data with multi-arm trials given as treatment differences with adjusted standard errors (Rücker and Schwarzer, 2014).

[a] Compare with 102 data points.

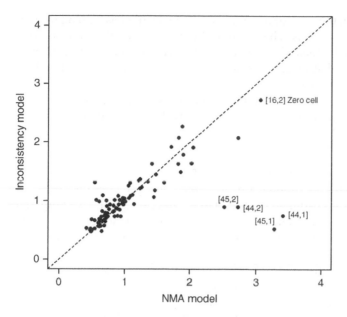

Figure 7.3 Full Thrombolytics example: plot of the individual data points' posterior mean deviance contributions for the network meta-analysis model (NMA) (horizontal axis) and the UME inconsistency model with original arm-level data (vertical axis), along with the line of equality. Points that have a better fit in the inconsistency model or correspond to a zero cell have been marked with the trial and arm number.

consistency model and 50,000 iterations on three chains after a burn-in of 20,000 for the inconsistency model.

Although the inconsistency model has a lower posterior mean of the residual deviance and hence is a better fit to the data, the DIC is very similar for both models – the difference in DIC is less than 3 (Chapter 3). This is because the inconsistency model has seven more parameters than the network meta-analysis model, as can be seen in the difference in the values of p_D (see Chapter 3). Note also that this difference in parameters corresponds exactly to the ICDF. A plot of each individual data points' posterior mean deviance contribution in each of the two models is presented in Figure 7.3. These contributions are obtained from WinBUGS by monitoring the node dev in each model, ensuring the studies are entered in exactly the same order in both models (so that the numbering corresponds). In this example, four data points show a much lower value of the posterior mean of the residual deviance in the inconsistency model, suggesting that a consistency model does not fit these points well. These four points correspond to the two arms of trials 44 and 45, which were the only two trials comparing Acc t-PA with ASPAC. Comparing the posterior estimates of the treatment effects of ASPAC versus Acc t-PA in Table 7.4, we can see that these differ between the consistency and inconsistency models with no overlap in the 95% credible intervals. The fact that the two trials on this contrast give similar results to each other, which are in conflict with the remaining evidence, supports the notion that there is a systematic inconsistency, rather than heterogeneity, within this contrast.

Consistency in this network has also been assessed using the node-split method, and the same comparison highlighted as showing evidence of inconsistency (Dias et al., 2010b).

7.4.2.1 Adjusted Standard Errors for Multi-Arm Trials

The UME model with the original arm-level data only provides results for 15 contrasts. There is no estimate of the relative effect of SK + t-PA compared with Acc t-PA for this model in Table 7.4 because these treatments have only ever been compared in a three-arm trial with SK as the reference arm. Therefore this comparison is not estimated in the standard UME inconsistency model. Since we are using a fixed effects model, we can adjust the data from the two multi-arm trials in this network so that all comparisons are considered (Rücker, 2012; Rücker and Schwarzer, 2014). For each of the 2 three-arm trials in this network, three log odds ratios are calculated and their standard errors adjusted, so that they can be included as if they originated from independent trials. These adjusted standard errors capture both the correlation and internal consistency between multiple estimates from the same study. For studies 1 and 6, the adjusted data are given in Table 7.5. Sample R code to obtain these adjusted estimates using the *netmeta* package (Rücker et al., 2015) is given in Appendix A.

A fixed effects UME inconsistency model with a binomial likelihood and logit link for all two-arm studies and a normal likelihood with identity link for the transformed three-arm studies was fitted, which estimated 16 independent

Table 7.5 Adjusted data for studies 1 and 6, given as treatment differences with adjusted standard errors.

t[,1]	t[,2]	y[,2]	se[,2]	Study
1	3	−0.147550	0.052643	1
1	4	−0.042067	0.049805	1
3	4	0.105483	0.075237	1
1	2	−0.023288	0.043316	6
1	9	−0.004314	0.042818	6
2	9	0.018974	0.043432	6

Obtained using *netmeta* in R.

mean treatment effects. The full code and results are given in *Ch7_FE_Bi_logit_incon2.odc*.

Results for the 16 contrasts estimated from the UME and consistency models are presented in Table 7.4, along with measures of model fit. A plot of individual data points' contributions to the residual deviance from this UME model and the consistency model is very similar to that presented in Figure 7.3. Conclusions based on these results are the same as using the model with the original arm-level data.

7.4.3 Example: Parkinson's

We will now consider the Parkinson's example, where data are presented as mean reduction in time off work for each arm of each study, with standard errors (Table 4.6 and Figure 4.6). The fixed effects network meta-analysis model with a normal likelihood was considered appropriate based on residual deviance and DIC (Table 4.7). However, we now need to check whether there is evidence of inconsistency in this network. As noted earlier, there are two loops that can show inconsistency: the loop formed by treatments 1, 3 and 4 and the four-way loop 1, 2, 4, 3, although note that there is a three-arm trial involved in two of the edges of the latter.

A fixed effects UME inconsistency model was fitted with a normal likelihood and identity link to obtain the direct estimates for the available comparisons and the measure of model fit (see Exercise 7.2). The output from WinBUGS is given in the following.

node	mean	sd	MC error	2.5%	median	97.5%	start	sample
d[1,2]	−1.833	0.3389	0.001558	−2.496	−1.833	−1.167	20001	150000
d[1,3]	−0.3045	0.6688	0.00332	−1.609	−0.306	1.01	20001	150000
d[1,4]	−0.669	0.6219	0.002554	−1.889	−0.6697	0.5522	20001	150000
d[1,5]	−0.3351	99.79	0.2531	−195.5	−0.5624	195.8	20001	150000
d[2,3]	−0.1878	99.87	0.2539	−196.6	0.08105	195.3	20001	150000
d[2,4]	−0.214	100.1	0.2718	−196.5	−0.04386	195.5	20001	150000
d[2,5]	0.1975	100.1	0.2541	−196.6	0.4017	196.2	20001	150000
d[3,4]	9.49E−04	0.3463	0.001275	−0.6787	6.72E−04	0.6816	20001	150000
d[3,5]	0.09132	99.88	0.2649	−197	0.02499	195.8	20001	150000
d[4,5]	−0.2992	0.2082	8.70E−04	−0.7069	−0.2991	0.11	20001	150000
totresdev	14.18	4.912	0.01492	6.597	13.52	25.51	20001	150000

Note that there is no posterior updating for comparisons of treatments 1 and 5, 2 and 3, 2 and 4, 2 and 5 and 3 and 5, which have a posterior distribution equal to the Normal$(0,100^2)$ prior distribution because there are no studies directly comparing treatments 1 and 5, 2 and 3, 2 and 5 or 3 and 5. Although there is a study comparing treatments 2 and 4, this is a three-arm study comparing treatments 1, 2 and 4 so the inconsistency model only uses the information on the relative effects of 2 versus 1 and 4 versus 1, ignoring the information on 4 versus 2, as explained earlier.

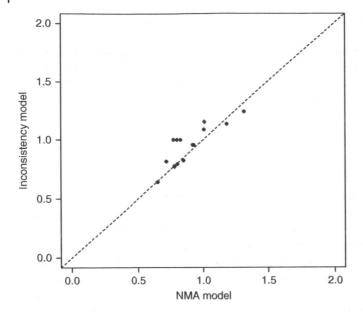

Figure 7.4 Parkinson's example: plot of each data points' contribution to the residual deviance for the network meta-analysis (NMA) with consistency and UME inconsistency models.

We can compare the posterior mean of the residual deviance (totres-dev=14.15) with the value for the fixed effects model in Table 4.7, which was 13.3. Thus the fit of the model is comparable and does not suggest evidence of inconsistency. In effect both the network meta-analysis and inconsistency models fit the data well. Comparing each individual data points' contribution to the residual deviance for both the network meta-analysis and inconsistency models, we can see that no data points have noticeable improvement in fit when the consistency assumption is relaxed (Figure 7.4). Thus there is no improvement in model fit when using the inconsistency model, suggesting that there is no evidence of inconsistency.

The DIC for the inconsistency model can be calculated as 26.15 by noting that $p_D = 12$, since there are seven study-specific baselines mu[i] to estimate and five comparisons d (note that d[2,4] is not estimated). We have added one parameter relative to the consistency model (Table 4.7) but obtained no improvement in residual deviance.

We can use the posterior mean and standard deviation from the UME inconsistency model as the 'direct' estimates to calculate the p-values for inconsistency in the (1, 3, 4) loop using Bucher's method (equation (7.1)) by calculating the inconsistency estimates in WinBUGS and comparing them with the direct estimates to obtain a probability of conflict. This is done by including the following code in the UME inconsistency model:

```
# Inconsistency assessment for loop (1,3,4)
# Indirect estimate
dInd.34 <- d[1,4]-d[1,3]
# difference between direct and indirect
diff.34 <- dInd.34-d[3,4]
# p-value
p.34 <- step(diff.34)
```

where dInd.34 calculates the indirect estimate for the comparison of treatment 4 to 3 according to equation (1.1), diff.34 calculates the inconsistency estimate as the difference between direct and indirect evidence (equation (7.1)) and p.34 counts how many times this difference is positive. Averaging p.34 over the number of iterations will give the probability that the difference is positive. For a two-sided Bayesian *p*-value, we use equation (7.3). Calculating the *p*-value in this way has the advantage of not requiring the assumption of approximate normality of the estimates since the full posterior distribution of the direct, indirect and difference estimates are used. It does not matter whether we calculate the inconsistency *p*-value for the comparison of treatment 4 to 3, treatment 3 to 1 or treatment 4 to 1, since they are part of the same loop the absolute value of the inconsistency estimate will be the same (Section 7.3.1.1), as will the Bayesian *p*-value. To confirm this we can add the following code, which calculates the indirect estimates, the difference between the direct and indirect estimates and the probability that this difference is positive for the remaining contrasts in the loop. By repeated application of equations (1.1) and (7.1), we extend the code to

```
# Inconsistency assessment for loop (1,3,4)
# Indirect estimates
dInd.13 <- d[1,4]-d[3,4]
dInd.14 <- d[1,3]+d[3,4]
# differences between direct and indirect
diff.13 <- dInd.13-d[1,3]
diff.14 <- dInd.14-d[1,4]
# p-values
p.13 <- step(diff.13)
p.14 <- step(diff.14)
```

Full WinBUGS code and results are given in *Ch7_FE_Normal_id_inconBucher.odc* (see also Exercise 7.2). Table 7.6 gives indirect estimates for all three comparisons in this loop. There is no evidence of conflict between the direct

Table 7.6 Parkinson's example: assessment of inconsistency for a single loop using Bucher's method in WinBUGS.

		Direct		Indirect		Difference			
X	Y	Mean	sd	Mean	sd	Mean	sd	Probability	*p*-Value
1	3	−0.3045	0.6688	−0.6699	0.7105	−0.3654	0.9750	0.354	0.71
1	4	−0.6690	0.6219	−0.3036	0.7527	0.3654	0.9750	0.646	0.71
3	4	0.0009	0.3463	−0.3645	0.9128	−0.3654	0.9750	0.354	0.71

and indirect estimates in this loop, and we can confirm that the differences between the direct and indirect estimates for all edges of the loop are the same, apart from their sign, which leads to the same Bayesian p-value of 0.71, far from suggesting inconsistency (Table 7.6).

Unfortunately we cannot use this approach directly to assess inconsistency in the four-way loop since no direct estimate is calculated for the comparison of treatments 2 and 4. See Exercise 7.2 for a solution to this.

7.4.4   Example: Diabetes

We will now consider the Diabetes example, where data are the number of new cases of diabetes over the trial duration period (Table 4.4 and Figure 4.3). Fixed and random effects network meta-analysis models with a binomial likelihood and cloglog link were fitted and the random effects model was considered the most appropriate based on the DIC (Table 4.5). However, we now need to check whether there is evidence of inconsistency in this network. This network has several non-independent loops and we will therefore only look at a global assessment of inconsistency. A random effects UME inconsistency model was fitted with a binomial likelihood and cloglog link to obtain the direct estimates for the available comparisons and the measures of heterogeneity and model fit. WinBUGS code and full results are available in *Ch7_RE_Bi_cloglog_incon.odc*. The posterior mean of the residual deviance for the inconsistency model is 50.7, which is slightly lower than that obtained for the network meta-analysis consistency model (Table 4.5). Comparing each individual data points' contribution to the residual deviance for both the consistency and inconsistency models, we can see that there are two data points that have some improvement in fit when the consistency assumption is relaxed and one which has a slight worse fit (Figure 7.5). These points correspond to trials 1 and 8, which have one of the longest and shortest follow-up times in the network, respectively. However, all these points have a relatively poor fit in both models, so no strong conclusions can be made. The posterior distribution of the between-study

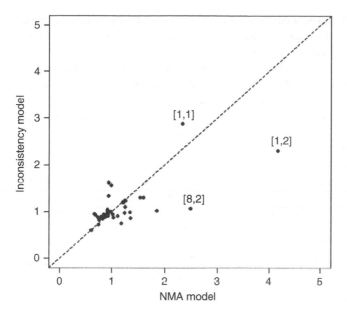

Figure 7.5 Diabetes example: plot of the individual data points' posterior mean deviance contributions for the network meta-analysis model with consistency (NMA) (horizontal axis) and the UME inconsistency model (vertical axis) along with the line of equality. Points that have a better fit in one of the models have been marked with the trial and arm number, respectively.

heterogeneity has median 0.12 and 95% CrI (0.04, 0.26) in the inconsistency model, which is very similar to that seen in the consistency model (Table 4.5). We therefore conclude that, overall, there is no evidence of inconsistency in this network.

7.5 Response to Inconsistency

Having found evidence of inconsistency in a network, the reliability of the estimates and any recommendations based on them is put in question.

There has been little work so far on how to respond to inconsistency when it is detected in a network, and it is likely that the best approach will depend on the specific characteristics of each problem. A reasonable principle is that decisions should be based on models that are internally coherent, that is, models in which the consistency assumptions are imposed, *and* that these models should fit the data. If the data cannot be fitted with a coherent model, then some kind of adjustment must be made. However, any adjustment in response to inconsistency is *post hoc*, which emphasises the importance of identifying

potential causes of heterogeneity of effect at the time of defining the trial inclusion and exclusion criteria and evaluating potential internal or external biases
in the evidence, in advance of synthesis.

It is important to appreciate that once inconsistency is detected, there is little
that statistical science can offer as remediation. For example, it might be that
the statistical analysis shows up one particular trial, or set of trials, as having a
bad fit within a consistency model. Very likely the poor statistical fit might
disappear if this trial, or trials, is removed or if the observed treatment effects
in the trials are adjusted. However, it is highly likely that a consistent network
of evidence can *also* be obtained by removing or making adjustments to *other*
trials. Worse, each different adjustment might be equally effective in reducing
inconsistency, but each represents a very different interpretation of the evidence and produces very different estimates, as has been shown in multi-
parameter evidence synthesis in epidemiology applications (Goubar et al.,
2008; Presanis et al., 2008). The essential point is that inconsistency is not a
property of individual studies, but of loops of evidence, and it may not always
be possible to isolate which loop is 'responsible' for the detected inconsistency,
let alone which edge of that loop (Lu and Ades, 2006).

The decision of how to address inconsistency cannot therefore be determined
by statistical methods. A thorough review of the entire evidence base by clinical
epidemiologists is required. This may result in the identification of one or more
trials that are 'different' in some way, where, for example, a treatment-modifying
covariate (e.g., dose, disease duration or bias) is present, or suspected.
Investigators must then make a series of decisions: how might these factors
relate to the target population for the decision? Are there specific trials that
should not be included in the evidence base? Should the treatment effects
observed in some trials be regarded as 'biased' and adjusted in some way, and if
so, what data are available on which this adjustment can be based (see Chapters
8 and 9)? The final option, of course, if there seems to be no explanation for an
apparent inconsistency or heterogeneity, is to consider it a chance finding and
assess the sensitivity of conclusions to different estimates (Caldwell et al., 2016).

Another possible cause of inconsistency is a poor choice of scale of measurement, which can also lead to increased heterogeneity (Deeks, 2002). Choosing
a different scale might be a solution, although it is not always obvious whether
to model treatment effects on a risk difference, logit or complementary log log
scale (Chapter 4). In Chapter 2 we noted that the choice of which scale was
most appropriate is essentially an empirical one, although there is seldom
enough evidence to decide on the basis of goodness of fit. However, it is often
the case that measurement scales that lead to higher heterogeneity also tend to
show more inconsistency and this should be investigated.

It is important to note that inconsistency in one part of the network does not
necessarily imply that the entire body of evidence is to be considered suspect.
In general, it is not possible to identify which contrast is 'deviant', but in some

cases the network structure may provide an indication of which estimates are contributing to the disagreement. In the Enuresis example (Section 7.3.1.3), there are four different estimates of a single contrast. If three agree and the fourth is different, it might be considered that three estimates have been 'corroborated' and that the fourth is 'deviant'. However, in this particular instance, an examination of the different estimates (Table 7.3) does not suggest any such simple interpretation, and it would be necessary to review *all* the studies involved in estimating these contrasts (see Caldwell et al. (2010) for a thorough examination of this dataset). It would be advisable, in fact, to *always* reconsider the entire network if inconsistency is located in any part of it, as evidence of inconsistency throws doubt on the trial inclusion criteria and the potential presence of effect modifiers in the entire network.

To ensure conclusions based on indirect evidence are sound, we must attend to the direct evidence on which they are based. This states that if the direct estimates of the AB and AC effects are unbiased estimates of the treatment effects *in the target population*, the indirect estimate of the BC effect must be unbiased as well. Conversely, any bias in the direct estimates, for example, due to effect-modifying covariates arising from the patients not being drawn from the target population, will be passed on to the indirect estimates in equal measure (see Section 1.5). The term 'bias' in this context must be seen broadly, comprising both internal and external threats to validity. This implies that if direct evidence on AB is based on trials conducted on a different patient population, and that a treatment effect modifier is present, what some may regard as an 'incorrect generalisation' from the AB trials to draw inferences about the target population can be considered as (external) bias that will be inherited by any indirect estimates based on these data. These issues are discussed in further detail in Section 12.3. Thus, the question 'are conclusions based on indirect evidence reliable?' should be considered alongside the question 'are conclusions based on pairwise meta-analysis reliable?'

While it is essential to carry out tests for inconsistency, the issue should not be considered in an overly mechanical way. Detection of inconsistency, like the detection of any statistical interaction, requires far more data than what is needed to establish the presence of a treatment effect. Investigators will therefore nearly always fail to reject the null hypothesis of consistency. But this is not an indication that there is no inconsistency.

7.6 The Relationship between Heterogeneity and Inconsistency

Although we have characterised heterogeneity as between-trial variation *within* treatment contrasts, and inconsistency as variation *between* contrasts, the difference is subtle. Eventually, all heterogeneity in relative treatment effects is a reflection of an interaction between the treatment effect and a trial-level varia-ble (see Chapter 12). To put it in another way, heterogeneity reflects the pres-ence of effect modifiers and so does inconsistency (see Chapter 1). If we consider the case of a triangular network (Figure 1.1e), if the effect modifiers are present in the AB and AC trials, but not the BC trials, then we may observe 'inconsist-ency'. However, if the effect modifiers are more evenly balanced across the net-work, then it is more likely that we will not find inconsistency, but we may find substantial heterogeneity. In addition, although a high level of heterogeneity increases the risk of inconsistency, it also lowers the chances that it will be detected. For this reason, even when inconsistency is not detected, and when, as with indirect comparisons, it *cannot* be detected because *ICDF* = 0, the question that must always be asked is 'how reliable are conclusions based on indirect evidence or network meta-analysis?' This is discussed further in Chapter 12.

One might try to distinguish an imbalance occurring by chance from one due to an *inherent inconsistency*. Suppose, for example, patients in BC trials are inherently different because they cannot take treatment A, and this is in addition associated with a different treatment effect. These patients have perhaps already failed on A, or have had adverse side-effects, or they have markers that counter-indicate A. At the other extreme, it might be that the inconsistency was just the result of chance, in which the BC trials just happened to be in those patients with an effect modifier. Access to IPD may help to explore inconsistency as well as heterogeneity (see Chapter 8). Between these extremes one might imagine that the BC trials are just *more likely* to concern patients who could not take A. In either case, heterogeneity and inconsistency are both reflections of a treatment effect modifier. Inconsistency is thus a special case of heterogeneity where there is an association between the effect modifier and the set of treatment contrasts compared. There is an immediate implication that inconsistency due to a chance imbalance in the distribution of effect modifiers should become increasingly less likely to occur as the number of trials on each contrast increases.

Networks with few trials per comparison highlight the drawbacks of using random effects model when the between-study heterogeneity cannot be ade-quately estimated. This is the case for both Bayesian and frequentist approaches. In a Bayesian framework, the practice of using vague prior distributions for the between-trial variation, combined with a lack of data, will generate posterior distributions that allow an unrealistically high between-study variance. This, in turn, is likely to mask all but the most obvious signs of inconsistency. In the thrombolytic example, we used a fixed effects analysis, but very similar results were obtained with a random effects model (Lu and Ades, 2006; Dias et al., 2010b). However, in this dataset, the extent of between-trial variation is unusu-ally low, so that the inconsistency stands out (Figure 7.3). Informative prior distributions, based on expert opinion or meta-epidemiological data, should be considered particularly in networks where there are few studies per com-parison (see Chapters 2 and 4).

In general, the risk of inconsistency is greatly reduced if between-trial heterogeneity is low. Empirical assessment of heterogeneity can therefore provide some reassurance or can alert investigators to the risk of inconsistency. The posterior distribution of the between-study standard deviation from a network meta-analysis model with consistency or the multiple posterior distributions from separate pairwise meta-analysis models on each contrast can be inspected to provide an indication of the extent of heterogeneity in the whole network. Another useful indicator is the between-trial variation in the reference arm, in trials that compare the reference treatment. If a large number of trials include comparisons with a reference treatment, perhaps placebo or a standard, and these arms all have similar proportions of events, hazards and so on, then this suggests that the trial populations are relatively homogeneous and that there will be little heterogeneity in the treatment effects. If, on the other hand, these study-specific baselines are highly heterogeneous, while not meaning that the *relative* effects are also heterogeneous, it does at least constitute a warning that there is a potential risk of heterogeneity in the relative effects. Heterogeneity across specific treatment arms can be examined via a Bayesian synthesis (see Chapter 5). Chapter 12 discusses these issues in further detail.

Any step that can be taken to avoid between-trial heterogeneity will be effective in reducing the risk of drawing incorrect conclusions from pairwise meta-analysis, indirect comparisons and network meta-analysis. Fortunately, the decision-making context is likely to have already eliminated the great majority of potentially confounding factors. The most obvious sources of potential heterogeneity of effect, such as differences in dose, treatment definition or differences in co-therapies, will already have been eliminated when defining the scope of the literature review, which is likely to restrict the set of trials to specific doses and co-therapies.

The HIV example in Section 7.3.1.2 where evidence of inconsistency was detected is an example of heterogeneity in treatment definitions introduced by the practice of trying to combine evidence on disparate treatment doses or treatment combinations within meta-analyses, often termed 'lumping'. Here, there were substantial, and independently recognised, differences in efficacy between the treatment combinations appearing in the direct and indirect evidence. If this is addressed, the difference between indirect and direct evidence is no longer statistically significant (Caldwell et al., 2007).

7.7 Summary and Further Reading

Logically, inconsistency testing should come after an examination of heterogeneity and after adjustment for known causes of heterogeneity through meta-regression or bias (see Chapters 8 and 9). Decisions should be based on coherent models that fit the data. Careful examination of different sources of evidence may reveal that some estimates are 'corroborated', while others are not. If inconsistency is detected, the entire network of evidence should be reconsidered from a clinical epidemiology viewpoint with respect to the presence of potential effect modifiers.

In spite of all the limits on heterogeneity resulting from the narrow scope required to make a decision, there is still the potential for treatment effect modifiers to be present in trials, and unrecognised. Results may not have been broken down by confounding variables, and their distribution over the sample may not have been recorded. Among typical variables that frequently appear as effect modifiers are age, severity at baseline (i.e., at start of treatment or randomisation) and previous treatments, all of which may be further confounded with each other. Investigators should make themselves aware of potential confounders, both within the network of evidence, and in previous literature, and consider the potential role of bias adjustment and meta-regression (see Chapters 8 and 9), prior to synthesis and consistency checking.

The choice of method to evaluate consistency should be guided by the evidence structure. If it is possible to isolate independent loops and construct independent tests, then the Bucher test or its extensions to larger 'circuit' structures and to chi-square tests represent the most simple and complete approach. In more complex networks, a repeated application of the Bucher method to all the possible loops fails to account for the whole network structure and should not be used. The UME inconsistency model presented in Section 7.4, with its shared variance parameter when random effects are considered, can be used to provide direct estimates that can be used to implement a Bucher approach for networks with independent loops. For complex networks, the fit of a consistency model can be compared with the fit of an 'inconsistency' model, such as the UME model (Section 7.4.1). Analyses of residual deviance can provide an 'omnibus' test of global inconsistency and can also help locate it. Reductions in heterogeneity in an inconsistency model, compared with a consistency model, can also signal inconsistency. Node-splitting is another effective method for comparing direct evidence with indirect evidence in complex networks and should be used to locate and quantify inconsistent loops if any evidence of inconsistency is found using an inconsistency model.

Measures of inconsistency variance (Lu and Ades, 2006) or incoherence variance (Lumley, 2002) and the back-calculation method (Dias et al., 2010b) are *not* recommended as indicators of inconsistency.

Inconsistency checking is closely related to cross-validation for outlier detection, described in Chapter 3. However, in the presence of heterogeneity, cross-validation is based on the predictive distributions of effects, while the concept of inconsistency between 'direct' and 'indirect' evidence refers to inconsistency in expected (i.e., mean) effects and is therefore based on the posterior distributions of the mean effects. This will frequently result in a situation where, in a triangular loop in which one edge consists of a singleton trial, we may find inconsistency in the expected effects, while cross-validation fails to show that the singleton trial is an outlier. In fact, in loops where one of the contrasts is informed by a single trial and a random effects model is used, we would recommend cross-validation with careful assessment of changes in heterogeneity over any methods for detecting inconsistency as this will provide a better understanding of whether this trial disagrees with the remaining evidence or if it is just another (plausible) draw from the random effects distribution (Madan et al., 2011). Of course, technically there is no reason why inconsistency checks cannot be made on the predictive distributions of the treatment effects, and this may be desirable if inference is to be based on the predictive treatment effects from a network meta-analysis. For example, predictive node-splitting can be used, where the predictive distribution of the direct estimate is used to compare with the indirect estimate (see Chapters 3 and 8). This will account for large between-study heterogeneity that may affect what we expect to see in a 'new' comparison added to this network. However, such methods are unlikely to detect any but the most extreme of inconsistencies due to the added levels of uncertainty.

Inconsistency should also be checked in network meta-analysis models with multiple outcomes, such as the Schizophrenia example in Chapter 4 or the models in Chapters 10 and 11. Node-split or UME models can be adapted to check consistency in these networks, although calculation of Bayesian p-values is complicated by the multiple dimensions at which inconsistency could manifest. Different options could be considered to calculate relevant p-values for each application – see also notes on cross-validation with multi-arm trials in Chapter 3.

7.8 Exercises

7.1 The Smoking Cessation example has been introduced in Chapter 5 and the possibility of inconsistency discussed in this chapter. The data, random effects network meta-analysis model and results are given in *Ch5_ Smoking_CEA.odc*.

a) Adapt the random effects network meta-analysis code to fit a random effects UME inconsistency model, as described in Section 7.4. Note that you can delete the cost-effectiveness analysis code or ignore those results in the inconsistency model. Run the inconsistency model and compare the measures of model fit and heterogeneity with the network meta-analysis model (Table 5.2). What would you conclude about inconsistency in this network?

b) Using the direct estimates obtained from the UME inconsistency model in a), calculate the Bucher inconsistency estimates for all seven loops: the three triangles and four quadrilaterals. What do you conclude about inconsistency in this network?

7.2 In the Parkinson's example presented in Section 7.4.3, we could not assess consistency in a four-way loop (1, 2, 4, 3) since no direct estimate of d_{24} was available since this comparison is only informed by study 3, a three-arm trial comparing treatments 1, 2 and 4.

a) Add code to *Ch7_FE_Normal_id_inconBucher.odc* that will provide a direct estimate for d_{24} from study 3, with which an indirect estimate can be compared. Add also code to compare the direct and indirect estimates for this contrast as well as the Bayesian measure of conflict. Comment on this approach. Is there any evidence of inconsistency in this loop?

b) * Rücker and Schwarzer (2014) suggested adjusting the standard error of each comparison provided by multi-arm studies that allows them to be incorporated as multiple two-arm studies, while retaining their correct correlation structure (and hence internal consistency). Use the R package *netmeta* (Rücker et al., 2015) to calculate the adjusted standard errors and relative treatment effects for study 3 (see Appendix A) and incorporate this modified study in the fixed effects UME inconsistency model given in *Ch7_FE_Normal_id_inconBucher.odc*. Note that a shared parameter model will be required (see Chapter 4). Re-calculate the Bucher inconsistency estimates and compare with those in Table 7.6 and to those obtained earlier.

7.3 Data from a HTA report with 18 RCTs comparing four treatments for non-small cell lung cancer (Brown et al., 2013) was introduced in Exercise 4.3. The fixed effects network meta-analysis model assuming consistency was preferred. This network has several loops so we need to assess inconsistency.

a) Adapt the fixed effect network meta-analysis code from Exercise 4.3 to fit a fixed effects UME inconsistency model, as described in Section 7.4. Run the inconsistency model and compare the measures of model fit with the network meta-analysis model in Exercise 4.3. What would you conclude about inconsistency in this network?

b) * R package *gemtc* (van Valkenhoef and Kuiper, 2016) can be used to implement the node-splitting method described in Section 7.3.2.4. Using the original data, obtain the pairs of nodes to split and the direct and indirect estimates as well as p-values and model fit statistics for this example. What would you conclude about inconsistency in this network? Does it agree with the conclusions from a)?

8

Meta-Regression for Relative Treatment Effects

8.1 Introduction

Heterogeneity in relative treatment effects is an indication of the presence of effect-modifying covariates, in other words of *interactions* between the treatment effect and trial-level or patient-level variables whose distribution might vary across included trials. A distinction is usually made between (i) true clinical variability in treatment effects due to variation between patient populations, protocols or settings across trials and (ii) biases related to methodological flaws in the way in which trials were conducted.

Clinical variability in relative treatment effects is said to represent a threat to the *external validity* of trials (Rothman et al., 2012) and limits the extent to which one can generalise trial results from one situation to another. The trial may deliver an unbiased estimate of the treatment effect in a certain setting, but it may be 'biased' *with respect* to the target population in a specific decision problem (Chapter 1). Careful consideration of inclusion and exclusion criteria can help to minimise this type of bias, but often at the expense of having little or no evidence to base decisions on. That is, if inclusion criteria are too strict, the majority, or even all, of the evidence may be discarded as 'not relevant', leaving no synthesis option. On the other hand, inclusion criteria that are too broad risk pooling populations with very different relative treatment effects, thus inducing a large heterogeneity and making interpretation of results very difficult. Biases or interaction effects due to imperfections in trial conduct represent threats to the *internal validity* of the results from RCTs. Although some of the methods presented in this chapter can be used to adjust for bias due to lack of internal validity, more powerful methods are described in detail in Chapter 9.

In this chapter we focus on methods for meta-regression that can address the presence of heterogeneity caused by known, and observed, effect modifiers. Although regression is usually seen as a form of adjustment for differences in

Network Meta-Analysis for Decision-Making, First Edition. Sofia Dias, A. E. Ades, Nicky J. Welton, Jeroen P. Jansen and Alexander J. Sutton.
© 2018 John Wiley & Sons Ltd. Published 2018 by John Wiley & Sons Ltd.
Companion website: www.wiley.com/go/dias/networkmeta-analysis

covariates, here we consider it as a method for 'bias adjustment' since these covariates affect the 'external validity' of trials (Turner et al., 2009) that may lead to biased results for the target population. The aim is therefore to remove unwanted variability in relative treatment effects that can be explained by known, and measured, effect modifiers, which vary between studies. We focus particularly on the technical specification of models that can adjust for potential causes of heterogeneity and on the interpretation of such models in a decision context. In a network meta-analysis context, effect modifiers causing variability in relative treatment effects across studies can also induce inconsistency across pairwise comparisons, so these methods are also appropriate for dealing with inconsistency (see Chapter 7). Unless otherwise stated, when we refer to heterogeneity, this can be interpreted as heterogeneity and/ or inconsistency.

The term 'meta-regression' can cause confusion in network meta-analysis. Our basic network meta-analysis models are already 'regression' models in a technical sense, but the regression coefficients, which are the treatment effect parameters, have special properties conferred on them by randomisation. Most of this chapter is devoted to models in which additional interaction terms are introduced for covariates to which patients have not been randomised to (e.g. age or disease severity). However, we also discuss models where the covariate is part of the treatment definition (Section 8.6). These include dose–response models for treatment effects, which at first sight one might wish to include as a type of covariate adjustment. We see these models in a rather different way, partly because patients are randomised to different doses and partly because rather than adding interaction terms, these models reconstruct the treatment effects as functional parameters, derived from a different set of basic parameters (Section 2.2.2).

The chapter begins by setting out the fundamental concepts (Section 8.2): types of covariate (trial or patient level, within- and between-trial comparisons, continuous or categorical covariates, aggregate data vs. individual patient data), how heterogeneity can be measured and implications of heterogeneity for decision making, in particular the predictive distribution. In Section 8.3 a pairwise meta-analysis is used to illustrate the impact of large heterogeneity in decision making, how covariate adjustment can be used to reduce heterogeneity, how results should be interpreted and how it can impact decisions.

Section 8.4 sets out the algebra and WinBUGS code for a series of meta-regression models for network meta-analysis. Worked examples are presented throughout to illustrate the main points. In Section 8.5 we discuss meta-regression with IPD. In Section 8.6 we introduce treatment effects models for dose–response, combination of treatment components and class effects. Section 8.7 reviews the chapter from a decision-making perspective, where we draw attention to some pragmatic considerations in model choice.

In Chapters 2–4, the general network meta-analysis model and methods for model comparison and criticism were described. In all that follows, it is implicit that those ideas can be applied throughout this chapter.

8.2 Basic Concepts

Meta-regression is used to relate the size of a treatment effect obtained from a meta-analysis to certain numerical characteristics of the included trials, with the aim of explaining some, or all, of the observed between-trial heterogeneity in relative treatment effects. These characteristics can be due to specific features of the individual participants in the trial or can be directly due to the trial setting or conduct. Meta-regression can be based on aggregate (trial-level) outcomes and covariates, or IPD may be available. However, even if we restrict attention to RCT data, the study of effect modifiers, with the exception of those relating to intervention definition (Section 8.6), is inherently observational (Higgins and Green, 2008; Borenstein et al., 2009). This is because it is not possible to randomise patients to one covariate value or another. As a consequence, the meta-regression techniques described in Section 8.4 inherit all the difficulties of interpretation and inference that attach to non-randomised studies: confounding, correlation between covariates and, most importantly, *the inability to infer causality from association.* However, there are major differences in the quality of evidence from a meta-regression that depend on the nature of the outcome, the covariate in question and the available data.

8.2.1 Types of Covariate

We will define trial-level covariates as those that relate to trial or participant characteristics that have been aggregated at the trial level and for which IPD, or a suitable breakdown of results by characteristic, are not available. Patient-level covariates are defined as covariates that relate to patient attributes and can be attributed to specific patients in each trial, either because IPD are available or because a sufficient breakdown of results has been provided.

For **categorical covariates**, we can distinguish between the following scenarios:

A1. **Trial-level covariates that relate to trial characteristics:** This type of covariate relates to a *between-trial* treatment–covariate interaction and is often termed subgroup analyses. All patients in all arms of a trial share the same characteristic, and these only differ between trials, for example, trials that have been conducted in primary or secondary care settings. This is equivalent to A2(a) (see Section 8.4.4). Risk of bias indicators also fall under this category and are discussed in Chapter 9.

A2. **Trial-level covariates that relate to patient characteristics:** In this case, the covariate relates to a patient characteristic, but it is aggregated at the trial level. Examples of such covariates include sex (male/female) or treatment status, that is, patients who are treatment naïve (first-line therapy) versus those that have previously failed on another therapy (second-line therapy). Data may be reported in different formats:

a) Trials that have been conducted on patients with homogeneous characteristics only, for example, trials including only treatment-naïve patients and trials including only second-line therapy patients. In this case we can think of treatment status as a *between-trial* covariate, even though strictly it refers to a patient characteristic. This is equivalent to A1. An example is given in Section 8.4.4.

b) Trials that include patients with mixed characteristics and report only the proportions of patients with each characteristic in the trial. For example, trials may include both naïve and second-line therapy patients and report the proportion of each type of patient included. This proportion is sometimes taken as a *between-trial* continuous covariate, which is then equivalent to B2. See Section 8.4.3.1.

c) Trials that include patients with mixed characteristics but do not report proportions or a breakdown of outcomes by characteristic. No meta-regression can be carried out unless further assumptions are made.

A3. **Patient-level covariates:** In this case the covariate relates to a patient characteristic, and we can identify the outcome for each patient and covariate value. Data may be reported in different formats:

a) Trials that have IPD available for the outcome and covariate of interest or where sufficient statistics are reported, for example, cross-tabulations for categorical outcomes and covariates or full covariance matrices for continuous outcomes. In this case the covariate can be used to explore *within-trial* covariate effects, which can then be explored further in the meta-regression. Methods for this type of meta-regression are discussed in Section 8.5.

b) Trials that include patients with mixed characteristics but report the treatment effect with a measure of precision separately for each covariate values or sufficient data (e.g. in contingency tables) that would allow these to be derived. This is a *within-trial* effect and, for the purpose of meta-regression, is equivalent to having IPD on that characteristic, if only this covariate is being modelled. This is true whether binary or continuous outcomes are reported, but only applies to categorical covariates. This type of meta-regression is discussed in Section 8.5.

A similar set of distinctions can be drawn for **continuous covariates**:

B1. **Trial-level covariates that relate to trial characteristics:** For example, duration of the treatment, timing of intervention (e.g. how many hours after surgery) or trial setting (e.g. community/hospital care). An example is given in Section 8.4.3.1.

B2. **Trial-level covariates that relate to patient characteristics:** For example, the mean age or mean disease duration of patients recruited to the trial. Baseline risk of patients in a trial, defined as the log odds of an event of patients on the reference treatment or similar, can also be thought of as a trial-level covariate that relates to unmeasured patient characteristics (see Section 8.4.1). This is equivalent to B1 and examples are given in Sections 8.4.3.2 and 8.4.3.3.

B3. **Patient-level covariates:** With binary outcomes (e.g. death), if the mean for that characteristic and its variance are reported separately (or can be derived) for events and non-events, then, for the purpose of meta-regression, this is as good as having IPD with each patient's exact characteristic (e.g. age or disease duration) recorded. If the mean covariate values are not reported separately, then IPD would be needed to perform meta-regression. For continuous outcomes with continuous covariates, IPD are always required for meta-regression. This is discussed in Section 8.5.

When investigating an interaction between treatment and covariate, one is comparing the treatment efficacy at different covariate values for categorical covariates and the linear change in efficacy per unit change in continuous covariates. There are two key differences between within- and between-trial comparisons. With between-trial comparisons, a given covariate effect (i.e. interaction) will be harder to detect as it has to be distinguishable from the 'random noise' created by the between-trial variation. However, for within-trial comparisons, the between-trial variation is controlled for, and the interaction effect needs only to be distinguishable from sampling error. With between-trial comparisons, because the number of observations (trials) may be very low while the precision of each trial may be relatively high, it is quite possible to observe a highly statistically significant relation between the treatment effect and the covariate that is entirely spurious (Higgins and Thompson, 2004).

A second difference is that between-trial comparisons are vulnerable to ecological bias or ecological fallacy (Rothman et al., 2012). This is a phenomenon in which, for example, a linear regression coefficient of treatment effect against the covariate in the between-trial case can be entirely different to the coefficient for the within-trial data. It is perfectly possible, of course, to have both within-trial and between-trial information in the same evidence synthesis. Depending on the availability of IPD, it may be possible to fit a model that estimates *both* a between-trial coefficient based on the mean covariate value and a within-trial coefficient based on the individual variation of the covariate around the mean. With continuous covariates and IPD, not only does the within-trial comparison avoid ecological bias, but it also has far greater statistical power to detect a true covariate effect. This is because the variation in patient covariate values will be many times greater than the variation between the trial means and the precision in any estimated regression coefficient depends directly on the variance in covariate values. See Section 8.5 for further details.

Finally, in cases where the covariate does not interact with the treatment effect, but modifies the probability of an event or the mean on the reference treatment (baseline risk), the effect of pooling data over the covariate is to bias the estimated treatment effect towards the null effect. This is a form of ecological bias known as *aggregation bias* (Rothman et al., 2012), which does not affect strictly linear models. Usually it is significant only when both the covariate effect on baseline risk and the treatment effect are quite strong. It is a particular danger in survival analysis because the effect of covariates such as age on cancer risk can be particularly marked and because the log-linear models routinely used are highly non-linear. When covariates that affect risk are present, even if they do not modify the treatment effect, the analysis must be based on pooled estimates of treatment effects from a stratified analysis for categorical covariates and regression for continuous covariates and not on treatment effects estimated from pooled data (Govan et al., 2010). See Chapter 5 for further details.

8.3 Heterogeneity, Meta-Regression and Predictive Distributions

A number of standard methods for measuring between-trial heterogeneity have been proposed (Sidik and Jonkman, 2007; Higgins and Green, 2008; Borenstein et al., 2009). In this book, and in keeping with the Bayesian framework, we compare the residual deviance and DIC statistics from fixed and random effects models to choose the preferred model (Chapter 3). If a random effects model is chosen, we examine the estimated heterogeneity in the context of the estimated treatment effects. An advantage of the Bayesian approach is that it provides a posterior distribution of the between-trial variance and – perhaps easier to interpret – the between-trial standard deviation, which gives investigators some insight into the range of values that are compatible with the data (Spiegelhalter et al., 2004, table 5.2). It is also possible to obtain a measure of uncertainty for the between-trial variance using classical approaches (Higgins and Thompson, 2002), but this is not often done. However, as stated in Chapter 2, the posterior distribution for the between-trial standard deviation is likely to be extremely sensitive to the prior distribution, and in particular using vague prior distributions is likely to result in posterior distributions that allow for unrealistically high levels of heterogeneity whenever the number of trials on each comparison is small or when the majority of trials are small. Informative prior distributions based on expert opinion or on meta-epidemiological data are possible solutions. See Chapters 2 and 4 for further comments on choice of prior distribution for between-study heterogeneity parameters.

In the presence of large between-trial variability in treatment effects (heterogeneity), interpretation of results requires care, since the uncertainty around the posterior mean of the treatment effects will not reflect the true uncertainty around the likely values of a future roll-out of the intervention or a future trial. The predictive distribution (Chapter 3) provides some insight into the wider uncertainty due to both the finite sample size and the variability across effects from different studies. It is also informative to compare the size of the heterogeneity to the largest estimated relative treatment effect. If the heterogeneity indicates that the between-trial variability is of the same order of magnitude as the observed effects, this will cast doubts on the suitability of the results for decision making. Again, the predictive distribution will be a better guide to the true uncertainty around the relative treatment effects.

In this section we illustrate the implications of substantial heterogeneity on the interpretation of results and decision making using a worked example. We return to this example in Section 8.4.3.2 where the model, its fit and interpretation of results are described in detail.

8.3.1 Worked Example: BCG Vaccine

A meta-analysis of trials evaluating the efficacy of a BCG vaccine for preventing tuberculosis (TB) showed large between-study heterogeneity (Berkey et al., 1995; Welton et al., 2012). Data were available on the number of vaccinated and unvaccinated patients and the number of patients diagnosed with TB during the study follow-up period for each group as well as the absolute latitude at which the trial was conducted (Table 8.1).

Assuming a binomial distribution for the number of cases of diagnosed TB in arm k of trial i, fixed and random effects meta-analyses of the number of events (TB diagnosis) in vaccinated and unvaccinated individuals were conducted using the core model presented in Chapter 2 (binomial likelihood with logit link). The code for these analyses is given in *Ch8_BCG_Bi_logit_FE.odc* and *Ch8_BCG_Bi_logit_RE.odc*, respectively.

The fixed effects model had a very poor fit to the data (posterior mean of the residual deviance of 191 compared with 26 data points, $DIC = 205$), so the random effects model was preferred (Table 8.2). However, a large between-study heterogeneity was estimated (posterior median 0.65 with 95% CrI 0.39–1.17), comparable in size to the pooled log odds ratio (OR) of −0.76 (Table 8.2).

Table 8.1 BCG vaccine example: number of patients diagnosed with TB, r, out of the total number of patients, n, in the vaccinated and unvaccinated groups and the absolute latitude at which the trial was conducted, x. Adapted from Berkey et al. 1995.

	Not vaccinated		Vaccinated		Absolute degrees latitude
	Number diagnosed with TB	Total number of patients	Number diagnosed with TB	Total number of patients	
Trial number	r_{i1}	n_{i1}	r_{i2}	n_{i2}	x_i
1	11	139	4	123	44
2	29	303	6	306	55
3	11	220	3	231	42
4	248	12,867	62	13,598	52
5	47	5,808	33	5,069	13
6	372	1,451	180	1,541	44
7	10	629	8	2,545	19
8	499	88,391	505	88,391	13
9	45	7,277	29	7,499	27
10	65	1,665	17	1,716	42
11	141	27,338	186	50,634	18
12	3	2,341	5	2,498	33
13	29	17,854	27	16,913	33

Table 8.2 BCG vaccine example: results from the random effects meta-analyses with and without the covariate absolute distance from the equator.

	No covariate		Model with covariate[a]	
	Median	95% CrI	Median	95% CrI
b	–	–	−0.03	(−0.05, −0.01)
log OR	−0.76	(−1.21, −0.33)	−0.76	(−1.03, −0.52)
OR	0.47	(0.30, 0.72)	0.47	(0.36, 0.59)
σ	0.65	(0.39, 1.17)	0.27	(0.03, 0.74)
Model fit				
resdev[b]	26.1		30.6	
p_D	23.6		21.4	
DIC	49.7		52.0	

Posterior median and 95% CrI of the log OR, OR, interaction estimate (b) and posterior median between-trial heterogeneity (standard deviation, σ) for the number of patients diagnosed with TB (log OR < 0 and OR < 1 favour vaccination) and measures of model fit (posterior mean of the residual deviance, resdev, number of parameters, p_D and DIC).
[a] Treatment effects are at the mean value of the covariate: latitude = 33.46°.
[b] Compare to 26 data points.

The observed log ORs and their 95% confidence intervals (CI) are presented in Figure 8.1 along with the study-specific (shrunken) estimates, the pooled mean log OR (based on the posterior distribution) and the predictive log OR (based on the predictive distribution) and their 95% CrI, obtained from the random effects models (Table 8.2). Focusing only on the solid lines in Figure 8.1, the first thing to note is that there is substantial between-study variability: some studies have shown very positive effects, while others have shown no effect, even suggesting harmful effects. This impacts the width of the 95% CrI for the mean and the predictive effects, with the latter being extremely wide, due to the observed heterogeneity (Figure 8.1).

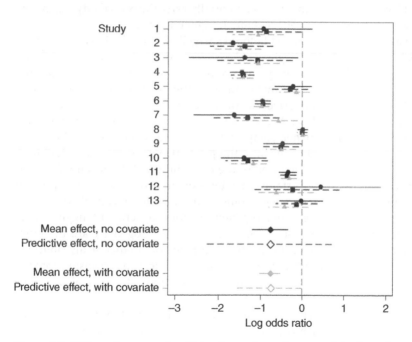

Figure 8.1 BCG vaccine example: effect of covariate adjustment (absolute distance from the equator). Observed log ORs with 95% CI (black circles, solid lines); posterior median with 95% CrI of the trial-specific log ORs (the 'shrunken' estimates) for the random effects models with no covariate (black squares, black dashed lines) and with covariate (grey triangles, grey dashed lines); median with 95% CrI of the posterior (black diamond, solid line) and predictive distribution (open diamond, dashed line) of the pooled treatment effect for the random effects model with no covariates; and median with 95% CrI of the posterior (grey diamond, grey solid line) and predictive distribution (grey open diamond, grey dashed line) of the pooled treatment effect at the mean covariate value for the random effects model with covariate absolute distance from the equator.

8.3.2   Implications of Heterogeneity in Decision Making

In the presence of high levels of heterogeneity, it is critical to consider its impact on decisions. In particular the size of the treatment effect should be interpreted in the context of the estimated between-trial variation. Figure 8.1 portrays a situation where a random effects model has been fitted and the mean effect is clearly different from zero with 95% CrI (−1.21, −0.33). However, given the large heterogeneity, median 0.65 95% CrI (0.39, 1.17) (Table 8.2), what is a reasonable CrI for our prediction of the outcome of a future trial of infinite size? The predictive distribution shown in Figure 8.1 gives the answer: in a model with no covariate adjustment, the 95% predictive interval for a future trial ranges from −2.27 to 0.72, spanning no effect and including a range of harmful effects. This means that while the probability that the vaccine is harmful based on the mean effect is essentially zero, the probability that a new trial would show a harmful effect is much higher at 14%.

This issue has been discussed before (Spiegelhalter et al., 2004; Ades et al., 2005; Welton et al., 2007; Higgins et al., 2009), and it has been proposed that, in the presence of heterogeneity, the predictive distribution, rather than the distribution of the mean treatment effect, better represents our uncertainty about the comparative effectiveness of treatments in a future 'roll-out' of a particular intervention. In an MCMC setting, a predictive distribution is easily obtained by drawing further samples from the distribution of effects, as described in Chapter 3. The mean of the predictive distribution, on its linear scale, will be the same as the mean of the distribution of the mean effect. But the implications on the uncertainty in a decision, in cases where there are high levels of unexplained heterogeneity, could be quite profound, and it is therefore important that the degree of heterogeneity is not exaggerated (Higgins et al., 2009). Methods to adjust for factors that cause heterogeneity are therefore important. See Section 5.6.2 for further comments on predictive distributions and alternative characterisation of the treatment effect in a decision-making context and Chapter 9 for bias adjustment methods that can also reduce heterogeneity.

For this example, it has been suggested that the absolute latitude, or distance from the equator, at which the trials were conducted might influence vaccine efficacy (Berkey et al., 1995). The crude ORs obtained from Table 8.1 are plotted (on a log scale) against distance from the equator in Figure 8.2 where, for each study, the size of the plotted bubble is proportional to its precision so that larger, more precise studies have larger diameters. It seems plausible that the effect of the vaccine may differ at varying latitudes according to a linear relationship (on the log OR scale).

If instead we consider the model with the covariate distance from the equator (for details, see Section 8.4.3.1), we can see that much of the between-study variability is explained (Table 8.2). The posterior and predictive distributions

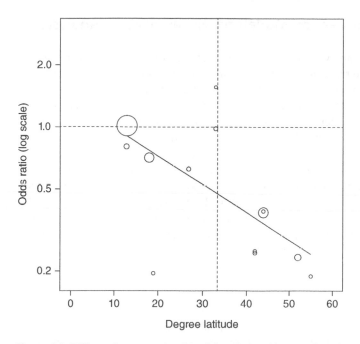

Figure 8.2 BCG vaccine example: plot of the crude odds ratios (on a log scale) against absolute distance from the equator in degrees latitude. The size of the circles is proportional to the studies' precisions, the horizontal line (dashed) represents no treatment effect, the vertical line (dashed) is at the mean covariate value (33.46° latitude), and the solid line is the regression line estimated by the random effects model including degrees latitude as a continuous covariate. Odds ratios below 1 favour the vaccine.

at the mean covariate value (latitude = 33.46°) for the model with the covariate are represented by the grey lines in Figure 8.1. The predictive probability that a new trial, *carried out at the mean covariate value*, will show harmful effects is now only 3%. However, note that this probability will differ for different values of the covariate. For example, the predictive probability that a new trial conducted at the equator (latitude = 0°) will show harmful effects is 80%; if the trial is instead conducted at 13° or 50° latitude, the predictive probabilities of a harmful effect are 35 and 0.6%, respectively. The estimated relative treatment effects and the width of the credible intervals around them will also differ with different covariate values. There is now a continuum of treatment effects for different covariate values, and this can have major implications when deciding whether or not to recommend the intervention. So while the vaccine seems effective at the mean latitude (Figure 8.1) and far from the equator (Figure 8.2), it has a very small effect at lower latitudes, and this needs to be taken into account by the decision maker who will be interested in a target population at a particular latitude. See Section 8.4 for further details.

8.4   Meta-Regression Models for Network Meta-Analysis

In network meta-analysis, a trial-level covariate can be seen as a variable that interacts with the treatment, but these interactions may be different for every treatment. The hypothesis is that the size of the treatment is different for different values of the covariate and that the relationship is linear on the chosen scale. To model this, we introduce as many interaction terms as there are basic treatment effects, $\beta_{12}, \beta_{13}, ..., \beta_{1S}$. Each of these added terms represents the *additional* (interaction) treatment effect per unit increase in the covariate value in comparisons of treatments 2, 3, ..., S to *treatment* 1. These terms are exactly parallel to the main effects $d_{12}, d_{13}, ..., d_{1S}$. As with the main effects, for trials comparing, say, treatments 3 and 4, the interaction term would be the difference between the interaction terms on the effects relative to treatment 1, so that $\beta_{34} = \beta_{14} - \beta_{13}$. The generic network meta-analysis model in Chapter 4 with interactions can then be written as

$$\theta_{ik} = \mu_i + \delta_{ik} + \beta_{t_{i1}, t_{ik}} x_i = \mu_i + \delta_{ik} + \left(\beta_{1,t_{ik}} - \beta_{1,t_{i1}}\right) x_i \tag{8.1}$$

with t_{ik} representing the treatment in arm k of trial i; x_i the trial-level covariate for trial i, which can represent a subgroup, a continuous covariate or baseline risk; and β_{ck} the regression coefficient for the covariate effect in comparisons of treatment k to c, which can be written as the difference in interactions with the reference treatment $(\beta_{1k} - \beta_{1c})$. We set $\delta_{i1} = \beta_{11} = 0$ so that $\theta_{i1} = \mu_i$ and note that the treatment and covariate interaction effects (δ and β) only act on the treatment arm, not on the control. In this model δ_{ik} represent the relative effect of the treatment in arm k compared with the treatment in arm 1 of trial i when the covariate value is zero. Similarly, the pooled effects d_{1k} will be the relative effects of treatments $k = 2, ..., S$ compared with the reference treatment *when the covariate is zero*.

The special case of a pairwise (two-treatment) meta-analysis has $t_{i1} = 1$ and $t_{ik} = 2$ for all included trials, and therefore only one regression coefficient, β_{12}, and one relative treatment effect, d_{12}, are estimated.

This model can be used to fit categorical or continuous covariates, although for continuous covariates it is generally advisable to centre the covariate to improve convergence (Draper and Smith, 1998; Welton et al., 2012). Therefore the model becomes

$$\theta_{ik} = \mu_i + \delta_{ik} + \left(\beta_{1,t_{ik}} - \beta_{1,t_{i1}}\right)\left(x_i - m_x\right) \tag{8.2}$$

with $\delta_{i1} = \beta_{11} = 0$ as before. Now δ_{ik} represent the relative effect of the treatment in arm k compared with the treatment in arm 1 of trial i at the centring value m_x, which is usually the mean covariate value $\bar{x}$, but could be some other value that aids interpretation such as the mean in the decision population of interest.

The pooled effects d_{1k} will be the relative effects of treatments $k = 2, ..., S$ compared with the reference treatment *at the centring value* m_x. Note that the model in equation (8.1) is exactly equivalent to the model in equation (8.2) when $m_x = 0$.

The treatment effects can be un-centred and transformed to produce treatment effect estimates at any covariate value. The mean treatment effect for treatment c compared with treatment k at covariate value z is

$$d_{ck} + \beta_{ck}(z - m_x). \tag{8.3}$$

For a random effects model, the trial-specific treatment effects δ_{ik} in equations (8.1) and (8.2) come from a common distribution as described in Chapter 4 with a common variance. However, we could instead assume that the between-study variances depend on the covariate values, for example, in subgroup analyses, provided there was enough data to inform them or informative prior distributions were used.

For a fixed effects model, we replace δ_{ik} in equations (8.1) and (8.2) with $d_{t_{i1}t_{ik}} = d_{1,t_{ik}} - d_{1,t_{i1}}$. In all models we also set $d_{11} = 0$. The likelihood, link function and prior distributions for μ_i, d_{1k} and the between-study heterogeneity are chosen taking into account the scale of analysis, as detailed in Chapters 2 and 4.

However, in a network meta-analysis context, there are a very large number of models that can be proposed for the interaction terms, β, each with very different implications. Three general meta-regression models can be defined, describing different assumptions for the interactions in a multiple-treatment context:

1) **Independent, treatment-specific interactions:** We assume that there is an interaction effect between the covariate and the treatment, but the interactions are different for every treatment. To model this, we introduce as many interaction terms as there are basic treatment effects, that is, $S - 1$, and consider them to be entirely unrelated. Thus the interaction effects are given unrelated non-informative prior distributions, such that for treatment $k = 2, ..., S$

$$\beta_{1k} \sim \text{Normal}(0, 100^2) \tag{8.4}$$

The interaction effects β_{1k} represent the change in relative effect d_{1k} for each unit increase in the covariate x_i. Interaction effects for other treatments can be obtained by subtraction.

2) **Exchangeable, related, treatment-specific interactions:** In this model we assume that the interaction effects for each treatment are exchangeable, that is, they are similar, but not equal. This has the same number of parameters as the previous model, but now the $(S - 1)$ 'basic' interaction terms are

drawn from a random distribution with a common mean and between-treatment variance, so for treatment $k = 2, ..., S$

$$\beta_{1k} \sim \text{Normal}(b, \sigma_b^2).\tag{8.5}$$

The mean interaction effect and its variance are estimated from the data. They are given non-informative prior distributions, for example, $b \sim \text{Normal}(0,100^2)$ and σ_b could be given a uniform prior distribution with lower limit at zero and a suitable chosen upper limit, depending on the outcome scale used (Chapters 2 and 4). Informative prior distributions, which limit how similar or different the interaction terms are, could also be used.

3) **The same interaction effect for all treatments:** In this more restrictive model, there is a single interaction term b that applies to relative effects of all the treatments relative to treatment 1. An important point to note is that the assumption of a common regression term b allows the interaction parameter to be estimated even for comparisons in the network which only have one trial and therefore do not provide information on a regression slope. The model can be expressed, and coded for computer implementation, in many ways. We have chosen to retain the treatment-specific interaction effects but set them all equal to b. This guarantees that the terms cancel out in comparisons of the non-reference treatments against each other. Thus, for all treatments $k = 2, ..., S$, we set

$$\beta_{1k} = b.\tag{8.6}$$

A non-informative prior distribution can be given to b, for example, $b \sim \text{Normal}(0,100^2)$, or an informative prior distribution can be used instead. In this model, the assumption is that the change in treatment effects *relative to treatment* 1, $d_{12}, d_{13}, ..., d_{1S}$, all increase or decrease by the same amount b, for each unit increase in the covariate x_i. However, the effects of treatments 2, 3, ..., S *relative to each other* are exactly the same regardless of the covariate value, because the interaction terms now cancel out. This means that the choice of reference treatment 1 becomes important and the results for models with covariates are sensitive to this choice. In fact, it will only make sense to use this type of model if the reference treatment is somehow different from the others, such as a placebo, an older treatment or 'standard care'. It is also important to ensure that the data are set up so that treatments are coded in ascending order by arm, as described in Chapter 2, to ensure the desired assumptions are implemented. Readers should be aware of the interpretation of parameters when coding all models, but it is particularly important for models including covariates.

Readers will note that it would be relatively straightforward to build models that incorporated a combination of these assumptions for different treatment comparisons by taking advantage of the modularity of the WinBUGS code.

This would be suitable if, for example, treatments belonging to different classes are included in the network and different assumptions apply to different classes of treatments (see also Section 8.7). For example, in a six-treatment network with placebo as the reference, if treatments 2–4 belong to the same class, they may be assumed to have the same interaction effect (equation (8.6)). If treatments 5 and 6 are different from the others and sufficient data are available, a separate interaction effect can be assumed for comparisons of each of these treatments to placebo. The consistency equations on the interaction terms will ensure that the correct relative effects are used each time.

8.4.1 Baseline Risk

A special kind of covariate is the average baseline risk in the trial, defined as the outcome on the reference treatment in that trial, on the chosen scale. This is usually the log odds, probability or rate of an event on the reference treatment, when the outcome is binary, but it can also be the mean outcome on the reference treatment. The baseline risk is often chosen as a proxy for underlying, often unmeasured or unknown, patient-level covariates that are thought to modify the treatment effect, but which cannot be accounted for directly in the model.

The meta-regression model on baseline risk is the same as in equation (8.2), but now $x_i = \mu_i$, the trial-specific baseline for the control arm in each trial. Thus, when arm-based data are available, no additional data are required to consider baseline risk in the meta-regression model, since this is already estimated for each trial. An important property of this Bayesian formulation is that it takes the 'true' trial-specific baseline (as estimated by the model) as the covariate and automatically takes the uncertainty in each μ_i into account (McIntosh, 1996; Thompson et al., 1997; Welton et al., 2012). Naïve approaches that regress against the observed baseline risk fail to take into account the correlation between the treatment effect and baseline risk and the consequent regression to the mean phenomenon.

It is important to note that the covariate value μ_i is on the same scale as the linear predictor (e.g. the logit, log or identity scales – see Chapter 4), and therefore the mean covariate value for centring needs to be on this scale too. For example, when using a logit link function, the covariate should be centred by subtracting the mean of the log odds in the trial-specific baseline arms ($k = 1$) of each trial that compares treatment 1 from μ_i. In a network meta-analysis context, the treatment in arm 1 will not always be treatment 1 (the reference treatment). However, for the model in equation (8.6), which assumes the same interaction effect for all treatments compared with treatment 1, the regression terms will cancel for all other comparisons, so no baseline risk adjustment is performed for trials that do not include treatment 1. Fitting one of the other models, care should be taken to ensure that the risk being adjusted for refers to the estimated risk for the reference treatment (treatment 1) that may not have been compared in every trial. This can be done by augmenting the data so that all trials have treatment 1 in arm 1 with missing data when this treatment was not compared. For example, for binary outcomes, an extra arm with treatment 1 should be added to studies that do not already have it, where the number of observed events is zero and the number of individuals is arbitrary, for example, 1. WinBUGS will then generate a prediction for the baseline risk on treatment 1 in those studies based on the model (Dias et al., 2011d) accounting for the full uncertainty in the missing baseline risk. For continuous data a missing observation with arbitrary variance (e.g. 1) should be added (Achana et al., 2013). An alternative is to ensure that the baseline risk considered reflects the expected outcome on treatment 1. This can be done by subtracting the effect of the treatment in arm 1 in study i from μ_i, ensuring that the baseline risk is corrected using the mean relative effect. See Exercise 8.3 for an illustration and comparison of the two methods.

When considering the different models that allow for effect modification, one of the factors that can influence choice of model is the amount of data available. If a fixed treatment effects model is being considered with a binary covariate (subgroup), the unrelated interaction model, equation (8.4), requires *two* connected networks, one for each subgroup, including all the treatments, that is, with at least $(S-1)$ trials in each. With random treatment effects, even more data are required to estimate the common between-trial variance: at least one, but preferably more, treatment comparison has to have multiple trials at the same covariate level if we assume a common between-study heterogeneity across covariate values. If the heterogeneity values are allowed to differ, a lot more data will be required. If a continuous covariate is considered, data need to include a suitable range of covariate values across all treatment comparisons, unless external data, in the form of informative prior distributions, are used.

It may be possible to estimate the exchangeable interaction model in equation (8.5) with less data. However, to use this model, we need to have a clear rationale for exchangeability, and we would suggest that this model is most useful for model checking, for example, to check whether there is evidence that the assumption of a common interaction may not hold. See Section 8.7 for a further discussion of the issues with the different models.

8.4.2 WinBUGS Implementation

The generic WinBUGS code for a random effects network meta-analysis presented in Chapter 4 can be extended to implement the meta-regression model in equation (8.2) simply by adding the extra term containing the regression coefficients multiplied by the centred covariate values to the linear predictor, so that for an appropriate link function (see Chapter 4), we write

```
# model for linear predictor, covariate effect relative
# to treat in arm 1
link function <- mu[i] + delta[i,k] + (beta[t[i,k]]-
                                    beta[t[i,1]]) * (x[i]-mx)
```

Values for the covariate vector x and the centring constant mx are given as data. If no centring is required, we set mx=0.

The rest of the generic code remains the same, including the multi-arm adjustment. However, further code needs to be added to implement the chosen assumptions for the interaction terms beta. To implement the assumption of **independent, treatment-specific interactions** (equation (8.4)), the following code should be added before the final closing brace:

```
beta[1] <- 0                    # covariate effect is zero
                                for reference treatment
for (k in 2:nt){                # LOOP THROUGH TREATMENTS
   beta[k] <- dnorm(0, 0.0001)  # independent covariate
                                effects
}
```

To implement the assumption of **exchangeable interactions** (equation (8.5)), the following code should be added before the final closing brace:

```
beta[1] <- 0                    # covariate effect is zero
                                for reference treatment
for (k in 2:nt){                # LOOP THROUGH TREATMENTS
    beta[k] ~ dnorm(B, tau.B)   # exchangeable covariate
                                effects
}
B ~ dnorm(0,0.0001)   # Prior for mean of distribution
                        of regression parameters
sd.B ~ dunif(0, appropriate upper bound) # Prior for sd
                                    of distr of
                                    regression
                                    parameters
tau.B <- pow(sd.B,-2)           # precision of regres-
                                sion parameters
```

To implement the assumption of **equal interactions** (equation (8.6)), we instead add the following code before the final closing brace:

```
beta[1] <- 0                    # covariate effect is zero for
                                reference treatment
```

```
for (k in 2:nt){          # LOOP THROUGH TREATMENTS
  beta[k] <- B            # common covariate effects
}
B ~ dnorm(0,0.0001)       # Prior for common regression
                          parameter
```

To implement models with adjustment for **baseline risk**, x[i] is replaced with mu[i] in the definitions of the linear predictor, and the centring constant mx should be on appropriate scale.

Fixed effects models with covariates can also be fitted by adapting the generic code for fixed effects network meta-analysis given in Chapter 4 in the same way.

In addition, code can be added before the final closing brace to estimate the treatment effects relative to the reference treatment for covariate values in vector z of length nz, given as data:

```
for (k in 1:nt){        # LOOP THROUGH TREATMENTS
  for (j in 1:nz) {     # LOOP THROUGH COVARIATE VALUES
    dz[j,k] <- d[k] + (beta[k]-beta[1])*(z[j]-mx)
             # treatment effect when covariate = z[j]
  }
}
```

For example, for a meta-regression with two subgroups, a vector z of length two would be added to the list data statement: list(z=c(0,1), nz=2).

Further code can be added to estimate all relative effects and to produce estimates of absolute effects on all treatments, given additional information on the absolute treatment effect on one of the treatments, for given covariate values. For example, if we had chosen to use a logit link, the following code would be added to calculate all relative effects for covariate values in z:

```
# pairwise ORs and LORs for all possible pairwise
# comparisons at covariate=z[j]
for (c in 1:(nt-1)) {
   for (k in (c+1):nt)  {
      for (j in 1:nz) {
         orz[j,c,k] <- exp(dz[j,k] - dz[j,c])
         lorz[j,c,k] <- dz[j,k]-dz[j,c]
      }
   }
}
```

If a different link function had been used, the code would be adjusted as described in Chapter 4. The absolute effects of each treatment, `Tz []`, for each treatment can be calculated using the following code:

```
for (k in 1:nt) {      # LOOP THROUGH TREATMENTS
   for (j in 1:nz){    # LOOP THROUGH COVARIATE VALUES
      logit(Tz[j,k]) <- A + dz[j,k]  # logit link function used
      }
   }
```

Note that the only difference in the main expression for T is that we now use the relative effects for each covariate value, given in `dz`, and have to include a double loop through all the treatments and then through the vector with the covariate values of interest.

Depending on the scale of analysis, the link function may need to be replaced as described in Chapter 4.

Deviance calculations, DIC and predictive distributions at the centring value can be calculated as described in Chapters 3 and 4. Predictive distributions at different values of the covariate can be calculated by adapting the code to implement equation (8.3) for the predictive relative treatment effects.

8.4.3   Meta-Regression with a Continuous Covariate

We will assume that there is a trial-level covariate defined on a continuous scale, which is given in the data as a vector with each term, x_i, representing the covariate value for trial i. This covariate could be an average patient characteristic such as the proportion of males in the trial (B2 in Section 8.2.1), or it could be a genuine trial characteristic (B1 in Section 8.2.1), such as distance from the equator, in absolute degrees latitude (Section 8.3.1).

We return to the meta-analysis of trials evaluating the efficacy of a BCG vaccine for preventing TB described in Section 8.3.1, to provide details on the model and interpretation of results, and then present an example with a continuous covariate in a network meta-analysis context.

8.4.3.1   BCG Vaccine Example: Pairwise Meta-Regression
with a Continuous Covariate

Assuming a binomial distribution for the number of cases of diagnosed TB in arm k of trial i, and letting x_i be the continuous covariate representing absolute degrees latitude, the meta-regression model in equation (8.2) was fitted to the data, with centring at the mean covariate value $\bar{x} = 33.46°$ latitude. Both fixed and, random treatment effects were considered, but note that since this analysis only involves two treatments, there is only one interaction term. Therefore the models in equations (8.4) and (8.6) are equivalent, and the hierarchical model in equation (8.5) is not relevant.

A random treatment effects model with covariate was fitted using the implementation in equation (8.6). The data structure is similar to that presented in Chapter 2, but now we add a column containing the value of covariate x for each trial and the centring value (the mean of the covariate) mx to the list data for centring. In addition we want to calculate all the pairwise log ORs and ORs for covariate values 0, 13 and 50, so a vector z of length three is also added to the data. Initial values will need to be given for B, which requires a single number, as well as for the other parameters, as described in Chapter 2. See *Ch8_BCG_Bi_logit_RE-x1.odc* for details on the code, data and initial values for this example (results are based on 100,000 iterations from three independent chains after a burn-in of 40,000).

The results of fitting a random effects model with and without the covariate 'absolute degrees latitude' are presented in Table 8.2. Note that the treatment effect for the model with covariate adjustment is interpreted as the log OR *at the mean value of the covariate* (33.46° latitude). The estimated log ORs at different degrees latitude are represented by the solid line in Figure 8.2. See also Section 8.3 for a further discussion of the results.

Comparing the values of the DIC (Table 8.2), it would appear that the models with and without the covariate are not very different; differences of less than 3 are not considered important (Chapter 3). Although the model without covariates has a smaller posterior mean of the residual deviance, the model with the covariate allows for more shrinkage of the random treatment effects, resulting in a smaller effective number of parameters (p_D). We can however see that the heterogeneity is considerably reduced when we add the covariate: the posterior medians are 0.65 for the model with no covariate and 0.27 for the model with covariate, and the 95% CrI for the interaction term *b* does not include zero (Table 8.2). Note also that the model with the covariate has the effect of 'shrinking' the study-specific estimates further towards the regression line (Figure 8.1), particularly for studies which were farther from it (Figure 8.2), and the 95% CrI for its predictive distribution no longer crosses the line of no effect. We might now ask whether the covariate has explained *all* the heterogeneity, in effect allowing us to fit a fixed effects model with the covariate. See Exercise 8.1.

It is important to note that when deciding whether a covariate should be included in a random effects model, the posterior mean of the regression coefficient and the posterior between-trial standard deviation (heterogeneity) should be looked at. Reductions in heterogeneity and a 95% CrI for the regression coefficient that does not include zero are signs that the model with the covariate should be preferred. However model fit may not differ by very much between random effects models with and without a covariate, because random effects models usually fit the data well, at the expense of higher between-trial variation.

8.4.3.2 Certolizumab Example: Network Meta-Regression with Continuous Covariate

A review of trials of certolizumab pegol (CZP) for the treatment of rheumatoid arthritis (RA) in patients who had failed on disease-modifying anti-rheumatic drugs (DMARDs), including methotrexate (MTX), was conducted for a single technology appraisal at NICE (National Institute for Health and Clinical Excellence, 2010). Twelve MTX controlled trials were identified, comparing seven different treatments – placebo plus MTX (coded 1), CZP plus MTX (coded 2), adalimumab plus MTX (coded 3), etanercept plus MTX (coded 4), infliximab plus MTX (coded 5), rituximab plus MTX (coded 6) and tocilizumab plus MTX (coded 7), forming the network presented in Figure 8.3. This type of network, where comparisons are all relative to one common treatment and there are no loops, is often called a 'star network'.

Table 8.3 shows the number of patients achieving ACR-50 at 6 months, that is, the number of patients who have improved by at least 50% on the ACR scale (ACR-50 at 3 months was used when this was not available), r_{ik}, out of all included patients, n_{ik}, for each arm of the included trials, along with the mean disease duration in years for patients in each trial, x_i ($i = 1, ..., 12; k = 1, 2$).

It is thought that mean disease duration can affect relative treatment. The crude ORs from Table 8.3 are plotted (on a log scale) against mean disease duration in Figure 8.4, with the numbers 2–7 representing the OR of that treatment relative to placebo plus MTX (chosen as the reference treatment). Note that due

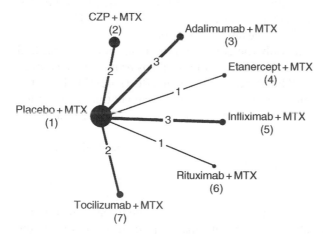

Figure 8.3 Certolizumab example: treatment network. Each circle represents a treatment, and connecting lines indicate pairs of treatments that have been directly compared in randomised trials. The numbers on the lines indicate the numbers of trials making that comparison, and the numbers by the treatment names are the treatment codes used in the modelling. Line thickness is proportional to the number of trials making that comparison, and the width of the circles is proportional to the number of patients randomised to that treatment.

Table 8.3 Certolizumab example: number of patients achieving ACR-50 at 6 months, *r*, out of the total number of patients, *n*, in the arms 1 and 2 of the 12 trials and mean disease duration (in years) for patients in trial i, x_i.

Study name	t_{i1}	t_{i2}	r_{i1}	n_{i1}	r_{i2}	n_{i2}	x_i
RAPID 1	Placebo	CZP	15	199	146	393	6.15
RAPID 2	Placebo	CZP	4	127	80	246	5.85
Kim 2007	Placebo	Adalimumab	9	63	28	65	6.85
DE019	Placebo	Adalimumab	19	200	81	207	10.95
ARMADA	Placebo	Adalimumab	5	62	37	67	11.65
Weinblatt 1999	Placebo	Etanercept	1	30	23	59	13
START	Placebo	Infliximab	33	363	110	360	8.1
ATTEST	Placebo	Infliximab	22	110	61	165	7.85
Abe 2006[a]	Placebo	Infliximab	0	47	15	49	8.3
Strand 2006	Placebo	Rituximab	5	40	5	40	11.25
CHARISMA[a]	Placebo	Tocilizumab	14	49	26	50	0.915
OPTION	Placebo	Tocilizumab	22	204	90	205	7.65

All trial arms had MTX in addition to the placebo or active treatment.
[a] ACR-50 at 3 months.

to zero events in one of the treatment arms, for plotting purposes, the crude OR for the Abe 2006 study was calculated by adding 0.5 to each cell. The original zero cell was used in the analysis (see also Chapter 6).

Due to the paucity of data, only the common interaction model described in equation (8.6) will be fitted. The disease duration covariate will be centred at its mean $\bar{x} = 8.21$ years. The relative treatment effects obtained are the estimated log ORs at the mean covariate value (8.21 years in this case), which can be transformed to produce the estimate at any covariate value of interest, as described in Section 8.4.2.

In this network, the generic random effects model with covariate disease duration and a Uniform(0,5) prior distribution for the between-study heterogeneity σ is not identifiable (see Exercise 8.2). This is because there is a trial with a zero cell, not many replicates of each comparison and no indirect evidence on any contrast. Due to the paucity of information from which the between-trial variation can be estimated, in the absence of an informative prior distribution for the between-study heterogeneity, the relative treatment effects for this trial will tend towards infinity. We have therefore used an informative half-normal prior distribution with mean 0.26, which ensures stable computation:

$$\sigma \sim \text{Half-Normal}\left(0, 0.32^2\right) \tag{8.7}$$

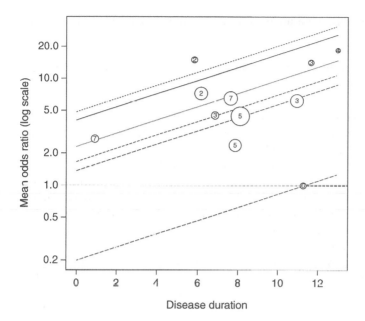

Figure 8.4 Certolizumab example: plot of the crude odds ratios (on a log scale) of the six active treatments relative to placebo plus MTX against mean disease duration (in years). The plotted numbers refer to the treatment being compared with placebo plus MTX, the blobs around the numbers are proportional to the precision of the study, and the lines represent the relative effects of the following treatments (from top to bottom) compared with placebo plus MTX based on a random effects meta-regression model: etanercept plus MTX (treatment 4, dotted line), CZP plus MTX (treatment 2, solid line), tocilizumab plus MTX (treatment 7, short long dash line), adalimumab plus MTX (treatment 3, dashed line), infliximab plus MTX (treatment 5, dot-dashed line) and rituximab plus MTX (treatment 6, long-dashed line). Odds ratios above 1 favour the plotted treatment, and the horizontal line (thin dashed) represents no treatment effect.

This prior distribution was chosen to ensure that, a priori, 95% of the trial-specific ORs lie approximately within a factor of 2 from the median OR for each comparison – for details, see Appendix B.

This prior distribution should not be used unthinkingly. It should be adapted to ensure it suitably reflects likely values of the heterogeneity for each example. Informative prior distributions allowing wider or narrower ranges of values can be used by changing the value of prec in the previous code. Alternatively, empirically based prior distributions (Turner et al., 2015b) could be used (see Exercise 8.2).

To fit the random effects meta-regression model with the prior distribution in equation (8.7), the line of code annotated as 'vague prior for between-trial SD' in the generic network meta-analysis code should be replaced with the following two lines:

```
sd ~ dnorm(0,prec)I(0,)    # prior for between-trial SD
prec <- pow(0.32,-2)
```

The WinBUGS code for the fixed and random effects meta-regression model with covariate disease duration is given in *Ch8_CZP_Bi_logit_FE-x1.odc* and *Ch8_CZP_Bi_logit_RE-x1prior.odc*, respectively.

In this example, the posterior distribution obtained for σ differs slightly from the half-normal prior distribution, suggesting there has been some updating based on the data. The range of plausible values for σ does not change much, but the probability of values very close to zero is smaller than that suggested by the prior distribution (Figure 8.5).

Table 8.4 shows the results of fitting fixed and random treatment effects network meta-analysis models with and without the covariate disease duration (results are based on 100,000 iterations from three independent chains after a burn-in of 50,000). The WinBUGS code for the fixed and random effects network meta-analysis models without covariate is given in *Ch8_CZP_Bi_logit_FE.odc* and *Ch8_CZP_Bi_logit_RE.odc*, respectively.

The estimated ORs for different durations of disease are represented by the parallel lines in Figure 8.4. The assumption of a common regression term implies that the interaction parameter is estimated even for the comparison of rituximab plus MTX (treatment 6) with placebo plus MTX that only has one trial. The model assumptions imply that a line parallel to the others is drawn through this point (Figure 8.4). This analysis also suggests that adding rituximab to MTX may be of much less benefit to patients than the other treatments and predicts, perhaps implausibly, that it can be harmful for patients with a shorter disease duration.

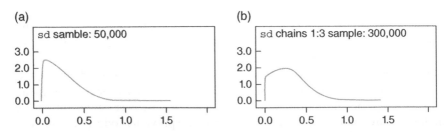

Figure 8.5 Certolizumab example: probability density function of (a) Half-Normal(0,0.32²) prior distribution simulated in WinBUGS and (b) the posterior distribution for the between-study heterogeneity for the meta-regression model with informative Half-Normal(0,0.32²) prior distribution – from WinBUGS.

Table 8.4 Certolizumab example: results from the fixed and random effects models with and without the covariate 'disease duration'.

	No covariate				Covariate 'disease duration'			
	Fixed effects		Random effects[a]		Fixed effects		Random effects[a]	
	Median	95% CrI	Median	95% CrI	Median	95% CrI	Median	95% CrI
b	–	–	–	–	0.14	(0.01, 0.26)	0.14	(−0.03, 0.32)
d_{12}	2.20	(1.73, 2.72)	2.27	(1.53, 3.10)	2.50	(1.96, 3.08)	2.55	(1.79, 3.44)
d_{13}	1.93	(1.52, 2.37)	1.96	(1.33, 2.64)	1.66	(1.19, 2.16)	1.70	(1.04, 2.41)
d_{14}	3.26	(1.45, 6.74)	3.28	(1.26, 6.63)	2.64	(0.71, 5.96)	2.61	(0.42, 6.01)
d_{15}	1.38	(1.06, 1.72)	1.46	(0.90, 2.21)	1.40	(1.08, 1.74)	1.46	(0.94, 2.16)
d_{16}	0.00	(−1.40, 1.39)	0.01	(−1.61, 1.63)	−0.42	(−1.86, 1.04)	−0.43	(−2.09, 1.21)
d_{17}	1.65	(1.22, 2.10)	1.57	(0.77, 2.28)	1.93	(1.45, 2.53)	1.99	(1.11, 2.93)
σ	–	–	0.34	(0.03, 0.77)	–	–	0.28	(0.02, 0.73)
resdev[b]	37.6		30.9		33.8		30.2	
p_D	18.0		21.1		19.0		21.2	
DIC	55.6		52.0		52.8		51.3	

Posterior median and 95% CrI interaction estimate (b), log ORs (d_{XY}) of treatment Y relative to treatment X and between-trial heterogeneity (σ) for the number of patients achieving ACR-50 ($d_{XY} < 0$ favours the reference treatment) and measures of model fit (posterior mean of the residual deviance, resdev, number of parameters, p_D and DIC). Treatment codes are given in Figure 8.3.
[a] Using informative prior distribution for sd.
[b] Compare to 24 data points.

The DIC and posterior means of the residual deviances for the models in Table 8.4 do not decisively favour a single model. Comparing only the fixed effects models, we can see that the fit is improved by including the covariate interaction term b that also has a 95% CrI, which does not include zero. Looking at the random effects models, although the model with covariate reduces the heterogeneity compared with the model with no covariate (Table 8.4), the 95% CrI for the interaction parameter b includes zero. Thus, the meta-regression models appear reasonable but not strongly supported by the evidence. Nevertheless the finding of smaller treatment effects with a shorter disease duration has been reported with larger sets of studies (Nixon et al., 2007), and the implications of this for the decision model need to be considered. The issue is whether or not the use of biologics should be confined to patients whose disease duration was above a certain threshold. This is not an unreasonable idea, but it would be difficult to determine this threshold on the basis of the regression in Figure 8.4 alone. The slope is largely determined by treatments 3 and 7 (adalimumab and tocilizumab), which are the only treatments trialled at more than one disease duration and which appear to have different effects at each duration. Furthermore, the linearity of relationships is highly questionable, and the prediction of negative effects for treatment 6 (rituximab) is not really credible. This suggests that the meta-regression model used is not plausible and other explorations of the causes of heterogeneity should be undertaken (see also Section 8.4.3.3).

8.4.3.3 Certolizumab Example: Network Meta-Regression on Baseline Risk

Figure 8.6 shows the crude OR obtained from Table 8.3 plotted against the baseline odds of ACR-50 (on a log scale) for the certolizumab example. Numbers 2–7 represent the OR of that treatment relative to placebo plus MTX (chosen as the reference treatment). Due to a zero cell in one arm, for plotting purposes, the crude OR for the Abe 2006 study was calculated by adding 0.5 to each cell, and the baseline log odds were assumed to be 0.01. Figure 8.6 seems to suggest a strong linear relationship between the treatment effect and the baseline risk (on the log scale). The model in equation (8.6) assumes that parallel regression lines are fitted to the points in Figure 8.6, where the differences between the lines represent the true mean treatment effects adjusted for baseline risk.

Both fixed and random treatment effects models with a common interaction term were fitted. The basic parameters d_{1k} and b are given non-informative normal prior distributions Normal$(0,100^2)$; the prior distributions for the μ_i were Normal$(0,1000)$, which have a slightly reduced variance to avoid numerical errors, and $\sigma \sim$ Uniform$(0,5)$. The WinBUGS code for meta-regression on baseline risk is given in *Ch8_CZP_Bi_logit_FE-xbase.odc* and *Ch8_CZP_Bi_logit_RE-xbase.odc*.

The analysis used centred covariate values, achieved by subtracting the mean of the observed log odds on treatment 1, $\bar{x} = -2.421$, from each of the estimated μ_i.

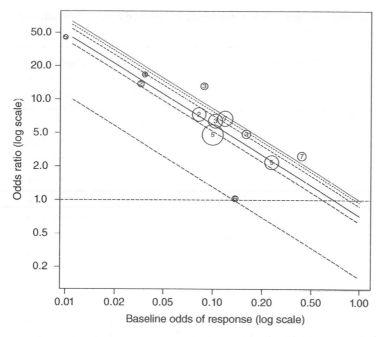

Figure 8.6 Certolizumab example: plot of the crude odds ratios of the six active treatments relative to placebo plus MTX against odds of baseline response on a log scale. The plotted numbers refer to the treatment being compared with placebo plus MTX, and the lines represent the relative effects of the following treatments (from top to bottom) compared with placebo plus MTX based on a random effects meta-regression model: tocilizumab plus MTX (7, short–long dash line), adalimumab plus MTX (3, dashed line), etanercept plus MTX (4, dotted line), CZP plus MTX (2, solid line), infliximab plus MTX (5, dot-dashed line) and rituximab plus MTX (6, long-dashed line). Odds ratios above 1 favour the plotted treatment, and the horizontal line (dashed) represents no treatment effect.

The treatment effects for models with covariate adjustment are interpreted as the effects for patients with a baseline logit probability of ACR-50 of −2.421 that can be converted to a baseline probability of ACR-50 of 0.082 using the inverse logit function (Table 4.1). These treatment effects can be un-centred and transformed to produce estimates at any value of baseline risk, as described in Section 8.4.2.

Table 8.5 shows the results of the interaction models with fixed and random treatment effects, with baseline risk as the covariate (results are based on 100,000 iterations from three independent chains after a burn-in of 60,000).

Both the fixed and random effects models with covariate have a credible region for the interaction term that is far from zero, suggesting a strong interaction effect between the baseline risk and the treatment effects. The estimated ORs for different durations for the random effects model with baseline risk interaction are represented by the different parallel lines in Figure 8.6. The DIC

Table 8.5 Certolizumab example: results from the fixed and random effects models with and without the covariate 'baseline risk'.

	Fixed effects		Random effects	
	Mean	95% CrI	Median	95% CrI
b	−0.96	(−1.03, −0.69)	−0.98	(−1.10, −0.70)
d_{12}	1.85	(1.67, 2.06)	1.83	(1.35, 2.29)
d_{13}	2.13	(1.90, 2.35)	2.18	(1.79, 2.63)
d_{14}	2.06	(1.47, 2.80)	2.03	(1.19, 2.94)
d_{15}	1.68	(1.49, 1.86)	1.71	(1.30, 2.16)
d_{16}	0.39	(−0.72, 1.26)	0.39	(−0.86, 1.45)
d_{17}	2.20	(1.92, 2.46)	2.24	(1.75, 2.79)
σ	–	–	0.19	(0.01, 0.70)
resdev[a]	27.4		24.2	
p_D[b]	19.0		21.6	
DIC	46.4		45.8	

Posterior median and 95% CrI interaction estimate (b), log ORs d_{XY} of treatment Y relative to treatment X and between-trial heterogeneity (σ) for the number of patients achieving ACR-50 ($d_{XY} < 0$ favours the reference treatment) and measures of model fit (posterior mean of the residual deviance, resdev, number of parameters, p_D and DIC). Treatment codes are given in Figure 8.3.
[a] Compare to 24 data points.
[b] p_D calculated outside WinBUGS.

statistics and the posterior means of the residual deviance favour the models with this covariate over the models without covariate or with disease duration as the covariate (Table 8.4). In fact, we might argue that baseline risk explains all the heterogeneity as a fixed effects model with this covariate is now appropriate.

As noted in Section 8.4.3.2, the assumption of a common regression term b allows the interaction parameter to be estimated for comparisons that only have one trial and gives estimates of treatment effects at values of the baseline risk outside the ranges measured for some comparisons. Again looking at rituximab plus MTX (treatment 6) with placebo plus MTX, the assumptions of parallel lines (common b) (Figure 8.6) predict, perhaps implausibly, that adding rituximab to MTX can be harmful if baseline risk is above 0.15.

The striking support in Figure 8.6 for a single interaction term for all treatments, except maybe treatment 6, has several implications for decision making and for synthesis in practice. Firstly, it clearly suggests a relation between efficacy and baseline risk that needs to be incorporated into cost-effectiveness analysis (CEA) models. Secondly, Figure 8.6 illustrates how variation in effect size due to a covariate will, if not controlled for, introduce severe heterogeneity in pairwise meta-analysis and potential inconsistency in network synthesis. It is clear that *both* the differences between trials (within treatments) *and* the differences between drugs are minimal once baseline risk is accounted for.

8.4.4 Subgroup Effects

In the context of treatment effects in RCTs, a subgroup effect can be understood as a categorical trial-level covariate that interacts with the treatment. The hypothesis would be that the size of treatment effect is different in, for example, male and female patients, or that it depends on age group, previous treatment, etc. The simplest way of analysing such data is to carry out *separate analyses* for each group, using the models described in Chapters 2 and 4, and then examine the estimates of the relative treatment effects. However, this approach has two disadvantages. First, if the models have random treatment effects, having separate analyses means having different estimates of between-trial variation. As there is seldom enough data to estimate the between-trial variation, it may make more sense to assume that it is the same for all subgroups. A second problem is that running separate analyses does not immediately produce a credible interval for the interaction effect. If this credible interval does not cross the value of no interaction (usually zero, depending on scale), it lends statistical support to the inclusion of the covariate when considered in conjunction with the reduction in the between-trial heterogeneity and model fit. The alternative is to run a single integrated analysis with a shared between-trial heterogeneity parameter and an interaction term, β, introduced on the treatment effect, as described in equation (8.1). Different assumptions for the interaction effects can then be used, as described previously.

The WinBUGS code is as described in Section 8.4.2, but care needs to be taken to specify the covariate vector so that it reflects the subgroups under consideration. For a model with two subgroups, these will be coded in a vector, where each element x_i will hold the value 0 or 1, depending on which subgroup the trial was conducted on. For a model with a single interaction term β, this will be interpreted as the change in relative effect of all treatments compared with the reference for patients in the subgroup coded 1 over the patients in subgroup coded zero.

These ideas extend naturally, but not necessarily easily, from binary effect modifiers to multiple categories. For example, for trials on patients categorised as mild, moderate and severe, two interaction terms can be introduced: one for moderate compared with mild and the second for severe compared with mild. Alternatively, disease severity can be examined as a continuous covariate (see Section 8.4.3) or as regression on baseline risk (see Section 8.4.3.3).

8.4.4.1 Statins Example: Pairwise Meta-Analysis with Subgroups

A meta-analysis of 19 trials of statins for cholesterol lowering against placebo or usual care (Sutton, 2002; Welton et al., 2012) included some trials on which the aim was primary prevention (patients included had no previous heart disease) and other trials on which the aim was secondary prevention (patients had previous heart disease). Note that in this case the subgroup indicator is a patient-level covariate that applies to all patients in the trials and can thus be considered a trial-level covariate (scenario A2(a) that is also equivalent to A1 (Section 8.2.1)). The outcome of interest is all-cause mortality and the data are presented in Table 8.6. The potential effect modifier, primary versus secondary prevention study, can be considered a subgroup in a pairwise meta-analysis of all the data using the model in equation (8.1), or two separate meta-analyses can be conducted on the two types of study.

Table 8.6 Statins example: data on statins and placebo for cholesterol lowering in patients with and without previous heart disease (Sutton, 2002) – number of deaths due to all-cause mortality in the control and statin arms of 19 RCTs.

Trial ID	r_{i1}	n_{i1}	r_{i2}	n_{i2}	x_i
1	256	2223	182	2221	Secondary
2	4	125	1	129	Secondary
3	0	52	1	94	Secondary
4	2	166	2	165	Secondary
5	77	3301	80	3304	Primary
6	3	1663	33	6582	Primary
7	8	459	1	460	Secondary
8	3	155	3	145	Secondary
9	0	42	1	83	Secondary
10	4	223	3	224	Primary
11	633	4520	498	4512	Secondary
12	1	124	2	123	Secondary
13	11	188	4	193	Secondary
14	5	78	4	79	Secondary
15	6	202	4	206	Secondary
16	3	532	0	530	Primary
17	4	178	2	187	Secondary
18	1	201	3	203	Secondary
19	135	3293	106	3305	Primary

The number of deaths in arm k of trial i, r_{ik}, is assumed to have a binomial likelihood $i = 1, \ldots, 19$; $k = 1, 2$. Defining x_i as the trial-level subgroup indicator such that

$$x_i = \begin{cases} 0 & \text{if study } i \text{ is a primary prevention study} \\ 1 & \text{if study } i \text{ is a secondary prevention study} \end{cases}$$

our interaction model is given in equation (8.1) where the linear predictor is $\theta_{ik} = \text{logit}(p_{ik})$ (Chapter 4). Note that since there are only two treatments, there is only one interaction effect, so we will use the model for a common interaction effect in equation (8.6). In this set-up, μ_i and δ_{i2} represent the log odds of the outcome in the reference treatment (i.e. the treatment indexed 1) and the trial-specific log ORs of success on the treatment group compared with the reference for primary prevention studies, respectively.

The WinBUGS code for the fixed and random effects subgroup meta-regression models is given in *Ch8_Statins_Bi_logit_FE-group.odc* and *Ch8_Statins_Bi_logit_RE-group.odc*, respectively.

The results of the two separate analyses and the single analysis using the interaction model for fixed and random treatment effects models are shown in Table 8.7 (results are based on 100,000 iterations from three independent chains after a burn-in of 50,000). Note that in a fixed effects context, the two analyses deliver exactly the same results for the treatment effects in the two groups, while in the random effects analysis, due to the shared variance, treatment effects are not quite the same: they are more precise in the single analysis, particularly for the primary prevention subgroup where there was less evidence available to inform the variance parameter, leading to very wide 95% CrI for all estimates in the separate random effects meta-analysis. However, only the joint analyses offer a 95% CrI for the interaction term β, which, in both cases, includes the possibility of no interaction, although the point estimate is negative, suggesting that statins might be more effective in secondary prevention patients.

8.5 Individual Patient Data in Meta-Regression

IPD meta-analyses have been described as the gold standard (Stewart and Clarke, 1995), and they clearly enjoy certain advantages over syntheses conducted on summary data, including the possibility of standardising analysis methods (Riley et al., 2010). When *patient-level* covariates are of interest, using the IPD to regress individual patient characteristics on individual patient outcomes will produce a more powerful and reliable analysis (Berlin et al., 2002; Lambert et al., 2002) compared with the use of aggregate outcome and covariate data and can avoid the potential ecological biases (see Section 8.2.1).

Table 8.7 Statins example: results from the fixed and random effects models for primary and secondary prevention groups.

Separate analyses

| | Fixed effects | | | | Random effects | | | |
| | Primary prevention | | Secondary prevention | | Primary prevention | | Secondary prevention | |
	Median	95% CrI	Median	95% CrI	Median	95% CrI	Median	95% CrI
log OR	−0.11	(−0.30, 0.09)	−0.31	(−0.42, −0.21)	−0.10	(−2.01, 1.12)	−0.34	(−0.72, −0.07)
OR	0.90	(0.74, 1.09)	0.73	(0.66, 0.81)	0.91	(0.13, 3.07)	0.71	(0.48, 0.94)
σ	–	–	–	–	0.79	(0.06, 3.90)	0.16	(0.01, 0.86)
resdev	16.9^a		29.0^b		11.9^a		28.3^b	
p_D	6.0		15.0		8.7		17.7	
DIC	22.9		44.0		20.6		46.0	

Single analysis

| | Fixed effects | | | | Random effects | | | |
| | Primary prevention | | Secondary prevention | | Primary prevention | | Secondary prevention | |
	Median	95% CrI	Median	95% CrI	Median	95% CrI	Median	95% CrI
log OR	−0.11	(−0.30, 0.09)	−0.31	(−0.42, −0.21)	−0.08	(−0.48, 0.36)	−0.35	(−0.72, −0.07)
OR	0.90	(0.74, 1.09)	0.73	(0.66, 0.81)	0.92	(0.62, 1.43)	0.71	(0.49, 0.94)
β	−0.21	(−0.42, 0.01)			−0.27	(−0.86, 0.20)		
σ	–	–			0.19	(0.01, 0.76)		
resdev^c	45.9				42.6			
p_D	21.0				25.0			
DIC	66.9				67.6			

Posterior summaries, mean, standard deviation (sd) and 95% CrI of the log OR, OR and posterior median, sd and 95% CrI between-trial heterogeneity (σ) of all-cause mortality when using statins (log OR < 0 and OR < 1 favour statins) and measures of model fit (posterior mean of the residual deviance, resdev, number of parameters (p_D) and (DIC).

[a] Compare to 10 data points.

[b] Compare to 28 data points.

[c] Compare to 38 data points.

Furthermore an IPD meta-regression analysis is essential when dealing with a continuous covariate and a continuous outcome.

In meta-analysis of IPD, historically, two broad approaches have been considered, the one- and two-stage approaches (Simmonds et al., 2005). In a two-stage approach, the analyst estimates the effect size(s) of interest from each study, together with a measure of uncertainty (e.g. standard error) in a first step, and then in a second step conducts a meta-analysis in the standard way using this summary data. In the context of exploring heterogeneity, the effect size could relate to a treatment by covariate interaction (Simmonds and Higgins, 2007). In some circumstances, it may be possible to carry out such an IPD analysis even if the analyst does not have access to all the IPD, that is, owners of the data may be willing to calculate and supply such interaction effects when they are not willing to supply the whole IPD dataset. However, such an approach becomes cumbersome/infeasible if multiple covariates are to be considered simultaneously.

IPD random effects pairwise meta-analysis models have been developed for continuous (Goldstein et al., 2000; Higgins et al., 2001), binary (Turner et al., 2000), survival (Tudor Smith et al., 2005) and ordinal (Whitehead et al., 2001) variables, and all allow the inclusion of patient-level covariates. Although most of the models are presented in the single pairwise comparison context, it is possible to extend them to a network meta-analysis context (Higgins et al., 2001; Tudor Smith et al., 2007; Cope et al., 2012). Simmonds and Higgins (2007) consider simple criteria for determining the potential benefits of IPD to assess patient-level covariates, and their work is recommended reading.

Treatment by covariate interactions can be estimated exclusively using between-study information when only summary data are available (meta-regression) and exclusively using within-study (variability) information if IPD are available. However, a subtlety when using IPD is that both between- *and* within-study coefficients can be estimated (Higgins et al., 2001). This can be achieved by including two covariates: the mean covariate value in that study (i.e. each individual in a study gets the same value – which is the value that would be used if an aggregate meta-regression analysis were being conducted) and a second covariate that is the individual patient response minus the mean value in that study (Riley and Steyerberg, 2010). Note that this applies most naturally to continuous covariates, but it can also be applied to binary covariates (e.g. if the binary covariate is sex, the between-study covariate would be the proportion of women).

There are a number of ways in which these dual effect (within- and between-study interaction) models can be used. The most appealing option is to use the interaction estimate derived exclusively from the within-trial variability, since this is free from ecological/aggregation biases and other potential sources of confounding between studies. Potentially, power could be gained by including the information in the between-trial variability by having the same parameter for within and between covariates. This, of course, comes at the cost of potentially inducing bias. It has been suggested (Riley and Steyerberg, 2010) that a statistical test of the difference between the two estimates could be carried out and the decision of whether to have the same interaction effect for within and between covariates could be based on this test. However, we suspect this test will have very low power in many situations, and further investigation of this approach is required before it can be recommended.

There may be situations where IPD are available from a number of, but not all, relevant studies. In this case, there are three potential options available for exploring heterogeneity. The first is to exclude all trials for which IPD are not available. This keeps the analysis simple, and can be based exclusively on within-study comparisons, but has the obvious disadvantage of not including all of the relevant trials. Furthermore, the analysis could potentially be biased if the reason for not providing IPD is related to the treatment effect. The second is to carry out a meta-regression on the aggregate data. This would potentially mean all trials could be included, but the benefits of having some IPD would be forgone. Finally, it is quite conceivable that IPD may be available for all trials of some comparisons, while none may be available for others. This may be particularly true for single technology appraisals done by industry where a company may have complete access to trial data for their own products, but only aggregate data on competitors' products (Signorovitch et al., 2012; Phillippo et al., 2016). Models have been developed that allow the incorporation of IPD where available and aggregate data where not (Riley et al., 2007b; Jansen, 2012; Saramago et al., 2012, 2014; Donegan et al., 2013). This approach allows all the data to be included at the most detailed level available from all the studies, but as for an IPD-only analysis, a decision has to be made on whether between-study variability is to be included in the estimation of effects. The difference between the effects using between- and within-study variability can be assessed and used to decide which approach to take, noting that in many contexts there will not be enough data to do this reliably. Models that allow the incorporation of IPD and aggregate data have been described for binary (Riley et al., 2007b; Sutton et al., 2008) and continuous (Riley et al., 2008) outcomes.

As described in Section 8.4, a decision has to be made on whether interaction effects with placebo/usual care are assumed to be the same, exchangeable or different across treatments. Although we have suggested a single interaction parameter for all treatments within the same class, models for all these possibilities can be constructed. Extensions to the dual within- and between-covariate models are possible, and there have been initial explorations of this (Saramago et al., 2011). The availability of IPD for several different treatments would allow a much more thorough investigation of whether patient-level interactions are the same across treatments, as well as linearity of interaction effects for continuous covariates (Donegan et al., 2012; Saramago et al., 2012, 2014).

8.6   Models with Treatment-Level Covariates

When patients are randomised to treatments at different doses, to treatments that belong to classes or to combinations of treatments, we may wish to express the relationship between relative treatment effects through a dose–response model, a class model or a treatment combination model. These models can be particularly useful when networks are sparse as they have fewer basic parameters than the standard network meta-analysis models described in Chapters 2 and 4, in which relative effects of every dose and every treatment and treatment combination compared with the reference are represented as a different basic parameter. This can have a large impact on network connectivity, since reducing the number of basic parameters may lead to a more connected network when data are sparse (Chapter 6) (Soares et al., 2012; Welton et al., 2015). However, as we stress in the following sections, the assumptions being made are strong and generally hard to check statistically and therefore require a high degree of clinical and empirical plausibility.

These types of model have previously been considered as meta-regression problems (Soares et al., 2012; Del Giovani et al., 2013; Fu et al., 2013; Thorlund et al., 2014; Welton et al., 2015), where dose has been added as a covariate. The basic parameters represent the treatment effects at zero or mean dose, requiring care in interpretation.

It should be noted that although similar in structure, these models are different from standard meta-regression models since patients *are* randomised to treatment with different doses, classes or combinations.

8.6.1   Accounting for Dose

We have previously noted that lumping across drug doses is to be avoided, as it may cause heterogeneity and make the results of any analysis difficult to interpret for decision making (Chapter 1). If each drug/dose combination is considered a different treatment in the network, that is, given its own treatment code, the standard network meta-analysis model on the appropriate scale can be used, as described in Chapters 2 and 4 (Cope et al., 2013; Naci et al., 2013a, 2013b; Alfirevic et al., 2015). This model does not make any assumption about the relationship between doses of the same treatment and is perhaps the least restrictive way of analysing data when patients are randomised to treatments at different doses. We suggest that this should usually be the base-case model. However, separating all the doses can lead to very sparse networks with limited information on the relative effects of each drug/dose combination.

The alternative is to make assumptions about the dose–response relationship within the same drug, thereby estimating fewer parameters, potentially resulting in more precise estimates (Mawdsley et al., 2016). We can think of this model as a model with a special kind of covariate, dose of drug, to which patients have been randomised to.

We begin by defining a treatment to be the actual drugs (compounds) or placebo, without reference to dose, and code these from 1 to S. As for the usual data set-up for the observed outcomes described in Chapter 4, the data structure describing the drug/dose combinations compared in each arm of each trial will consist of a treatment matrix **t** with elements t_{ik} holding the code for the drug compared in arm k of trial i and a dose matrix **x** with elements x_{ik} holding the dose of drug t_{ik} in arm k of trial i defined on some continuous scale (e.g. log dose or 1, 2, 3, etc.) that will be assumed linear, although the model could be extended to incorporate other functional forms. If a particular 'drug' is a placebo, then its dose should be set to zero (on the appropriate scale) to reflect the effect at zero dose of every drug. For no treatment arms, or for arms comparing treatments that do not really have doses (e.g. psychotherapy, regular monitoring, provide a leaflet), the corresponding element of the dose matrix should be set to a fixed number, say, 1, without loss of generality.

The generic network meta-analysis model described in equation (4.1) is the same, but now the basic parameters are the relative dose effects β_k, ($k = 1, ..., S$), reflecting the change in efficacy for a unit increase in the dose of treatment k when compared with placebo (i.e. the treatment at dose zero).

Thus for a linear dose model with random treatment effects, we use equation (4.2) where

$$d_{t_{i1}, t_{ik}} = \beta_{t_{ik}} x_{ik} - \beta_{t_{i1}} x_{i1}$$

and $\delta_{i1} = 0$ as before. For a fixed effects model, we write (equation (4.3))

$$\theta_{ik} = \mu_i + \beta_{t_{ik}} x_{ik} - \beta_{t_{i1}} x_{i1}$$

The dose effects are given non-informative prior distributions:

$$\beta_k \sim \text{Normal}(0,100^2)$$

The generic WinBUGS code for fixed and random effects introduced in Chapter 4 needs to be changed so that wherever we had `d[t[i,k]] - d[t[i,1]]` we now write

```
beta[t[i,k]]*x[i,k] - beta[t[i,1]]*x[i,1]
```

including in the code to correctly account for the correlation in multi-arm trials (see Chapter 2). Relative effects of treatments at different doses can be obtained by noting that the relative effect of drug b at dose X_b compared with the relative effect of treatment c at dose X_c can be written as $\beta_b X_b - \beta_c X_c$. Thus if we wanted the relative effect of drug 3 at dose 100 mg compared with placebo (dose set to zero), we would have $100 \times \beta_3$.

Note that whenever the reference drug (coded 1) is not a placebo, β_1 will reflect changes per unit increase in dose of drug 1 and will need to be estimated. Thus using this formulation, we estimate (at most) S relative effect parameters, whereas previously we set one of the effects to zero.

In general, to inform the parameters of this type of model, trials comparing several dose/treatment combinations are needed. In addition, trials comparing multiple doses of the same treatment are particularly informative as they provide information on the dose-relationship without being subject to between-study heterogeneity.

Note also that while the dose model proposed here estimates one parameter per drug, the standard network meta-analysis model would estimate one relative effect per dose of drug, compared with the reference treatment, leading to potentially many more parameters to estimate.

Using the dose effects model may greatly reduce the parameter space, which can result in considerable gains in precision of the estimates. However, this comes at the expense of an assumption of linearity (or other functional form) of the dose effects for each treatment on the chosen scale, which should always be validated clinically and empirically where possible (Mawdsley et al., 2016). Assessment of model fit (Chapter 3) may also bring insights into the suitability of the model.

Alternative models could be used, including those that allow for a non-zero intercept. Care is then required when interpreting the results since the intercept would no longer represent the expected effect at zero dose. Such models may be useful when approximate linearity is expected within the range of observed doses, but results cannot then be extrapolated beyond the range of doses observed (viz. to placebo or dose zero).

8.6.2 Class Effects Models

Another type of model with treatment-level classification that appears similar to regression models is a class model (Dominici et al., 1999; Dakin et al., 2011; Haas et al., 2012; Kew et al., 2014; Mayo-Wilson et al., 2014; Soares et al., 2014; Warren et al., 2014). This is where we may have a network with S treatments, but the treatments fall into classes with similar modes of action, making it reasonable to assume that there is a relationship between the effects of treatments in the same class. The extent of this relationship can be defined in several ways. For example, we may assume that the treatments belonging to a class have identical relative effects when compared with treatments in other classes or that the relative effects of treatments with a class are exchangeable, that is, they come from a common distribution. These models allow borrowing of strength across treatments in the same class. Alternatively we may assume that there are in fact no class effects and a standard network meta-analysis model on each separate treatment should be used. It should be noted that these three different assumptions about the class model are very similar to the possible assumptions on the regression parameters described in Section 8.4.

Defining D_k as the class to which treatment k belongs, the generic network meta-analysis model described in equations (4.1) and (4.2) is the same, but now instead of giving non-informative prior distributions to the basic parameters (equation (2.10)), different assumptions are made.

For an exchangeable class effects model, the basic parameters are assumed to come from a distribution with a common mean and variance, if they belong to the same class:

$$d_{1k} \sim \text{Normal}\left(m_{D_k}, \tau_k^2\right)$$

The within-class standard deviations τ_k should be given suitable prior distributions for the scale under consideration. When fitting an exchangeable class effects model, comparisons of treatments within the same class are particularly valuable as they can inform the within-class variability. Sparse networks with no within-class comparisons and few loops may require informative prior distributions for the within-class variability τ_k, or further assumptions, for example, that the within-class variability is the same or exchangeable across some or all classes. All these different assumptions can be implemented in WinBUGS by making small changes to the generic network meta-analysis code presented in Chapter 4.

For the model where all treatments in a class are assumed to have the same effects (fixed class effects model), the basic parameters are assumed equal for all treatments in a class $d_{1k} = m_{D_k}$.

The within-class mean treatment effects are given vague prior distributions $m_k \sim \text{Normal}(0, 100^2)$, $k = 1, ..., C$, where C is the number of classes.

Class models can be useful when the main decision is which class of treatment to recommend, or when data are sparse, as they allow different levels of borrowing of strength within classes, depending on the assumptions being made. These models are particularly useful when the number of classes, C, is much lower than the number of treatments, S. Assessment of model fit and model comparison techniques (Chapter 3) should be used to compare models and assess the suitability of assumptions.

8.6.3 Treatment Combination Models

Another type of treatment-level structure that may be encountered is the case where W discrete treatment components A, B, C, D, ... are defined, but some trials compare combinations of these, for example, B + D or B + C + D (Melendez-Torre et al., 2015). Examples include psychological intervention with multiple components (Welton et al., 2009b), combinations of drug treatments with different modes of action, designed to supplement each other, in chronic obstructive pulmonary disease (Riemsma et al., 2011; Mills et al., 2012; National Institute for Health and Care Excellence, 2012), interventions to encourage safe behaviours in the home (Cooper et al., 2012; Achana et al., 2015) and smoking cessation (Madan et al., 2014b).

Once again, if each component and combination is considered a different treatment in the network, that is, given its own treatment code, the standard network meta-analysis model on the appropriate scale can be used, as described in Chapters 2 and 4. Thus, if we had c combination interventions, the network would consist of $S - W + c$ treatments. This does not make any assumption on the relationship between combinations of the same intervention but in some cases may lead to very sparse networks with limited information on the relative effects of each treatment.

The alternative is to make an assumption about the relationship between the relative treatment effects of single and combination interventions. A simple assumption might be that they are additive on the linear predictor scale, so that the effect of combining elements B + D compared with the reference treatment is the sum of the individual effects of B and D relative to the reference on a suitable scale (Welton et al., 2009b; Riemsma et al., 2011). The form of this relationship could be extended to include other forms, for example, multiplicative or proportional, although note that additivity on a logistic or log scale implies a multiplicative relationship on the original (e.g. probability) scale.

Defining treatment components 1, 2, ..., S as the unique coding of interventions (single or combinations) where the first W treatments are single and the remaining are combinations of the single interventions, the generic network meta-analysis model described in equations (4.1) and (4.2) is the same, but now the non-informative prior distributions for the basic parameters (equation (2.10)) apply only to the first W treatments. The remaining relative effects implement the assumptions about the combination treatments. For example, if treatment $W + 1$ was a combination of treatments 2 and 4 ($W \geq 4$), to implement an additive assumption, we could state that

$$d_{1,W+1} = d_{12} + d_{14}$$

and similar assumptions could be stated for the remaining combination treatments. These assumptions can be easily coded in WinBUGS by adding suitable expressions to the generic code in Chapter 4 (see Exercise 8.4).

Fitting these models can result in a large reduction in the number of relative effect parameters to estimate (from $S - 1$ to $W - 1$), leading to stronger inferences, particularly when there are few comparisons between the S treatments. However, we stress that the assumptions being made are quite strong and need to have clinical and empirical plausibility. Assessment of model fit and model comparison techniques (Chapter 3) should also be used.

8.7 Implications of Meta-Regression for Decision Making

The implications of using the meta-regression models proposed in Section 8.4 for decision making can be quite profound. In practice, there is seldom enough data to fit the independent, treatment-specific interaction models, although related and exchangeable interactions might seem at first sight to offer an attractive approach. The difficulty is that even with ample data, using either of these models in clinical practice or in decision making could lead to recommendations that are counter-intuitive and difficult to defend. The claim made by these models is that there are *real* differences between the relative efficacies of the treatments included in the synthesis at different covariate values. If the models for interactions in equations (8.4) and (8.5) were used as a basis for treatment recommendation, a strict application of incremental CEA could lead to different treatments being recommended for different values of the covariate. This might be considered perverse, unless the hypothesis of different interaction effects was shown to be statistically robust, which will usually require very large amount of data. More importantly, this hypothesis needs to be clinically plausible, for example, decisions made based on a measure of severity or subgroup (e.g. screened positive for some marker) for which it is clinically plausible that the effects will differ. In addition, definition of cut-points of continuous covariates for changing decisions could be controversial. For these reasons, it has been recommended that only models implementing equation (8.6), which assumes an *identical* interaction effect across all treatments with respect to the reference treatment, are used (Dias et al., 2011a, 2013b).

However, we do not completely rule out the alternative models with different or exchangeable interaction effects (Nixon et al., 2007; Cooper et al., 2009) as they can have an important role in exploratory analyses or hypothesis-forming exercises. One rationale for departing from the identical interaction effects model could be to allow for the same covariate effect for different treatments within the same *class*, but different covariate effects across classes (which may or may not be considered exchangeable). So, if treatment 1 is a standard or placebo treatment while the other treatments belong to 'classes', and can therefore be assumed similar within a class, we would have different interaction effects for elements of different classes relative to the reference treatment. For example, one might imagine one set of equal interaction terms for aspirin-based treatments for atrial fibrillation relative to placebo and another set of interactions for warfarin-based treatments relative to placebo (Cooper et al., 2009) and a further set of interactions for novel oral anticoagulants relative to placebo.

There are however situations where it is reasonable to propose the more restricted model. Rather than a single interaction term for all active treatments within a class, we could simply have a single interaction term for *all* active treatments, regardless of class. For example, some treatments are so effective that they can virtually eliminate symptoms. In this case it is almost inevitable that there will be an 'interaction' between severity and treatment efficacy, because the extent of improvement is inevitably greater in more severely affected patients. Note however that in such circumstances, choice of analysis scale can be important. In some circumstances use of a log scale may eliminate the need for a covariate (Button et al., 2015). Potential examples might be different classes of biologic therapy for inflammatory arthritis or perhaps certain treatments for pain relief. In these cases the 'interaction' may reflect a property of the scale of measurement rather than the pharmacological effects of the treatment. Informed clinical and scientific input to model formulation is, as ever, critical. Model fit should be assessed as described in Chapter 3, and if several candidate models are considered, the preferred model choice can also be chosen according to the methods described in Chapter 3.

Using baseline risk as a covariate also has implications for interpretability as well as decision making. Often it will not be possible to accurately determine the baseline risk attaching to a particular patient, so unless baseline risk can be quantified by measureable patient attributes, these regression models are not very useful for making decisions for individuals. However, a guideline or reimbursement agency, interested in making decisions for a particular population, may have access to information (e.g. from patient registries or hospital statistics) on the baseline risk (e.g. log odds of an event) of the population of interest on the current treatment. If this is the case, and the statistical analysis is sufficiently robust and convincing, results could be used to guide decisions on which treatment to recommend.

When considering models with subgroups, ideally, we would want to include clinically meaningful subgroup terms whether they had a 95% CrI for the interaction term that included zero or not, possibly using informative prior distributions elicited from clinical experts. However, the NICE Methods Guide (National Institute for Health and Clinical Excellence, 2008b) suggests that subgroup effects should be statistically robust if they are to be considered in a cost-effectiveness model, as well as having some a priori justification. In practice, it would be difficult to sustain an argument that a treatment should be accepted or rejected based on a statistically weak interaction; thus models allowing for subgroups should be interpreted with care (see Section 8.4.4).

Models that incorporate treatment-level covariates allow decision makers to appraise treatments across the values of the treatment-defining covariate. For example, dose–response models (Section 8.6.1) allow decision makers to identify the optimal (e.g. most cost-effective) dose, although care has to be taken not to extrapolate beyond the range of doses in the included RCTs, and to consider all outcomes (including adverse events). Some doses may not be licensed, and therefore do not form part of the decision set of treatment options, but can still form part of the evidence set that contributes to the estimation of relative treatment effects.

Class effects models (Section 8.6.2) are appealing when we might expect treatment effects to be similar within class. However, cost effectiveness may differ between treatments within a class, so decisions are usually made at the treatment rather than class level. This can be achieved by using the shrunken treatment effect within class to inform decision models. If there is little variability in costs and adverse event profiles between treatments within a class, then decisions may be made to recommend a class of treatments, leaving it to local commissioners to identify the lowest cost option within a class.

Models for components of complex interventions are helpful for understanding the 'active ingredients' that increase efficacy, which can be helpful in the development of new interventions. However, policy decisions need to be for 'whole' interventions that can be recommended and rolled out in practice. Treatment combination models (Section 8.6.3) can be used to estimate the overall efficacy of 'whole' interventions for use in decision models, although care should be taken to ensure any assumptions (e.g. additivity) are justified.

8.8 Summary and Further Reading

In this chapter we have outlined the basic concepts: trial-level versus patient-level characteristics, continuous versus discrete covariates, subgroups, baseline risk as a covariate and the reasons why IPD is far more valuable than aggregate data for studying interactions. We have shown how the generic network meta-analysis code presented in Chapter 4 should be adapted to perform network meta-regression and which extra modelling assumptions are required.

We have identified models for the treatment effect itself – dose–response models, class models and treatment combination models – as special forms of 'meta-regression' with slightly different properties. Rather than add a further regression 'slope' coefficient, representing an interaction between treatment effect and the covariate, to a treatment effect parameter that represents the 'intercept', these models all represent different assumptions imposed on the treatment effect terms.

Although we have shown that all the models are easy to code in WinBUGS, requiring only small changes to the generic network meta-analysis code presented in Chapter 4, it is nonetheless extremely important to ensure models are coded accurately since small changes in coding and even choice of reference treatment can lead to major changes in the model assumptions and interpretation of results. It is also important to take care to interpret the parameters adequately, noting, for example, whether the covariate was centred and which is the reference treatment.

As usual, the decision-making context has been emphasised. From a statistician's perspective, models in which slope coefficients associated with different drugs are drawn from a random effects distribution elegantly express the very reasonable idea that similar products should have similar regression terms. However, for decision making in practice, this could create serious anomalies: a drug that is estimated to be the most effective in patients aged 75 might not be the most effective at age 60, and a third product might be best at age 45. One suspects that neither clinicians nor manufacturers would accept recommendations of this sort, unless of course they were supported by strong statistical evidence of real differences in regression slopes and strong a priori clinical plausibility.

Although the majority of work on meta-regression has been devoted to aggregate data, there is little doubt that IPD meta-analysis offers far greater insights. We would recommend to readers a careful look at some of the very fine applied work using IPD meta-analysis to throw light on the existence, or not, of covariate effects in pairwise meta-analyses (Collins et al., 1990; Berlin et al., 2002; Boutitie et al., 2002; Cholesterol Treatment Trialists' Collaboration, 2010). A listing of pairwise IPD meta-analyses can be found at http://ipdmamg. cochrane.org/ipd-meta-analyses. There is not, as yet, a large body of applied work using IPD meta-regression with network meta-analysis, although there is a review of this area (Veroniki et al., 2015, 2016).

Some RCTs may report results by subgroups or for combinations of subgroups (collapsed categories). The methods described in Section 5.3.4 can be used to address this.

8.9   Exercises

8.1   Fit the fixed effects network meta-analysis model with covariate 'absolute distance from the equator' to the BCG example. Compare the fit of this model to the random effects models with and without covariate (Table 8.2). Note that the data are given in *Ch8_BCG_Bi_logit_RE-x1.odc*.

8.2   Consider the CZP example presented in Section 8.4.3.2 and the results displayed in Table 8.4. Note that the full data and code are given in *Ch8_ CZP_Bi_logit_RE-x1prior.odc*.

a)   Fit the network meta-analysis model with covariate disease duration to the CZP example using a Uniform(0,5) prior distribution for the between-study standard deviation. Note the 95% CrI for sd and compare it to the prior bounds. Note also the posterior density for sd.

b)   Now fit the same model using an empirically based log-normal prior distribution, as suggested in table IV of Turner et al. (2015b). In this example the interventions are pharmacological compared with control, and we will assume the outcome is best categorised as 'signs/symptoms reflecting continuation/end of condition' (Turner et al., 2015b). Thus the suggested prior distribution for the between-study **variance** is log-Normal(-2.06, 1.51^2). [If we instead wanted to classify the outcome as 'general physical health indicators', we would use log-Normal(-2.29, 1.53^2) (Turner et al., 2015b).]

Note the 95% CrI and posterior density for sd, model fit statistics and estimated treatment effects and regression coefficient and compare them to the results in a).

8.3 Achana et al. (2013) fitted network meta-analysis models including baseline risk as a covariate to a dataset comprising 56 studies comparing four analgesics to reduce post-operative morphine consumption following major surgery. The outcome is the amount of morphine consumed over a 24 h period (in milligrams). The treatment network is in Figure 8.7 and the data are given in *Ch8_Ex3_AchanaPaindata-original.txt*. The augmented data, where treatment 1 arms with missing data are added to the two studies that did not include it, are given in *Ch8_Ex3_AchanaPaindata-augmented.txt*.

a) Adapt the code for the random effects network meta-regression model accounting for baseline risk to incorporate a continuous outcome, which can be assumed to have a normal likelihood. Fit the three interaction models detailed in Section 8.4 (common interaction, exchangeable and independent interactions) to this dataset, ensuring that you predict the missing baseline risk for studies that do not include treatment 1. See if you can replicate the results in table III of Achana et al. (2013), namely, the results referring to models A1, B1 and C1 (Achana et al., 2013). Check that you can interpret the output correctly and that you know how 'baseline risk' is defined in this case.

b) For comparison, now fit the model with the non-augmented data, ensuring that the baseline risk is corrected using the relative effect of the treatment in arm 1. Compare the results.

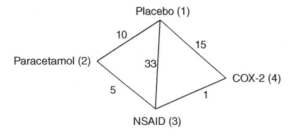

Figure 8.7 Pain data example (data from Achana et al., 2013): treatment network. Connecting lines indicate pairs of treatments that have been directly compared in randomised trials. The numbers on the lines indicate the numbers of trials making that comparison, and the numbers by the treatment names are the treatment codes used in the modelling.

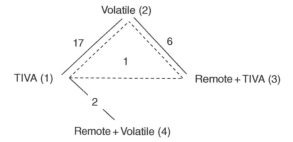

Figure 8.8 Mortality after cardiac surgery example (Zangrillo et al., 2015): treatment network. Connecting lines indicate pairs of treatments that have been directly compared in randomised trials. Solid lines represent two-arm studies, and the connected dotted lines represent a three-arm study. The numbers on the lines indicate the numbers of trials making each comparison, and the numbers by the treatment names are the treatment codes used in the modelling. http://journals.plos.org/plosone/article?id=10.1371/journal.pone.0134264. Licensed under CC BY 4.0. https://creativecommons.org/licenses/by/4.0/

8.4 Zangrillo et al. (2015) considered the hypothesis that the combination of total intravenous anaesthesia (TIVA) and volatile agents (Volatile) with the application of remote ischaemic preconditioning (Remote) would have an additive effect on survival after cardiac surgery. The outcome of interest was mortality at the longest available follow-up, and there were 26 studies where at least one event occurred, comparing four treatment strategies involving these three interventions. These are presented in the treatment network shown in Figure 8.8. Standard fixed and random effects network meta-analysis models were fitted, and the fixed effects model was preferred on the basis of the DIC. The code to fit this model, with the data included, is given in *Ch8_Mort_FE_Bi_logit.odc*. Adapt the code to fit a (fixed effects) treatment combination model with the assumption that the log odds ratios of TIVA + Remote and Remote + Volatile compared with TIVA are the sum of the log odds ratios of each component compared with TIVA (see Section 8.6.3). Compare results to those for the standard network meta-analysis code, noting model fit, relative effect estimates and their precision across the two models.

9

Bias Adjustment Methods

9.1   Introduction

Ideally all RCTs included in a meta-analysis or network meta-analysis will have been conducted with a high standard of methodological rigour on exactly the target population we are interested in making treatment recommendations and decisions for. In practise, however, this is unlikely to be the case, and the results of individual trials may provide *biased* estimates of the relative treatment effects that we would expect to see in the target population of interest. Bias arises as a result of effect-modifying mechanisms, in other words *interactions* between relative treatment effects and trial-level variables that have not been accounted for.

As in Chapter 8, we distinguish between two types of interaction effect, those that threaten the *external* validity of a trial and those that threaten the *internal* validity. Where effect modifiers have been measured and reported, covariate adjustment using meta-regression techniques may be used to adjust for bias caused by issues to do with *external* validity (Chapter 8). In this chapter we focus on interaction effects due to deficiencies in the way the trial was conducted or reported, which threaten *internal validity* (Rothman et al., 2012). Here, the trial delivers a biased estimate of the treatment effects in the target population for the trial, defined by that trial's inclusion/exclusion criteria. Typically, biases due to lack of internal validity are considered to vary randomly in size over trials, but do not necessarily have a zero mean. The clearest examples are the biases associated with markers of methodological rigour such as lack of allocation concealment of the randomisation process, or lack of double-blinding; these have been shown to be associated with larger treatment effects than trials without such markers (Schulz et al., 1995; Wood et al., 2008). Other examples include publication bias, where

Network Meta-Analysis for Decision-Making, First Edition. Sofia Dias, A. E. Ades, Nicky J. Welton, Jeroen P. Jansen and Alexander J. Sutton.
© 2018 John Wiley & Sons Ltd. Published 2018 by John Wiley & Sons Ltd.
Companion website: www.wiley.com/go/dias/networkmeta-analysis

the chance of publication depends on the effect estimate and its 'statistical significance', and missing data where the response on treatment is linked to whether an individual is lost to follow-up.

Note that pairwise or network meta-analyses can suffer from biases due to problems with *both* internal and external validity, where the trial delivers a biased estimate of the treatment effect in the target population for the trial, which may not be the same as the target population of interest for the decision. Where there are loops in a network meta-analysis, biases may manifest as inconsistency, and methods for bias adjustment may resolve issues with inconsistency (Chapter 7).

Confronted by trial evidence where there is some doubt about the internal validity of some of the trials, investigators have had two options: they can restrict attention to studies at low risk of bias or they can include all trials, at both low and high risk of bias, in a single analysis. Both options have disadvantages: the first ignores what may be a substantial proportion of the evidence, while the second risks delivering a biased estimate of the treatment effect. Recently, methods have been developed to allow a third option: bias adjustment.

The aim of bias adjustment is to transform estimates of treatment effect that are thought to be biased relative to the desired effect in the target population, into unbiased estimates. It is therefore essential to have in mind a target population for decision-making, which should have been specified in advance of conducting literature searches to identify evidence for the meta-analysis or network meta-analysis. Bias adjustment is appropriate when the evidence available, or at least some of the evidence, provides potentially biased estimates of the target parameter, due to issues with internal validity, external validity or both.

We present three different methods that have been proposed: adjustment and down-weighting of evidence at risk of bias based on external data to account for internal biases; estimation of bias associated with markers of risk of internal bias within a network meta-analysis; and adjustment for internal and/or external biases based on expert opinion or other evidence. It is necessary in all cases to take into account the uncertainty in external data or prior opinions that are used. The fourth approach to bias adjustment is to use meta-regression methods for covariate adjustment to address issues with external validity. Meta-regression methods have already been presented in Chapter 8, and we refer the reader to that chapter for details on covariate adjustment. The four different bias adjustment methods are summarized in Box 9.1.

All the ideas presented in this chapter can be applied to the general network meta-analysis models (Chapters 2 and 4), and methods for model criticism (Chapter 3) can be applied directly.

Box 9.1 Different Approaches to Bias Adjustment

Bias Adjustment Methods

1) **Meta-regression** (Chapter 8): A decision is required for a specific target population and specific treatments, but much of the evidence involves other populations or other (similar) treatments. This approach is suitable for pairwise meta-analysis, indirect comparisons and network meta-analysis of RCTs and works better with larger datasets, where there is enough evidence on the range of covariate values across the treatment comparisons in the included RCTs.
2) **Adjustment for potential bias associated with trial-level markers based on meta-epidemiological data** (Welton et al., 2009a): The evidence base contains some studies with markers of potential bias, and a prior distribution for this bias can be estimated from external meta-epidemiological data. This approach is suitable for pairwise meta-analysis, indirect comparisons and network meta-analysis with RCTs of mixed 'quality', particularly for small datasets, including the analysis of single trials. However its acceptability depends on the relevance of the meta-epidemiological data used.
3) **Estimation and adjustment of bias associated with trial-level markers in network meta-analysis** (Moreno et al., 2009b; Dias et al., 2010c; Salanti et al., 2010; Chaimani and Salanti, 2012; Mavridis et al., 2013, 2014; Turner et al., 2015a): The extent of the bias can be estimated internally from the existing trial evidence. This approach is suitable for network meta-analysis of RCTs of mixed 'quality'. It works better with larger datasets and can be combined with external prior distributions from meta-epidemiological data (method 2).
4) **Elicitation of internal and external bias distributions from experts** (Turner et al., 2009): This can be applied on its own or simultaneously in conjunction with methods 1–3 outlined previously, therefore it is suitable for pairwise meta-analysis, indirect comparisons, network meta-analysis of RCTs and/or observational studies (Thompson et al., 2011). This approach is good for small datasets, including single studies, but can be very time-consuming.

9.2 Adjustment for Bias Based on Meta-Epidemiological Data

Schulz et al. (1995) compared the results from RCTs rated at 'low risk of bias' with RCTs rated at 'high risk of bias' according to certain indicators of lack of internal validity: lack of allocation concealment or lack of double-blinding. Their dataset included over 30 meta-analyses, in which both high and low risk of bias trials were present (known as a meta-epidemiological database). Their results suggested the relative treatment effect in favour of the newer treatment was, on average, higher in the high risk of bias studies. The effect was large, with odds ratios in favour of the newer treatment on average about 1.6 times higher. Savovic et al. (2012a, 2012b) extended this work by combining evidence from seven such meta-epidemiological databases, removing overlaps, resulting in a dataset of 234 unique meta-analyses containing 1973 trials. They found a modest, yet significant, effect that odds ratio estimates seemed to be exaggerated in trials with inadequate or unclear (vs adequate) random sequence generation (ratio of odds ratios, 0.89, 95% CrI (0.82, 0.96); in trials with inadequate or unclear (vs adequate) allocation concealment (ratio of odds ratios, 0.93, 95% CrI (0.87, 0.99); and in trials with lack of or unclear double-blinding (vs double-blinding) (ratio of odds ratios, 0.87, 95% CrI (0.79–0.96)). These effects were mainly driven by trials with 'subjective' outcomes, where effects were stronger than for mortality and other objective outcomes.

Welton et al. (2009a) suggest an approach that uses all the data, but simultaneously adjusts and down-weights the evidence from studies assessed as being at high risk of bias. For a pairwise meta-analysis, the model for the studies at low risk of bias is the standard model introduced in Chapter 4:

$$\theta_{ik} = \mu_i + \delta_{ik} \quad \text{for } k = 1,2, \quad \text{where} \quad \delta_{i1} = 0 \tag{9.1}$$

For the studies at high risk of bias, the assumption is that each trial provides information not on δ_{i2}, but on a 'biased' parameter $\delta_{i2} + \beta_{i2}$, where the trial-specific bias terms β_{i2} are drawn from a random effects distribution, with a mean b_0 representing the expected bias, and a between-trial variance κ^2. Thus, for the trials at high risk of bias:

$$\theta_{ik} = \mu_i + \left(\delta_{ik} + \beta_{ik}\right) \quad \text{for } k = 1,2, \quad \text{where} \quad \delta_{i1} = \beta_{i1} = 0$$
$$\beta_{i2} \sim \text{Normal}\left(b_0, \kappa^2\right) \tag{9.2}$$

Prior distributions for the mean bias, b_0, and between-trial variance in bias, κ^2, can be obtained from a Bayesian analysis of an external dataset, for example, from collections of meta-analyses (Schulz et al., 1995; Savovic et al., 2010; Savovic et al., 2012a, 2012b). The same model as set out by equations (9.1) and (9.2) is used for the meta-epidemiological analysis, but with an additional level of hierarchy to reflect between meta-analysis variations in mean bias. The resulting model estimates provide a prediction for b_0 and κ^2 in a new meta-analysis that is considered exchangeable with the meta-analyses in the meta-epidemiology database. In this way prior distributions can then be put on b_0 and κ^2 to form a prior distribution for study-specific biases in equation (9.2), which simultaneously adjusts for and down-weights treatment effects for risk of bias in the new meta-analysis. Savovic et al. (2012a, 2012b) fitted this model to their meta-epidemiological database to obtain the ratio of odds ratios given previously. They also provide prior distributions for the bias in a new trial and mean bias in a new meta-analysis by outcome type and by risk of bias indicator.

This analysis hinges critically on whether the study-specific biases in the dataset of interest can be considered exchangeable with those in the meta-epidemiological data used to provide the prior distributions used for adjustment and in particular whether they would be considered exchangeable by *all* the relevant stakeholders in the decision (Welton et al., 2009a). It is already clear that the degree of bias is dependent on the nature of the outcome measure, being greater with subjective (patient- or physician-reported) outcomes, and virtually undetectable with all-cause mortality and other objective outcome measures (Wood et al., 2008; Savovic et al., 2012a, 2012b). Furthermore, there is now evidence that there are differences between clinical areas in the magnitude of biases (Savovic et al., 2012a), suggesting that sets of prior distributions tailored for particular outcome types and disease conditions, as reported by Savovic et al. (2012a), are required. There has also been some work on how multiple indicators of risk of bias might interact (Savovic et al., 2012a), which suggest that the effects may be less than additive, although this result is very uncertain because it is difficult to obtain enough power to estimate interaction effects, even in very large datasets.

Despite concerns about exchangeability of a new meta-analysis with previous meta-analyses, one might take the view that any reasonable bias-adjusted analysis is likely to give a better reflection of the true parameters than an unadjusted analysis. Welton et al. (2009a) suggest that, even when there are doubts about a particular set of values for the bias distribution, investigators may wish to run a series of sensitivity analyses to show that the presence of studies at high risk of bias, with potentially overoptimistic results, is not having an impact on the decision.

In principle the same form of bias adjustment could be extended to other types of bias, such as novel agent effects, industry sponsor effects or small-study effects, or to mixtures of RCTs and observational studies. Each of these extensions, however, depends on detailed and far-ranging analyses of very large meta-epidemiological datasets, which have not yet been performed. There is no reason why prior distributions from meta-epidemiological studies cannot be applied to network as well as pairwise meta-analyses; however, a key challenge in doing so is in defining the direction in which a bias is expected to act, especially in studies comparing active treatments (see Section 9.3 for more details) as it may not be clear which treatment may be favoured in studies at high risk of bias. Savovic et al. (2012a) excluded trials where it was not clear in which direction the bias would act; otherwise biases between studies may 'cancel out', leading to an underestimate of the mean bias in a meta-analysis. Chaimani et al. (2013b) report results from a network meta-epidemiological study, where a collection of network meta-analyses are analysed to estimate bias resulting from indicators of risk of bias. The relationship between treatment effects and study precision (small-study effects, see Section 9.3.4) was also considered. However, they restrict attention to 'star networks' (i.e., those where all treatments have been compared with a common comparator) so that the direction in which the bias might act could be assumed to be against the common comparator. They found that imprecise studies were associated with large treatment effects.

9.3 Estimation and Adjustment for Bias in Networks of Trials

We turn next to a method that removes the difficulties associated with the strong 'exchangeability' assumptions required if using meta-epidemiological data to adjust for bias. Instead, the parameters of the bias distribution, b_0 and variance κ^2, are estimated *internally*, within the dataset of interest without recourse to external data. The method also sets out how the bias model in equations (9.1) and (9.2) can be extended from pairwise to network meta-analysis.

Imagine a set of trials, some of which are at 'high' and some at 'low' risk of bias according to an indicator of risk of bias due to issues with internal validity. In a pairwise meta-analysis, one can always use trials categorised in this way to estimate the size of bias and – with enough data – the variability in bias across studies. However, estimating the bias distribution adds nothing to our knowledge of the true treatment effect: the studies at high risk of bias provide information on the bias distribution, while those at low risk of bias provide information on the relative treatment effect. We might just as well have restricted the analysis to the low risk of bias studies alone in the first place. In other words in a pairwise meta-analysis there are not sufficient degrees of freedom to borrow strength from biased data to adjust treatment effects.

For indirect comparisons and network meta-analysis, if we assume that the mean and variance of the study-specific biases are the same for each treatment comparison, then it is possible to simultaneously estimate the treatment effects and the bias effects in a single analysis. This will produce treatment effects that are based on the entire body of data, including studies at both high and low risk of bias, and also adjusted for bias (Dias et al., 2010c). Furthermore, the consistency equations (equation (2.9)) provide further degrees of freedom to allow us to jointly estimate and adjust for bias. The model is exactly the same as in Section 9.2, where equations (9.1) and (9.2) are combined to give

$$\theta_{ik} = \mu_i + \left(\delta_{ik} + \beta_{ik} x_i \right), \quad \text{where} \quad \delta_{i1} = \beta_{i1} = 0 \tag{9.3}$$

with $x_i = 1$ if study i is considered to be at risk of bias and zero otherwise and β_{ik} is the trial-specific bias of the treatment in arm k relative to the treatment in arm 1 of trial i. Note that equation (9.3) is a more general form of the meta-regression model introduced in equation (8.1), without the requirement of consistency of the regression coefficients, β_{ik}. We assume that the study-specific biases are exchangeable:

$$\beta_{ik} \sim \text{Normal}\left(b_{t_{i1},t_{ik}}, \kappa^2\right) \tag{9.4}$$

with between-study variance in bias, κ^2, and where the mean bias $b_{t_{i1},t_{ik}}$ depends on the treatment comparison being made between arm k and arm 1 in study i. In order to be able to estimate the bias parameters, we will need to make some simplifying assumptions on the mean biases b_{k_1,k_2} for each pair of treatments k_1 and k_2. One possibility is that the mean bias is the same for all active treatments that are compared with a standard or placebo treatment ('active vs placebo' trials) so that $b_{1,k_2} = b_1$, for $k_2 = 2,\dots$. Note this is equivalent to the 'exchangeable, related, treatment-specific interactions' case introduced in equation (8.5).

It is less clear what to assume about bias in trials that make comparisons between active treatments. One approach might be to assume a mean bias of 0, based on the assumption that the mean bias against placebo is the same for the two active treatments, and so it should cancel out when compared head to head, based on the consistency of regression coefficients of covariates. However, such consistency may not be reasonable. For example, it may be the case that average bias is always in favour of the newer treatment ('optimism bias'), and we require a model to reflect this novel agent bias (Song et al., 2008; Salanti et al., 2010). Another approach might be to propose a separate mean bias term for active versus active comparisons (Dias et al., 2010c) so that $b_{k_1,k_2} = b_2$, for $k_1 \neq 1$ and $k_2 \neq 1$.

9.3.1 Worked Example: Fluoride Therapies for the Prevention of Caries in Children

Dias et al. (2010c) present bias adjustment models in a network meta-analysis of fluoride therapies to prevent the development of caries in children (Salanti et al., 2009). There are 130 trials of which one was a four-arm trial, three were three-arm trials, and the remaining were two-arm trials. The treatments are coded: 1 = No Treatment, 2 = Placebo, 3 = Toothpaste, 4 = Rinse, 5 = Gel and 6 = Varnish. The network is presented in Figure 9.1. The outcome available from each trial arm is the mean increase in number of caries, y_{ik}, for a given number of patients at risk, n_{ik}, from which we can derive the total number of additional caries, $r_{ik} = y_{ik} n_{ik}$. Follow-up time, $time_i$, varied between the trials.

The total additional number of caries has a Poisson likelihood with mean equal to the rate of development of caries, λ_{ik}, multiplied by the person time at risk, $E_{ik} = n_{ik} \times time_i$ (equation (4.4)). The network meta-analysis model is put on the log rate scale $\theta_{ik} = \log(\lambda_{ik})$, and the bias model is as given in equations (9.3) and (9.4). Dias et al. (2010c) explored two models for the bias. In the first (model 1), it is assumed that there is no bias in active versus active

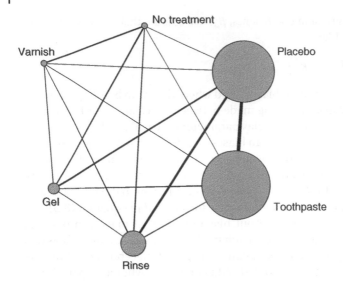

Figure 9.1 Fluoride example: network of treatment comparisons (drawn using R code from Salanti (2011)). The thickness of the lines is proportional to the number of trials making that comparison and the width of the bubbles is proportional to the number of patients randomised to each treatment (Salanti et al., 2009). Reproduced with permission of Elsevier.

comparisons, and the mean bias is the same for all of the active versus inactive comparisons (no treatment and placebo). Note that both treatments 1 and 2 are considered non-active treatments in this example, so that in model 1:

$$b_{k_1,k_2} = \begin{cases} b_1 & \text{for } k_1 = 1, 2 \quad \text{and} \quad k_2 = 3, 4, 5, 6 \\ 0 & \text{Otherwise} \end{cases}$$

In the second (model 2), it is assumed that there is bias in all treatment comparisons with a common mean bias for active versus active comparisons and a separate common mean bias for active versus inactive comparisons:

$$b_{k_1,k_2} = \begin{cases} b_1 & \text{for } k_1 = 1, 2 \quad \text{and} \quad k_2 = 3, 4, 5, 6 \\ b_2 & \text{Otherwise} \end{cases}$$

The code for model 1 is given in the following (available in *Ch9_RE_Po_Fluorbias1.odc* and the data in *Ch9_FluorData.xlsx*).

WinBUGS code for bias adjustment model 1, with common mean bias in active versus no treatment or placebo trials, and 0 mean bias in active versus active trials. Matrix C[i,k] indicates whether comparison is 1 = Placebo versus No Treatment, 2 = Active versus No Treatment, 3 = Active versus Placebo, 4 = Active versus Active. Vector bias[i] indicates a study at high risk of bias on a given marker.

The full code with data and initial values is presented in the online file *Ch9_RE_Po_Fluorbias1.odc*.

```
# Poisson likelihood, log link
# Random effects model for multi-arm trials, with bias adjustment (model 1)
model{
for(i in 1:ns){                                    # *** PROGRAM STARTS
                                                   # LOOP THROUGH STUDIES
    w[i,1] <- 0                                    # adjustment for multi-arm trials is zero for control arm
    beta[i,1] <- 0                                 # no bias term in baseline arm
    delta[i,1] <- 0                                # treatment effect is zero for control arm
    mu[i] ~ dnorm(0,.0001)                         # vague priors for all trial baselines
    for (k in 1:na[i]){                            # LOOP THROUGH ARMS
        r[i,k] <- n[i,k]*y[i,k]                    # total caries = (no. of patients) * (mean no. caries)
        theta[i,k] <- lambda[i,k]*n[i,k]*time[i]   # failure rate * exposure
        r[i,k] ~ dpois(theta[i,k])                 # Poisson likelihood
        log(lambda[i,k]) <- mu[i] + delta[i,k] + beta[i,k] * bias[i]    # linear predictor
# Deviance for Poisson
        dev[i,k] <- 2*((theta[i,k]-r[i,k]) + r[i,k]*log(r[i,k]/theta[i,k]))
    }
    resdev[i] <- sum(dev[i, 1:na[i]])              # summed residual deviance contribution for this trial
    for (k in 2:na[i]){                            # LOOP THROUGH ARMS
# model for bias parameter beta
        beta[i,k] ~ dnorm(mb[i,k], Pkappa)
        mb[i,k] <- A[C[i,k]]
# trial-specific RE distributions
        delta[i,k] ~ dnorm(md[i,k], taud[i,k])
        md[i,k] <- (d[t[i,k]] - d[t[i,1]]) + sw[i,k]    # mean of RE distributions (with multi-arm trial correction)
```

```
        taud[i,k] <- tau *2*(k-1)/k              # precision of RE distributions (with multi-arm trial
                                                 #   correction)
        w[i,k] <- delta[i,k] - d[t[i,k]] + d[t[i,1]]    # adjustment for multi-arm RCTs
        sw[i,k] <-sum(w[i,1:k-1])/(k-1)          # cumulative adjustment for multi-arm trials
        }
    }
totresdev <- sum(resdev[])                       # Total Residual Deviance
d[1] <- 0                                        # treatment effect is zero for reference treatment
for (k in 2:nt){d[k] ~ dnorm(0,.0001) }          # vague priors for basic parameters
sd.d ~ dunif(0,10)
var.d <- pow(sd.d,2)
tau <- 1/var.d                                   # vague prior for between-trial SD
# mean bias: assumptions
A[1] <- 0        # NT v Pl
A[2] <- b        # NT v A
A[3] <- b        # Pl v A
A[4] <- 0        # A v A
# bias model prior for variance
kappa ~ dunif(0,10)
kappa.sq <- pow(kappa,2)
Pkappa <- 1/kappa.sq
# bias model prior for mean
b ~ dnorm(0,.0001)
# all pairwise differences
for (c in 1:(nt-1)) {
    for (k in (c+1):nt) {
        lhr[c,k] <- d[k]-d[c]
        log(hr[c,k]) <- lhr[c,k]
    }
}
}

                # *** PROGRAM ENDS
```

For model 2 (available in *Ch9_RE_Po_Fluorbias2.odc*), the only changes required to the code are

```
# mean bias: assumptions (NT=no treatment, Pl=placebo,
# A=active)
A[1] <- 0         # NT v Pl
A[2] <- b[1]      # NT v A
A[3] <- b[1]      # Pl v A
A[4] <- b[2]      # A v A
# bias model prior for mean
for (j in 1:2){b[j] ~ dnorm(0,.0001)}
```

and the initial values for b need to be given as a vector with two values.

Different indicators of risk of bias, x_i, can be explored (Dias et al., 2010c). Here we consider $x_i = 1$ if allocation concealment is inadequate or unclear and $x_i = 0$ if allocation concealment is adequate. The first point to note is that convergence for these models is very slow, so long burn-in periods are necessary. Inference should be based on large samples post burn-in.

Table 9.1 shows the results from models 1 and 2. Both of these models give similar model fit according to the posterior mean residual deviance, which is also similar to that obtained from a model without any bias adjustment (278.3). Also, the estimated mean bias terms are close to 0. This suggests that there is no evidence that studies with inadequate or unclear allocation concealment produce results that are different to those with adequate allocation conceal-ment. However, there is some evidence of lack of fit, with a posterior mean residual deviance of approximately 278 compared with 270 data points. Dias et al. (2010c) explored this further using the studies where allocation conceal-ment was rated as unclear. Instead of assuming that these studies were at high risk of bias, they introduced a *probability* that each unclear study was at high risk of bias. This allows each unclear study to be classified as being either adequate or inadequate based on the predicted probability of being at risk of bias, rather than assuming all unclear studies to be at high risk of bias. See Dias et al. (2010c) for details on how to download the WinBUGS code. Fitting this model gave a posterior mean residual deviance of 274.6 for both models 1 and 2, and the between-study standard deviation falls to 0.12 with 95% CrI (0.10, 0.15). The estimated mean bias for active versus inactive comparisons (placebo or no treatment) was −0.19 with 95% CrI (−0.36, −0.02) for model 1, suggesting that trials with high risk of bias due to allocation concealment (when unclear studies are modelled to have a probability of being at risk of bias) have a tendency to overestimate treatment effects relative to placebo or no treatment. The parameter estimates are on a log rate ratio scale, which translates to an estimated rate ratio of $\exp(-0.19) = 0.83$ for studies at high risk of bias com-pared with those at low risk of bias. Model 2 gave very similar results, and the

Table 9.1 Fluoride example: posterior summaries for the bias model using allocation concealment rated as inadequate or unclear as an indicator of high risk of bias.

Parameter	Model 1, no active vs active bias	Model 2, including active vs active bias
Mean bias for active vs placebo or no treatment, b_1	−0.01 (−0.106, 0.094)	−0.01 (−0.114, 0.147)
Mean bias for active vs active, b_2	NA	0.39 (0.145, 0.614)
Between-study standard deviation in bias, κ	0.10 (0.012, 0.200)	0.11 (0.017, 0.196)
Placebo vs no treatment, d_2	−0.22 (−0.358, −0.087)	−0.18 (−0.320, −0.049)
Toothpaste vs no treatment, d_3	−0.50 (−0.670, −0.332)	−0.47 (−0.652, −0.291)
Rinse vs no treatment, d_4	−0.50 (−0.668, −0.328)	−0.46 (−0.652, −0.283)
Gel vs no treatment, d_5	−0.48 (−0.648, −0.321)	−0.47 (−0.652, −0.295)
Varnish vs no treatment, d_6	−0.61 (−0.807, −0.424)	−0.78 (−1.007, −0.556)
Between-study standard deviation in treatment effects, σ	0.19 (0.130, 0.245)	0.18 (0.124, 0.235)
Posterior mean residual deviance[a]	278.2	277.9

Results are shown for (i) model 1 where a common mean bias term is assumed for the active versus placebo or no-treatment comparisons and a zero mean bias for active versus active comparisons and (ii) model 2 where a common mean bias is assumed for active versus active comparisons that may be different to the common mean bias assumed for active versus placebo or no treatment. Results shown are posterior means and 95% credible intervals and treatment effects are interpreted as log rate ratios.

[a] Compare to 270 data points. Values much larger than this indicate some lack of fit. The posterior mean residual deviance in a model with no bias adjustment is 278.3.

estimated mean bias for active versus active comparisons was 0 with 95% CrI (−0.65, 0.57). The model with no active versus active bias (model 1) was therefore preferred on the basis of parameter estimates and model fit. Note that model 1, where the unclear studies have a probability of being at high risk of bias, estimates a mean bias with a credible interval that does not contain 0 (suggesting evidence of bias), whereas when all unclear studies are assumed to be at high risk of bias, we do not find evidence of bias. This is due to a small number of the unclear studies being classified at high risk of bias, whereas the majority of unclear studies are classified as being at low risk of bias by the model. When all are assumed high risk, then the bias effect is masked due to this apparent misclassification.

Figure 9.2 shows the estimated log rate ratios from the unadjusted network meta-analysis model (solid lines) and from model 1 where the unclear studies have a probability of being at risk of bias (dashed lines). It can be seen that the main impact of the bias adjustment is to move the treatment effect estimates

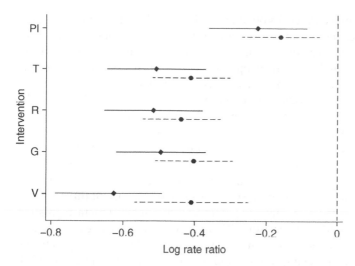

Figure 9.2 Fluoride example: estimated posterior means and 95% credible intervals for log-hazard ratios compared with no treatment for the following: Pl, placebo; T, toothpaste; R, rinse; G, gel; V, varnish. Results from a network meta-analysis model with no bias adjustment shown with diamonds and solid lines. Circles and dotted lines represent results from bias adjustment model 1 with common mean bias term for the active versus placebo or no-treatment comparisons, zero mean bias for active versus active comparisons, a probability of being at risk of bias in studies rated as unclear. The vertical dotted line represents no effect.

towards the null effect, suggesting that studies at high risk of bias slightly over-estimate treatment effects. This effect is especially strong for varnish, where there were only a few trials making comparisons with placebo or no treatment, all of which had inadequate or unclear allocation concealment. The bias-adjusted results also show that there is evidence of a placebo effect in these trials, likely to be due to the placebo involving brushing or other treatment of the teeth, albeit without fluoride. The bias-adjusted results show that all of the fluoride interventions are similar in effectiveness and are clearly better than no treatment and likely better than placebo. The bias-adjusted analysis also shows that there is no reason to believe varnish is more effective than the other fluoride interventions, as was found in the non-adjusted analysis.

9.3.2 Extensions

The method can in principle be extended to include syntheses that are mixtures of trials and observational studies, but this does not appear to have been attempted yet. It can also be extended to any form of 'internal' bias. A key advantage of the approach is that it adjusts for bias using only the randomised controlled trials that are included in the network; however there is still low power to estimate bias and results often have a high level of uncertainty. Because the underlying bias models in this section and the previous one are the same, it would be perfectly feasible to combine them to estimate the bias within the network meta-analysis, but using prior distributions based on external evidence. We are not aware that anyone has done this, which is an area for future work. Like the methods described in Section 9.2, these methods may be considered by some as semi-experimental. There is certainly a great need for further experience with applications, and there is a particular need for further meta-epidemiological data on the relationships between the many forms of internal bias that have been proposed (Dias et al., 2010a). However, they appear to represent reasonable and valid methods for bias adjustment and are likely to be superior to no bias adjustment in situations where data are of mixed methodological rigour. At the same time, the method is essentially a meta-regression based on 'between-study' comparisons. There is no direct evidence for a 'causal' link between the markers of study quality and the size of the effect. It is therefore important to avoid using the method for small datasets and to establish that the results are statistically robust and not dependent on a small number of studies.

In the following text we describe examples of estimation and adjustment for specific types of bias that have been developed.

9.3.3 Novel Agent Effects

Salanti et al. (2010) used a model similar to that described previously to explore the existence of novel agent effects, where relative effects of new treatments are overestimated due to optimism bias (Song et al., 2008), conflicts of interest or other mechanisms. Their model is the same as equation (9.3) but with the bias indicator x depending on treatment arm k as well as study i:

$$\theta_{ik} = \mu_i + \left(\delta_{ik} + \beta_{ik} x_{ik}\right), \quad \text{where} \quad \delta_{i1} = \beta_{i1} = 0,$$

where $x_{ik} = 1$ if the treatment on arm k is newer than the treatment on arm 1 in study i.

Salanti et al. (2010) applied their model to three network meta-analyses of chemotherapy and other non-hormonal systemic treatments for cancer (ovarian, colorectal, and breast cancer), further assuming that the novel agent bias was exchangeable across cancers. They found some evidence of novel agent effects, with hazard ratios for overall survival exaggerated by 6%, 95% CrI (2%, 11%) for newer treatments, although this had little effect on treatment rankings. Note that overall survival is an objective measure, and we would expect to see bigger effects on subjective outcomes (Savovic et al., 2012b). Salanti et al. (2010) assumed that the novel agent effect occurs regardless of other indicators of risk of bias. However, it is possible that effects are stronger in studies at high risk of bias according to other indicators. It would be interesting to extend this model to explore possible interactions.

9.3.4 Small-Study Effects

A particularly interesting application is 'small-study bias', where the idea is that the smaller the study, the greater the bias. One possible mechanism that might generate small-study effects is publication bias, where negative findings from small studies are less likely to get published than positive findings, but large studies are likely to get published regardless of the results. This mechanism leads to an overestimation of treatment effect in the smaller studies, due to the 'missing' small negative studies. The 'true' treatment effect can therefore be conceived as the effect that would be obtained in a study of infinite size. This, in turn, is taken to be the intercept in a regression of the treatment effect against the study variance. This model has the same form as equation (9.3) where x_{ik} is set equal to the variance of the relative effect estimate. Moreno et al. (2009a, 2009b) show that, in the context of antidepressants, the bias-adjusted estimate from this approach closely approximates the results found in a simple meta-analysis based on a register of prospectively reported data. Once again, in larger networks, some care would need to be exercised in how to code the direction of bias in 'active–active' studies. Chaimani et al. (2013b) found that in their collection of 32 star networks, imprecise studies were associated with larger treatment effects and tended to be those with inadequate conduct.

One explanation for small-study effects is 'publication bias', where studies that show statistically significant findings are more likely to be published and identified in a systematic review. Copas and Shi (2001) present a selection model to explore sensitivity of results to publication bias in pairwise meta-analysis, and these models have been extended to network meta-analysis (Chootrakool et al., 2011; Trinquart et al., 2012; Mavridis et al., 2013, 2014), including networks with loops (Mavridis et al., 2014). Trinquart et al. (2012) have compared the performance of regression and selection bias models applied to published trials on antidepressants, using the US Food and Drug Administration database as a gold standard representing the totality of the evidence. They found that both adjustment methods corrected for publication bias, but more successfully for some drugs than others.

9.3.5 Industry Sponsor Effects

It has been suggested that trials sponsored by industry tend to favour the product of the sponsor (Gartlehner and Fleg, 2010; Gartlehner et al., 2010; Flacco et al., 2015). Naci et al. (2014a) used a meta-regression approach to explore the effects of trials with and without industry sponsorship, in a network meta-analysis of statins for low-density lipoprotein (LDL) cholesterol reduction. They assumed a common mean bias term for all statins relative to placebo and found no evidence of industry sponsor effects in trials of statins for LDL cholesterol reduction. However, there were differences in effectiveness according to the dose of statin given, which may explain why the previous work, not accounting for dose variability, had found an association between treatment effect and industry sponsorship. It appears that the frequently reported tendency of industry trials to report results in favour of the sponsored product (Ioannidis, 2005; Bero, 2014) is due to the choice of comparator rather than bias due to trial conduct, analysis and reporting.

9.3.6 Accounting for Missing Data

Missing outcome data is common in RCTs and can occur for a variety of reasons, many of which can lead to biased effect estimates if not adjusted for (Little and Rubin, 2002). Methods to adjust for missing data in RCTs typically centre around imputation, where other observed variables are used to predict outcomes in those that are missing (Little and Rubin, 2002). The challenge in pairwise and network meta-analysis is that we usually only have summary-level data available and so cannot use imputation methods at the individual level, although methods have been described if individual patient data are available (Burgess et al., 2013). If all of the RCTs included in the meta-analysis have reported effect estimates adjusted for missing data, then these can be combined in meta-analysis; however adjustments need to be made if the missing proportion depends on effect size (Yuan and Little, 2009).

In most meta-analyses we do not have adjusted estimates from the RCTs and so must attempt to account for missing data within the meta-analysis. Higgins et al. (2008b) proposed the informative missingness odds ratio (IMOR) as a measure to adjust for bias arising from missing data on a binary outcome. The IMOR is defined as the odds of an event in the missing individuals divided by the odds of an event in the observed individuals and can depend on study and treatment arm. If the odds of an event is the same for the observed and the missing individuals, then the IMOR = 1 and the data are taken at face value. If the IMOR is not equal to 1, then it can be used to adjust the estimate for that study and arm. White et al. (2008a) suggest using prior information on the IMORs to reflect departures from 'missing at random' and explore robustness of results to the prior assumptions in a sensitivity analysis (Higgins et al., 2008b; White et al., 2008a). This approach has been extended to network meta-analysis (Spineli et al., 2013) and also to continuous outcome measures by defining the informative missingness difference of means and the informative missingness ratio of means (Mavridis et al., 2015).

In the absence of any prior information on the missingness mechanism, it is important to reflect the additional uncertainty in effect estimates as a result of the missing data. Gamble and Hollis (2005) proposed down-weighting studies where 'best-case' and 'worst-case' analyses give wide limits on treatment effect to reflect the uncertainty associated with missing data. Turner et al. (2015a) instead put a flat prior distribution on the probability of an event in the missing individuals to reflect our uncertainty on this.

There have been a few attempts to estimate missingness parameters within a meta-analysis. White et al. (2008b) proposed a 1-stage hierarchical model for the IMORs in an attempt to 'learn' about the IMORs from the observed data, although found there was limited ability to do so in a pairwise meta-analysis. Spineli et al. (2013) used a network meta-analysis model to estimate the IMOR but again found that the data were barely sufficient to identify the IMOR parameters. In random treatment effects models, there is almost complete confounding between the random treatment effect and the random IMOR elements. In fixed treatment effects models, particularly when some trials have only small amounts of missing data, then the data are sufficient to identify missingness parameters and 'learning' can take place. Turner et al. (2015a) present a general framework for a Bayesian 1-stage estimation for a given definition of the missingness parameter (e.g., IMOR, probability of an event in the missing individuals, etc.) that allows one to select the way missingness is parameterised, facilitating the use of informative prior distributions. They have only applied this to pairwise meta-analysis, but the extension to network meta-analysis is natural, which will provide a greater potential to learn about the missingness parameter, because of the 'spare' degrees of freedom generated by the consistency equations. This is an important area for further developments.

9.4 Elicitation of Internal and External Bias Distributions from Experts

Turner et al. (2009) proposed a method to elicit distributions for biases from experts for the purpose of bias adjustment. The method is conceptually the simplest of all bias adjustment methods, applicable to trials and observational studies alike. It is also the most difficult and time-consuming to carry out. One advantage may be that it can be used when the number of trials is insufficient for meta-regression approaches (Chapter 8). Readers are referred to the original publication for details, but the essential ideas are as follows. Each study is considered by several independent experts using a predetermined protocol. The protocol itemizes a series of potential internal and external biases, and each expert is asked to provide information that is used to develop a bias distribution. Among the internal biases that might be considered are selection biases (in observational studies), non-response bias, attrition bias and so on. A study can suffer from both internal *and* external biases. When this process is complete, the bias information on each study from each assessor is combined into a single bias distribution. The assessor distributions are then pooled mathematically. In the original publication the mean and variance of the bias distributions are statistically combined with the original study estimate and its variance to create what is effectively a new, adjusted estimate of the treatment effect in that study. The final stage is a conventional synthesis, in which the adjusted treatment effects from each study, and their variances, are treated as the data input for a standard pairwise meta-analysis, indirect comparison or network synthesis. The methods in Chapters 2 and 4 can then be applied to the adjusted study-specific estimates.

This methodology (Turner et al., 2009) in its full form requires considerable time and care to execute. The key idea of replacing a potentially biased study estimate with an adjusted estimate based on expert opinion regarding bias is one that can be carried out in many ways and with a degree of thoroughness that is commensurate with the sensitivity of the overall analysis to the parameters in question. There is an important conceptual difference between this approach and the others discussed in this chapter. As we noted in the previous section, under some circumstances data on treatment effects in unbiased studies can provide indirect information on the biases in other studies. However, this does not happen if the bias information is used in effect to 'adjust' the data. A slight modification of the Turner et al. (2009) method would be to combine the original data and the bias distributions in a single MCMC simulation under the consistency model. This is an exciting area for further development.

9.5 Summary and Further Reading

RCTs may produce biased estimates of treatment effect in the population we are interested in making treatment recommendations for, if there are interactions between treatment effects and trial characteristics that are unaccounted for. A distinction is made between bias resulting from issues with internal and external validity of an RCT. Methods to deal with issues with external validity include meta-regression to adjust for bias due to difference in covariates between trial and decision populations and adjustment for bias using prior distributions elicited from experts. Methods to deal with issues of internal validity include use of external information, such as collections of previous meta-analyses, to adjust and down-weigh evidence rated at high risk of bias, estimation of bias within a network meta-analysis and adjustment for bias using prior distributions elicited from experts or other sources. Markers of lack of internal validity that have been explored include issues with the randomisation procedure, whether there is adequate concealment of the allocation of individuals to treatments, lack of blinding or double-blinding, novel agent effects, small-study effects (including publication bias), industry sponsor effects and missing outcome data. Bias adjustment methods in meta-analysis are still evolving and should be considered as exploratory. However, we would argue that an attempt to adjust for bias will lead to more valid results than simply ignoring it. Bias adjustment models should at least be conducted as a sensitivity analysis, alongside analyses that omit studies at high risk of bias.

A general model for heterogeneity that encompasses bias due to both internal and external validities can be found in Higgins et al. (2009), but it is seldom possible to determine what the causes of heterogeneity are or how much is due to true variation in clinical factors and how much is due to other unknown causes of biases.

In theory, the methods described here for bias adjustment could be applied to generalised evidence synthesis (Prevost et al., 2000; Welton et al., 2012), where both RCT and observational evidence are pooled. In particular the method of eliciting prior distributions for bias due to issues with internal and external validity has been applied to a mixture of RCT and observational studies (Turner et al., 2009).

9.6   Exercises

9.1   For the fluoride example, fit the bias adjustment model 1 (zero bias in active vs active comparisons) and model 2 (common mean bias in active vs active trials) using the risk of bias indicator: $x_i = 1$ if allocation concealment is inadequate or the trial is not double blind, and $x_i = 0$ otherwise. The data are available in *Ch9_FlourData.xlsx* in the tab 'FluorData_allocORblind'. You can use the code given in *Ch9_RE_Po_Fluorbias1.odc* and *Ch9_RE_Po_Fluorbias2.odc*, making changes to the data. Compare the posterior mean residual deviance with that from the model with no bias adjustment (178.3) and look at estimates of mean bias. Is there any evidence of bias according to this indicator for

a) Active versus active comparisons?
b) Active versus placebo or no treatment?

9.2   *For the fluoride example, fit the bias adjustment model 1 (zero mean bias in active vs active comparisons) using the risk of bias indicator: $x_i = 1$ if allocation concealment is inadequate, and $x_i = 0$ otherwise, using an informative prior distribution for $\kappa \sim$ Gamma(10, 50), and $b \sim$ Normal($-0.16, 0.1^2$) (based on (Savovic et al., 2012a), assuming a subjective outcome). You will need to adjust the code available in *Ch9_RE_Po_Fluorbias1.odc* to add in the informative prior distribution (the data and initial values are unchanged). Remember that WinBUGS parameterises the normal distribution using the precision ($=1/(0.1^2)$ here). How does adding in the informative prior distributions change the results?

10

*Network Meta-Analysis of Survival Outcomes

10.1 Introduction

We are often interested in relative effectiveness of competing treatments on the time from treatment initiation to the occurrence of a particular event (time to event). For example, in oncology, nearly all studies report time-to-event data on overall survival (OS), where the event is death from any cause, and on progression-free survival (PFS), where the event is death from any cause or disease progression, whichever occurred first. Time-to-event outcomes have particular features that require different analytical techniques than those for other continuous outcomes. Firstly, the distribution of times to event tends to be skewed, so the normal likelihood is not usually appropriate. Secondly, it is usually the case that not every patient will have experienced the event (e.g. progression or death) during the follow-up period of the study, or they may have become lost to follow-up. Such patients provide censored observations. Censored observations need to be incorporated in the analysis because they provide information regarding the lowest possible value of the time to event for the individual. Statistical procedures for analysis of continuous outcomes that do not account for censoring will provide biased estimates (Collett, 2003).

The hazard and survival functions are central to the analysis of time-to-event data. Let U be a non-negative continuous variable reflecting the time to an event, with probability density function $f(u)$. The survival function is the probability that the event of interest, for example, death, has not yet occurred by time u and given by

$$S(u) = \Pr(U \geq u) = 1 - \int_0^u f(x)\,dx$$

Network Meta-Analysis for Decision-Making, First Edition. Sofia Dias, A. E. Ades, Nicky J. Welton, Jeroen P. Jansen and Alexander J. Sutton.
Companion website: www.wiley.com/go/dias/networkmeta-analysis

The hazard function represents the instantaneous probability of experiencing the event of interest (e.g. progression or death) at time u conditional on not having experienced the event prior to time u and is given by

$$h(u) = \frac{f(u)}{S(u)}$$

Pairwise and network meta-analysis of time-to-event outcomes is complicated because the hazard function varies over time, and unless a simplifying assumption, such as proportional hazards (PH), is made, the relative treatment effects may be multidimensional. Also, the shape of the hazard function may be different across studies. Furthermore, there are a variety of ways in which time-to-event outcomes can be reported, and different studies may report different summaries, such as median times, hazard ratios (HRs) and empirical survival curve plots. Finally, in a decision-making context, we are interested in making predictions in the longer term beyond the follow-up period of some or all of the included studies. Model selection can have a big impact on extrapolation to the longer term and should therefore be considered carefully.

In this chapter we begin by describing common ways in which survival data are presented. We then introduce survival models for data from a single trial before going on to describe network meta-analysis models in cases where the relative treatment effect is assumed to act on a single parameter. We then show some extensions to multidimensional treatment effects models. Having described the key models, we show how data in different formats can inform the models through the likelihood and linking functions. We discuss issues specific to network meta-analysis with time-to-event outcomes, such as model choice, meta-regression, presenting results, and use in decision making. The methods are illustrated with an example of treatments for multiple myeloma.

10.2 Time-to-Event Data

10.2.1 Individual Patient Data

Ideally we would like to have individual patient data (IPD) from each study to provide us the flexibility to fit a variety of survival models. For time-to-event outcomes such as PFS and OS, this will comprise an observation, u_{ijk}, for patient j in arm k of study i, which represents the time of event for those patients who experienced the event during the study follow-up and represents the length of follow-up available for each of the patients who are censored. In addition, there will be an indicator variable, z_{ijk}, that indicates whether patient j in arm k of study i experienced the event ($z_{ijk} = 1$) or was censored ($z_{ijk} = 0$). Patient-level covariates may also be available.

10.2.2 Reported Summary Data

In practice IPD are often not made available, and instead we need to rely on the summary statistics reported in the trial publications. Commonly reported summaries include median survival time, proportion of patients alive at certain time points (e.g. 1-year survival, 2-year survival), Kaplan–Meier estimated survival curves (see Section 10.2.3) and HR for comparisons of interest (see Section 10.4). In addition to the summary estimates, measures of uncertainty (e.g. 95% confidence intervals (CI), standard errors or p-values) are required. One may also see life table data where for multiple time-related intervals the population at risk is presented along with the number of patients who have had an event in that interval, but this is less common in the clinical literature. It is unusual to see the mean survival time reported (Guyot et al., 2011); however the restricted mean survival time, which is the mean time conditional on an event having occurred during the study follow-up, may sometimes be reported (Royston and Parmar, 2013).

10.2.3 Kaplan–Meier Estimate of the Survival Function

In reports of RCT it is standard practice to present the Kaplan–Meier estimates of the survival function, $S^{KM}(u_m)$, estimated from IPD summarised with a series of M time intervals defined by the event times so that at least one event occurs at the start of each interval. For each interval $m = 1, 2, ..., M$, the Kaplan–Meier data consist of the number of events that occur at the start of the interval, e_m; the number of individuals censored on the interval, c_m; and the number of patients still at risk just before the start of the interval, n_m (Collett, 2003), defined iteratively as

$$n_{m+1} = n_m - e_m - c_m$$

The Kaplan–Meier estimate of the survival function $S^{KM}(u_m)$ at time u at the start of interval m, u_m, can be obtained according to

$$S^{KM}(u_m) = \prod_{j=1}^{m} \frac{n_j - e_j}{n_j} = S^{KM}(u_{m-1}) \times \frac{n_m - e_m}{n_m}$$

The Kaplan–Meier method is a non-parametric method for estimating the survival function and does not require specific assumptions about the underlying distribution of the survival times.

10.3   Parametric Survival Functions

Models in which a specific probability distribution is assumed for the survival times are known as parametric models. Common parametric distributions for survival times are the exponential, Weibull, Gompertz, log-normal and log-logistic. Their survival and hazard functions are presented in Table 10.1. These can be reparametrised to obtain scale and shape parameters on which we might expect the treatment effects to be additive (see column 4 in Table 10.1) and can take a variety of shapes depending on the values of their parameters. Note that the exponential distribution is a special case of the Weibull distribution. The generalised gamma and F distributions provide more flexibility that can be used for constant, monotonically increasing or decreasing, arc-shaped and bathtub-shaped hazard functions. However these distributions may be hard to fit and still lack sufficient flexibility. More flexible models include cubic spline models (Royston and Parmar, 2002) and fractional polynomials (FP) (Royston and Altman, 1994).

Royston and Altman (1994) introduced FP models for determining the functional form of a continuous predictor, which are well suited for nonlinear data. These have been used in many applications including survival analysis (Berger et al., 2003; Bossard et al., 2003; Bagnardi et al., 2004; Sauerbrei et al., 2007). The first-order FP model can be written as

$$y = \alpha_0 + \alpha_1 u^p \tag{10.1}$$

where y is a continuous dependent variable and u is a continuous predictor. The power p is chosen from the following set: -2, -1, -0.5, 0, 0.5, 1, 2, 3 with $u^0 = \ln(u)$. α_0 and α_1 are regression parameters.

The second-order FP is defined as

$$y = \alpha_0 + \alpha_1 u^{p_1} + \alpha_2 u^{p_2} \tag{10.2}$$

When we set $p_1 = p_2 = p$ in equation (10.2), a 'repeated powers' model is obtained (Royston and Altman, 1994):

$$y = \alpha_0 + \alpha_1 u^p + \alpha_2 u^p \ln(u)$$

By varying p_1, p_2, α_0, α_1 and α_2, a wide range of monotonically increasing and decreasing, arc- and bathtub-shaped curves can be obtained (Royston and Altman, 1994). Using these models to describe the log-transformed hazard over time (i.e. $y = \ln(h(u))$), it is easy to see that first-order FP (equation (10.1)) with $p = 0$ and $p = 1$ correspond to the reparametrised functions of Weibull and Gompertz model, respectively (Table 10.1). As such, α_0 represents a scale parameter, and α_1 a shape parameter of the log hazard function. Adding a second time-related parameter (or second shape parameter α_2) makes changes

Table 10.1 Common distributions used for the analysis of time-to-event data along with the corresponding survival and hazard functions.

Distribution	Survival function	Hazard function	Transformation of $S(t)$ or $h(t)$ to a linear predictor function	Shapes of hazard functions
Exponential	$S(u) = \exp(-\lambda u)$	$h(u) = \lambda \quad \lambda > 0$	$\ln(h(u)) = \alpha_0$ with $\alpha_0 = \ln(\lambda)$	Constant
Weibull	$S(u) = \exp(-\lambda u^{\varphi})$	$h(u) = \lambda \varphi u^{(\varphi-1)} \quad \varphi > 0$ and $\lambda > 0$	$\ln(h(u)) = \alpha_0 + \alpha_1 \ln(u)$ with $\alpha_0 = \ln(\lambda \varphi)$ and $\alpha_1 = (\varphi - 1)$	Constant Monotonically in/decreasing
Gompertz	$S(u) = \exp\left(\dfrac{\lambda}{\theta}(1 - \exp(\theta u)) \right)$	$h(u) = \lambda \exp(\theta u) \quad \lambda > 0$	$\ln(h(u)) = \alpha_0 + \alpha_1 u$ with $\alpha_0 = \ln(\lambda)$ and $\alpha_1 = \theta$	Constant Monotonically in/decreasing
Log-normal	$S(u) = 1 - \Phi\left(\dfrac{\log(u) - \mu}{\sigma} \right)$ where Φ is the cumulative distribution function for the standard normal distribution and μ and σ are the mean and standard deviation of the variable's natural logarithm	$h(u) = \dfrac{f(u)}{S(u)}$ with $f(u) = \dfrac{1}{u\sigma\sqrt{2\pi}} \exp\left(-\dfrac{1}{2}\left(\dfrac{\log(u) - \mu}{\sigma} \right)^2 \right)$	$\Phi^{-1}(S(u)) = \alpha_0 + \alpha_1 \ln(u)$	Monotonically decreasing Arc shaped
Log-logistic	$S(u) = \dfrac{1}{1 + (\lambda u)^{\theta}}$	$h(u) = \dfrac{\lambda \theta (\lambda u)^{\theta-1}}{1 + (\lambda u)^{\theta}} \quad \lambda > 0$	$\ln\left(\dfrac{1 - S(u)}{S(u)} \right) = \alpha_0 + \alpha_1 \ln(u)$ with $\alpha_0 = \ln(\lambda)$ and $\alpha_1 = \theta$ $\dfrac{1 - S(u)}{S(u)}$ is the failure odds	Monotonically decreasing Arc shaped

in the direction of the hazard function a possibility, similar to those possible with log-logistic, log-normal and (generalised) gamma distributions. With the reparameterisation of the standard survival distributions (column 4 in Table 10.1), we can obtain nested models; the reparametrised exponential, Weibull and Gompertz log hazard functions are special cases of the second-order FP log hazard functions. This makes selection and comparison of competing models straightforward.

10.4 The Relative Treatment Effect

In order to assess the relative efficacy of treatments in an RCT, the time to event of interest of the different treatment groups are compared. Once the survival functions have been estimated, it is straightforward to obtain an estimate of the median survival time, which is the time beyond which 50% of the population is expected to survive. The difference in median survival time between the treatment groups is one estimate of the relative treatment effect. Alternatively, one can compare the proportion of patients alive at a certain time point among treatment groups, for example, with the odds ratio of survival at 1 year. However, these measures do not take into account the complete survival distribution, and conclusions can vary from one time point to the next.

Models to evaluate the effect of treatment using the complete survival distributions consist of a (baseline) function representing the survival times in the absence of treatment (or with the reference treatment) and relative effect parameters describing how the survival times vary according to treatment. Models where we assume a single parameter representing the effect of treatment can be roughly categorised as PH models or accelerated failure time (AFT) models.

A simple and frequently used model that focuses directly on the hazard function was introduced by Cox and Oakes (1984) where the hazard in one group at any time point is proportional to the hazard in the other group. The Cox model is considered a semi-parametric model because no parametric distribution is assumed for the baseline hazard. PH models can also be fitted assuming a parametric baseline hazard function. The relative treatment effect measure of interest obtained with PH models is the (constant) HR, and this is the most commonly reported outcome from RCT (Moher et al., 2010).

An AFT model focuses on the survival times (rather than the hazard function) and assumes a linear relationship between the log of survival time and treatment effect (Collett, 2003). More specifically, the treatment effect in the model represents a shift in the log survival time from the control group to the treatment group. In other words, the survival time with treatment is a multiplicative 'acceleration factor' of the baseline survival time. For example, if the acceleration factor is 1.5, that means that patients in the treatment group live 1.5 times as long as patients in the control group. Note that the AFT model does not imply PH regarding treatment effects, and vice versa the PH model does not imply treatment effects in terms of a constant acceleration factor.

The Weibull and Gompertz survival distribution can be parameterised as a PH model. Note that the application of a constant HR implies the assumption that the treatment only has an effect on the scale parameter (λ in Table 10.1) of these distributions. The log-logistic distribution provides the most commonly used AFT model. Other distributions suitable for AFT models include the Weibull, log-normal and gamma distributions.

In the aforementioned PH and AFT models, it is assumed that the effect of treatment is represented with one parameter and is not related to time. Given the modelling framework, the treatment effect is constant over time. Such an assumption may not always be supported by the data. For example, if the hazard functions of the treatment and control group cross or converge, the PH assumptions does not hold, as illustrated in Figure 10.1.

As an alternative to a constant HR, one can assume a time-varying HR. This requires that the parameters in the model representing the effect of treatment have a time-related component.

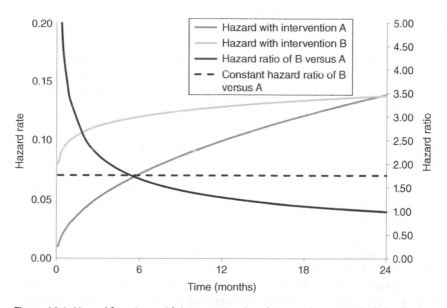

Figure 10.1 Hazard functions with intervention A and B according to Weibull distributions with different scale and shape parameters along with corresponding time-varying hazard ratio. The constant hazard ratio relying on the proportional hazard assumption is not supported by the data resulting in a biased estimate of the relative treatment effect.

10.5   Network Meta-Analysis of a Single Effect Measure per Study

10.5.1   Proportion Alive, Median Survival and Hazard Ratio as Reported Treatment Effects

A network meta-analysis of the proportion of patients alive at a specific time point, median survival or reported HR can be performed using the core model presented in Chapter 4 with appropriate likelihoods and link functions.

For the proportion of patients alive at a specific time point, a logit link function and binomial likelihood can be used, provided all studies report the number of events at that time point. The results of the analysis are odds ratios but can be transformed into survival probabilities for each treatment by applying the odds ratios to the odds of survival with a reference treatment (see Chapter 4). Alternatively, if an exponential survival model is considered realistic, the cloglog link function can be used to estimate the log HR when different studies report proportions at different follow-up times, or the Poisson likelihood with log link may be used if there is censoring and number of events for given person years at risk is reported (see Chapter 4).

A model with identity link and normal likelihood can be used for median survival, provided a 95% CI or standard error for the medians is reported (or can be estimated using, e.g. Greenwood's formula (Collett, 2003)) and the sample sizes are not too small.

After transforming reported HRs and their 95% CI into log HRs and their standard errors, we can use a normal likelihood and the network meta-analysis model for data in the format of treatment differences (Section 4.4.1). Results of this analysis will be the (pooled) log HRs for all contrasts in the network that need to be transformed back to the HR scale.

10.5.2   Network Meta-Analysis of Parametric Survival Curves: Single Treatment Effect

When IPD, that is, survival times for each individual in the study, are available, or Kaplan–Meier curves have been published that allow recreation of the IPD (see Section 10.7.4), we can assume a certain parametric survival function (e.g. Weibull, log-normal or generalised gamma) across the studies in the network and obtain, for each study, a constant HR or acceleration factor that subsequently can be incorporated in a network meta-analysis. For example, let us assume survival times in study i follow a Weibull distribution and treatment only has an effect on the scale of this distribution. For uncensored individuals the likelihood is $u_{ijk} \sim \text{Weibull}(\varphi_i, \lambda_{ik})$ where u_{ijk} is the observed survival time of subject j in study i for arm k, λ_{ik} is the scale parameter in study i for arm k and φ_i is the study-specific shape parameter. For censored individuals, the likelihood contribution is $S(u_{ijk}) = \exp(-\lambda_{ik}(u_{ijk})^{\varphi_i})$. The study-specific scale parameters λ_{ik} can be incorporated in a network meta-analysis with an arm-based likelihood using a log link function (Chapter 4). The results of the network meta-analysis are the differences in the log-scale parameters between treatments, which is the log HR.

Alternatively, we can reparameterise the study- and treatment-specific hazard function $h_{ik}(u) = \lambda_{ik}\varphi_i u^{\varphi_i-1}$, according to Table 10.1, and implement the network meta-analysis model assuming the treatment effects act on $\alpha_{0,ik} = \ln(\lambda_{ik}\varphi_i)$:

$$\ln\left(h_{ik}\left(u\right)\right) = \alpha_{0,ik} + \alpha_{1,i}\ln\left(u\right)$$
$$\alpha_{0,ik} = \mu_{0,i} + \delta_{0,ik} \tag{10.3}$$
$$\delta_{0,ik} \sim \text{Normal}\left(d_{0,1t_{ik}} - d_{0,1t_{i1}}, \sigma^2\right)$$

where $\alpha_{1,i} = (\varphi_i - 1)$. The study and treatment arm-specific parameters $\alpha_{0,ik}$ and $\alpha_{1,ik}$ describe the hazard over time. $\delta_{0,ik}$ are trial-specific treatment effects affecting the scale parameter α_0 of the treatment in arm k relative to the treatment in arm 1 of that trial (with $\delta_{0,i1} = 0$) and are drawn from a normal distribution with the mean effects expressed in terms of the overall reference treatment 1: $d_{0,1t_{ik}} - d_{0,1t_{i1}}$. The pooled results are log HRs. Note that we have introduced extra subscripts in the notation to facilitate the extension to the network meta-analysis models with time-varying treatment effects in Section 10.6.1.

10.5.3 Shared Parameter Models

Among the different studies to be included in the network meta-analysis, different data regarding relative treatment effects may be available. For example, for some studies IPD may be available or Kaplan–Meier curves may have been published. For other studies only medians or HRs may have been reported. As long as a parametric functional form is assumed, these different summaries can be combined in a network meta-analysis using a shared parameter model. For example, if an exponential model is assumed, then the median can be assumed to have a normal likelihood with mean equal to $2/\lambda_{ik}$, and the log HR can be assumed to have a normal likelihood with mean equal to $\log(\lambda_{i2}) - \log(\lambda_{i1})$. The IPD (recreated from the Kaplan–Meier curves) allow estimation of the λ_{ik} directly using an exponential likelihood with a log link. If two-parameter models are assumed (e.g. Weibull), then it is necessary to assume something about one of the parameters in order to estimate the other, for example, a constant or exchangeable shape parameter across studies (Welton et al., 2008b, 2010). See Chapter 4 for further details on shared parameter models.

10.5.4 Limitations

The main limitation of network meta-analysis of survival at a certain time point is that we only focus on the cumulative effect of treatment at that particular time point. We ignore any variation in treatment effects over time up to, as well as beyond, that time point. Similar limitations hold for analyses of median survival (Section 10.5.1). The HR summarises the treatment effect for the complete follow-up period of the trials (Sections 10.5.1 and 10.5.2). However, network meta-analysis of the constant HR relies on the PH assumption, which is implausible if the hazard functions of competing interventions do not run parallel. Assuming a single acceleration factor per treatment has similar limitations.

10.6 Network Meta-Analysis with Multivariate Treatment Effects

10.6.1 Multidimensional Network Meta-Analysis Model

As an alternative to the constant HR, which is a univariate treatment effect measure, we can also use a multivariate or multidimensional treatment effect measure that describes how the relative treatment effect (e.g. HR) changes over time.

Ouwens et al. (2010) and Jansen (2011) presented methods for (network) meta-analysis of survival data using a multidimensional or multivariate treatment effect as an alternative to the synthesis of one treatment effect (e.g. the constant HRs). The hazard functions of the interventions in a trial are modelled using known parametric survival functions, and the difference in the parameters are considered the multidimensional treatment effect, which are synthesised (and indirectly compared) across studies. With this approach the treatment effects are represented by multiple parameters rather than a single parameter. By incorporating additional parameters for the treatment effect, the PH assumption is relaxed and the network meta-analysis model can be fitted more closely to the available data.

In the next sections we present multidimensional models that relate to familiar survival distributions. These models can be considered special and simpler cases of the general multivariate models presented in Chapter 11, in particular the model represented with equation (11.7).

10.6.1.1 Weibull

In line with equation (10.3), a Weibull-based network meta-analysis model with a two-dimensional treatment effect can be defined according to

$$\ln\left(h_{ik}\left(u\right)\right) = \alpha_{0,ik} + \alpha_{1,ik}\ln\left(u\right)$$

$$\begin{pmatrix} \alpha_{0,ik} \\ \alpha_{1,ik} \end{pmatrix} = \begin{pmatrix} \mu_{0,i} \\ \mu_{1,i} \end{pmatrix} + \begin{pmatrix} \delta_{0,ik} \\ d_{1,1t_{ik}} - d_{1,1t_{i1}} \end{pmatrix} \tag{10.4}$$

$$\delta_{0,ik} \sim \text{Normal}\left(d_{0,1t_{ik}} - d_{0,1t_{i1}}, \sigma^2\right)$$

with study- and treatment arm-specific parameters $\alpha_{0,ik}$ and $\alpha_{1,ik}$ describing the hazard over time as defined in Table 10.1 (with $\alpha_{0,ik} = \ln(\lambda_{ik}\varphi_{ik})$ and $\alpha_{1,ik} = (\varphi_{ik} - 1)$, as before). The vectors $\begin{pmatrix} \mu_{0,i} \\ \mu_{1,i} \end{pmatrix}$ are trial specific and reflect the true underlying scale and shape parameters of the log hazard function for the treatment in arm 1; $\delta_{0,ik}$ are trial-specific treatment effects affecting the scale parameter α_0 of the treatment in arm k relative to the treatment in arm 1 of that trial (with $\delta_{0,i1} = 0$) and are drawn from a normal distribution with the mean effects expressed in terms of the overall reference treatment 1: $d_{0,1t_{ik}} - d_{0,1t_{i1}}$. The parameters $d_{0,1t_{ik}}$ correspond to the treatment effect under a PH assumption. A fixed effects model is assumed for the relative treatment effect regarding the shape parameter α_1 of the log hazard curve, expressed as $d_{1,1t_{ik}} - d_{1,1t_{i1}}$, reflecting the change in log HR over time or deviation from the PH assumption over time. In other words, for a PH model $d_{1,1t_{ik}}$ equals 0. By incorporating $d_{1,1t_{ik}}$ in addition to $d_{0,1t_{ik}}$, a multidimensional treatment effect is used. Consistency equations can be put on any parameter on any scale. However, the parameterisation needs to be such that treatment effects are additive. If they are not, that may manifest itself as heterogeneity or inconsistency. The reparameterisation of the Weibull hazard function according to Table 10.1 as used in equation (10.4) facilitates this. Variance σ^2 reflects the between-study heterogeneity in the difference in the scale parameters. A random effects model with a heterogeneity parameter only for the relative treatment effect in terms of scale α_0 implies that the between-study heterogeneity of the log HRs remains constant over time. For a fixed effects version of this model, $\delta_{0,ik}$ needs to be replaced with $d_{0,1t_{ik}} - d_{0,1t_{i1}}$. The random effects model represented by equation (10.4) treats multi-arm trials (>2 treatments) without taking account of the correlations between the trial-specific treatment effects. The multi-arm code introduced in Chapter 2 can be applied here as well. An additional heterogeneity parameter for treatment effects in terms of shape will have the flexibility to capture between-study variation regarding changes in the log HRs over time.

10.6.1.2   Gompertz
Replacing $\ln(h_{ik}(u)) = \alpha_{0,ik} + \alpha_{1,ik} \ln(u)$ in equation (10.4) with $\ln(h_{ik}(u)) = \alpha_{0,ik} + \alpha_{1,ik}u$, we obtain the model for multivariate network meta-analysis assuming Gompertz distributed survival times.

10.6.1.3   Log-Logistic and Log-Normal
If one wishes to apply network meta-analysis models with time-varying treatment effects in the context of log-logistic or log-normal distributed survival times, we can replace $\ln(h_{ik}(u))$ in equation (10.4) with the log failure odds $\ln((1 - S_{ik}(u))/S_{ik}(u))$ for a log-logistic model and with $\Phi^{-1}(S_{ik}(u))$ for the log-linear model, where $S_{ik}(u)$ is the survival proportion in study i for treatment k at time u.

10.6.1.4 Fractional Polynomial

If we replace the hazard function $\ln(h_{ik}(u)) = \alpha_{0,ik} + \alpha_{1,ik}\ln(u)$ in equation (10.4) with $\ln(h_{ik}(u)) = \alpha_{0,ik} + \alpha_{1,ik}u^p$ where $p = \{-2, -1, -0.5, 0, 0.5, 1, 2, 3\}$ and $u^0 = \ln(u)$, we obtain a first-order FP model with one scale and one shape parameter. (As mentioned before, with $p = 0$ we have a Weibull function and with $p = 1$ we have a Gompertz function.) For additional flexibility a second shape parameter can be added to obtain a second-order FP network meta-analysis model:

$$\ln\left(h_{ik}(u)\right) = \begin{cases} \alpha_{0,ik} + \alpha_{1,ik}u^{p_1} + \alpha_{2,ik}u^{p_2} & p_1 \neq p_2 \\ \alpha_{0,ik} + \alpha_{1,ik}u^p + \alpha_{2,ik}u^p \ln(u) & p = p_1 = p_2 \end{cases} \quad \text{with } u^0 = \ln(u)$$

$$\begin{pmatrix} \alpha_{0,ik} \\ \alpha_{1,ik} \\ \alpha_{2,ik} \end{pmatrix} = \begin{pmatrix} \mu_{0,i} \\ \mu_{1,i} \\ \mu_{2,i} \end{pmatrix} + \begin{pmatrix} \delta_{0,ik} \\ d_{1,1t_{ik}} - d_{1,1t_{i1}} \\ d_{2,1t_{ik}} - d_{2,1t_{i1}} \end{pmatrix} \tag{10.5}$$

$$\delta_{0,ik} \sim \text{Normal}\left(d_{0,1t_{ik}} - d_{0,1t_{i1}}, \sigma^2\right)$$

The relative treatment effect of the additional shape parameter α_2 is expressed as $d_{2,1t_{ik}} - d_{2,1t_{i1}}$. This model is a special case of the general FP model represented with equation (11.7) by assuming fixed treatment effects in terms of the shape parameters.

10.6.1.5 Splines

As an alternative to the FP, we can also use spline functions to describe the log hazard over time (Royston and Parmar, 2002; Royston and Lambert, 2011; Guyot et al., 2016). In order for the consistency assumption of the network meta-analysis to hold, the knots of the splines need to be in the same location for all arms of all trials when treatments effects are assumed to have an impact on the different segments of the spline functions (Vieira et al., 2013; Guyot, 2014).

10.6.2 Evaluation of Consistency

When one uses the proportion of patients alive, median survival or the constant HR as the effect measures of interest, the evaluation of consistency between direct and indirect evidence is in line with the approach described in Chapter 7. For a network meta-analysis using models with time-related treatment effects, the consistency between estimates based on direct and indirect evidence needs to be assessed for all treatment effect parameters, for example, both scale and shape (Jansen et al., 2015). In principle, the methods for evaluating consistency introduced for unidimensional treatment effects can also be applied to multidimensional effects. Relative treatment effect estimates in terms of scale and shape based on direct comparisons can be compared with the corresponding estimates based on indirect comparisons using node-splitting techniques. However, since the different dimensions are unlikely to be independent, additional work is needed to understand how this affects interpretation of inconsistency of treatment effect over time.

10.6.3 Meta-Regression

Network meta-analysis models can be extended to include treatment-by-covariate interactions in an attempt to improve consistency or to explore sources of heterogeneity, as extensively discussed in Chapter 8. The models presented in Section 10.6.1 have been extended with study-level covariates to explore treatment-by-covariate interactions regarding the scale parameters (Jansen and Cope, 2012) where it was assumed that treatment effects on shape parameters are unaffected by covariates, that is, the change in treatment effects over time is the same for the different levels of a covariate. Of course, there is no reason to believe that between-study heterogeneity only concerns the level of the treatment effect (the scale) and not the extent to which it changes over time. Further work is needed to develop meta-regression models where covariates interact with treatment effects in terms of both scale and shape parameters, although rich datasets (with long follow-up and low censoring rates) will be required to fit such models.

10.7 Data and Likelihood

10.7.1 Likelihood with Individual Patient Data

With IPD or reconstructed IPD available, such as those obtained with the algorithm by Guyot et al. (2012) (see Section 10.7.4), we can use the best fitting survival distribution to the data, for example, Weibull, Gompertz, lognormal, etc. The likelihood takes the form of the assumed distribution for uncensored individuals and the survival function at the censor time for censored individuals.

Subsequently, we set up a network meta-analysis model for the treatment effects acting on a single parameter (e.g. scale) or multiple parameters (e.g. scale and shape) of these distributions. In order for the assumptions of consistency of all treatment effects to hold, the likelihoods for the arms of the different trials included in the network need to be of the same family.

The most commonly used survival distributions presented in Table 10.1 have two parameters and offer a certain amount of variation in hazard functions, but may not be sufficient to capture all variation in the development of the hazard over time between treatments. The generalised gamma or F distributions capture many more different hazard functions over time, but it is not straightforward to implement these in the multivariate network meta-analysis framework where treatment effects act on several parameters.

If we wish to implement a likelihood representing the flexible FP or cubic splines in WinBUGS, we can use the 'zeros' trick (Lunn et al., 2013) where a dataset comprising entirely of zeros is given a Poisson distribution with its parameter defined equal to the negative log-likelihood (plus a sufficiently large constant). The log-likelihood function corresponding to the FP or spline is then written algebraically in the WinBUGS code (Lunn et al., 2013).

10.7.2 Discrete or Piecewise Constant Hazards as Approximate Likelihood

Rather than using a likelihood corresponding to the distribution of individual survival times in each arm of each trial, an alternative approach is to write the likelihood for incident events over multiple discrete time periods for each arm of each trial (Jansen and Cope, 2012). The form of the hazard function is incorporated in the model statements rather than directly in the likelihood statement. This approach avoids writing bespoke code to capture a likelihood that is not part of the standard list of distributions in WinBUGS. Another advantage of this approach with discrete hazards is that the computational time using WinBUGS is reduced.

Survival curves can be divided into M consecutive intervals over the follow-up period: $[u_1, u_2], (u_2, u_3], \ldots, (u_M, u_{M+1}]$ with $u_1 = 0$. For each time interval $m = 1, 2, 3, \ldots, M$, a binomial likelihood for the incident events can be described according to

$$r_{ikm} \sim \text{Binomial}\left(p_{ikm}, n_{ikm}\right) \tag{10.6}$$

where r_{ikm} is the observed number of events in the mth interval $(u_m, u_{m+1}]$ for treatment k in study i, n_{ikm} is the number of subjects at risk at the start of the mth interval adjusted for the subjects censored in the interval (see Section 10.7.5) and p_{ikm} is the corresponding underlying event probability or discrete hazard. When the time intervals are relatively short, the hazard rate h_{ikm} is assumed to be constant for any time point within the corresponding mth time interval. The hazard rate corresponding to the mth interval can be standardised by the unit of time (to overcome differences in length of time, e.g. months, from one interval to the next) and used for the analysis according to

$$h_{ikm} = \frac{-\ln\left(1 - p_{ikm}\right)}{\Delta u_{ikm}}$$

so

$$p_{ikm} = 1 - e^{-\Delta u_{ikm} h_{ikm}} \tag{10.7}$$

where Δu_{ikm} is the length of the interval. Note that this approach assumes piecewise constant hazards on each interval, regardless of the length of the interval, and is equivalent to using the cloglog link for each interval separately (see also Chapter 4). The development of the hazard h_{ikm} over time is defined by the survival model that is being fitted, for example, equation (10.4) or (10.5), evaluated at the time point at the end of each interval, u_{m+1}.

In order to use a likelihood based on discrete or piecewise constant hazards over time, data are required either in the Kaplan–Meier format, which in the absence of IPD can be obtained using the algorithm by Guyot et al. (2012) (see Section 10.7.4 and Appendix C), or as interval data using an algorithm proposed by Jansen and Cope (2012) (see Section 10.7.5 and Appendix C).

10.7.3 Conditional Survival Probabilities as Approximate Likelihood

If one wishes to perform network meta-analysis with time-varying treatment effects in the context of log-logistic or log-normal models (see Section 10.6.1.3), a similar approach can be taken as in the previous section. The incident events in the mth interval $(u_m, u_{m+1}]$ are assumed to have a binomial likelihood (equation (10.6)), but with probability of events

$$p_{ikm} = 1 - \frac{S\left(u_{ikm+1}\right)}{S\left(u_{ikm}\right)}$$

rather than equation (10.7). The network meta-analysis model may be put on either the failure odds $(1 - S(u_{ikm}))/S(u_{ikm})$ or probit, $\Phi^{-1}(S(u_{ikm}))$, at a given time point u_{ikm}.

10.7.4 Reconstructing Kaplan–Meier Data

Ideally we would like to have IPD for all trials included in a network meta-analysis. However, the reality is that for most, if not all, trials, there is no access to IPD and the synthesis is based on reported trial findings. Typically trials with survival outcomes (e.g. PFS and OS) as one of the main endpoints of interest do report Kaplan–Meier curves for the treatment groups, and data can be extracted from these to allow a variety of survival network meta-analysis models to be fitted.

Survival proportions over time can be obtained by digitally scanning the reported Kaplan–Meier curves. A variety of software is available to do this. The Kaplan–Meier curves are read into the software by defining the axes and clicking multiple time points of the curves. In principle, every step of the Kaplan–Meier curve should be captured, and it is important that the data points are captured accurately. Once extracted, the survival proportions may need some small adjustments to ensure that for increasing time points the survival proportions are the same or lower than at previous time points. Once the survival proportions have been recorded, an algorithm can be used to create a dataset that includes information about the population at risk over time, the occurrence of the events over time and censoring times.

Guyot et al. (2012) developed an algorithm to construct a dataset that produces a Kaplan–Meier curve that approximates the published curve. The method uses the survival proportions extracted from the published curves, together with reported numbers at risk under the survival curve, and, if reported, the total number of events and total number of individuals censored. A key feature of this approach is the use of iterative numerical methods to solve the inverted Kaplan–Meier equations, which is necessary to estimate the number of censored individuals on each interval, defined by the reported numbers at risk under the curves. The method assumes constant censoring within each interval, but censoring rates may differ between intervals. Full details of the method are given in Appendix C. Guyot et al. (2012) found, for the examples that they looked at, that their method gives a high level of accuracy for survival proportions and medians and a reasonable degree of accuracy for HRs as long as numbers at risk under the curve or total number of events are reported. The accuracy of the method relies on the quality of the published curve, and care taken in the 'clicking procedure'; however reproducibility between different 'clickers' was found to be high (Guyot et al., 2012). The dataset created with this algorithm allows the data to be analysed as if one had IPD, that is, time points of event or censoring for all subjects in the trial (Section 10.2.1). Note that this algorithm does not provide patient-level data for the covariates, and therefore the recreation of IPD survival times does not mean that we can perform subgroup analyses according to different values of patient-level covariates.

10.7.5 Constructing Interval Data

An approach to generate data when using likelihoods for discrete or piecewise constant hazards (see Sections 10.7.2 and 10.7.3) was presented by Jansen and Cope (2012). Their algorithm also makes use of the survival proportions extracted from the Kaplan–Meier curves and information on numbers at risk reported below the graphs whenever available. It provides a dataset with numbers at risk at the beginning of multiple time intervals adjusted for censoring during the interval and number of events that have occurred during the interval. Full details of the method are given in Appendix C.

10.8 Model Choice

In order to use a survival function for the network meta-analysis model that is appropriate given the data at hand, it is recommended to limit the possible competing options by first visually inspecting the development of the hazard over time for the different trials to be included in the analysis. Log-cumulative hazard plots versus log-time or time can be constructed for the trials of interest to inspect whether the hazard functions follow a Weibull distribution, exponential distribution or Gompertz distribution. Alternative plots can be used to assess the relevance of log-normal and log-logistic distributions (Latimer, 2011). Next, the fit of competing network meta-analysis models representing different survival functions over time, including the FP, can be compared using DIC (Chapter 3). This is more straightforward when a binomial likelihood is used for discrete hazards (and the development of hazards over time is captured at the 'model level'; see Section 10.7.2) than with survival distribution specific likelihoods (Section 10.7.1).

The second-order FP may pick up extreme (chance) fluctuations in the observed hazard towards the end of follow-up when the sample size of the at-risk population becomes small. This may result in relative treatment effects that, when extrapolated beyond the available time point, are unrealistic. When extrapolation is of interest, rather than solely relying on model fit criteria, it is recommended to also assess whether the estimated relative treatment effects towards the end of follow-up seem reasonable.

10.9 Presentation of Results

The relative treatment effect parameter estimates in terms of scale and shape obtained with a model assuming survival times follow a Weibull distribution can be transformed into the HR of treatment k relative to c at time u according to

$$\exp\left(d_{0,1k} - d_{0,1c} + \left(d_{1,1k} - d_{1,1c}\right)\ln\left(u\right)\right) \tag{10.8}$$

For a Gompertz model, $\ln(u)$ is replaced with u. For the second-order FP model, parameters related to d_2 need to be incorporated as well. The obtained time-varying HRs along with measures of uncertainty can be graphically plotted for the comparisons of interest. In addition, it may be useful to present the HRs of all contrasts in the network for a few key time points in a cross table.

One can translate the obtained HR functions over time into survival functions by treatment (using the hazard function of one treatment in the network as an 'anchor' or other external evidence, as discussed in Chapter 5) to help facilitate interpretation. This is also useful to inspect whether the tails make sense from a clinical or epidemiological perspective and help guide selection of the model for the relative treatment effect. This is pertinent for cost-effectiveness analysis where extrapolation of OS is frequently needed (see Chapter 5 and Section 10.11).

Rankograms for a single effect measure in the context of survival analysis, such as median survival, proportion alive at a specific time point or constant HR, can be easily presented (see Figure 2.6). A disadvantage of rank probabilities based on these one-dimensional effect measures is that they may be sensitive to the choice of effect measure. Given the use of network meta-analysis models where we try to capture the complete survival distribution allowing for the possibility of time-varying treatment effects, we may want to use rank probabilities based on the complete survival distribution or distribution of treatment effects over time (Cope and Jansen, 2013). Rank probabilities based on survival proportions over time reflect the cumulative effect of treatments over time and may be considered the most intuitive; however a baseline survival curve is needed to transform the relative treatment effects obtained with the network meta-analysis into survival curves for all competing interventions in the network. Rank probabilities based on the HR over time provide information of the treatment effect at each time point ignoring effects at previous time points, which result in over-interpreting the findings near the tails of the survival function. Rank probabilities based on the mean survival (i.e. area under the curve) at each time point gives more weight to the treatment effects when a greater proportion of patients are still alive.

10.10   Illustrative Example

By means of a network meta-analysis, we aim to compare the efficacy of be melphalan–prednisolone (MP), melphalan–prednisone–bortezomib (MPV), melphalan–prednisone–thalidomide (MPT) and cyclophosphamide–thalidomide–dexamethasone attenuated (CTDa) regimen for the first-line treatment of multiple myeloma in patients not eligible for high-dose chemotherapy and stem cell transplantation (HDT-SCT) based on RCT evidence identified with a systematic literature search. The outcome of interest was OS. The evidence base consisted of eight relevant RCT (for details see Appendix D). The network is presented in Figure 10.2.

The Kaplan–Meier curves for OS with all of the interventions in the eight trials were digitised and a dataset for interval data was created as outlined in Appendix C. We considered the following competing models for the network meta-analysis: fixed and random effects Weibull models, fixed effects and random effects Gompertz models and fixed and random effects second-order FP with power $p_1 = 0$, 1 and power $p_2 = 1$, all with the binomial likelihood for the discrete hazard over time (see Section 10.7.2). The corresponding WinBUGS code along with data and initial values is provided in *Ch10_FE_1st_order_model.odc, Ch10_RE_1st_order_model.odc, Ch10_FE_2nd_order_model.odc* and *Ch10_RE_2nd_order_model.odc*. An example of the implementation of the random effects second-FP model in WinBUGS is shown as follows. The syntax builds upon the WinBUGS code introduced in Chapter 4.

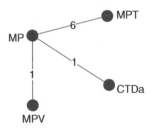

Figure 10.2 Evidence network of randomised controlled trials comparing melphalan–prednisone–bortezomib (MPV), melphalan–prednisone–thalidomide (MPT) and cyclophosphamide–thalidomide–dexamethasone attenuated (CTDa) regimen for the first-line treatment of multiple myeloma in patients not eligible for HDT-SCT.

WinBUGS code for random effects second order fractional polynomial model

```
model{
for (j in 1:N) {                          # *** PROGRAM STARTS
                                          # LOOP THROUGH EVENTS

# time in months transformed according to powers P1 and P2
timen[j]<-(time[j]
timen1[j]<-(equals(P1,0)*log(timen[j])+(1-equals(P1,0))*pow(timen[j],P1) )
timen2[j]<-((1-equals(P2,P1))*(equals(P2,0)*log(timen[j])+(1-equals(P2,0))*pow(timen[j],P2))
+equals(P2,P1)*(equals(P2,0)*log(timen[j])+(1-equals(P2,0))*pow(timen[j],P2)
*log(timen[j])))

r[j]~dbin(p[j], z[j])                     # likelihood according to eq. (10.6)
p[j]<-1-exp(-h[j]*dt[j])     # hazard rate in each interval standardized by
                                          unit of time according eq.(10.7)

#Random effects model
log(h[j])<-Alpha[s[j],a[j],1]+Alpha[s[j],a[j],2]*timen1[j]+Alpha[s[j],a[j],3]*time
                              n2[j]  # hazard over time according to FP
}

for (i in 1:ns) {                         # LOOP THROUGH STUDIES
w[i,1]<-0               # adjustment for multi-arm trials is
                                          zero for control arm

delta[i,1]<-0           # treatment effect is zero for control arm

for (k in 1:na[i]) {                      # LOOP THROUGH ARMS
Alpha[i,k,1]<-mu[i,1]+delta[i,k]     # model for linear predictor of alpha_0
Alpha[i,k,2]<-mu[i,2]+d[t[i,k],2]-d[t[i,1],2]   # model for linear predictor of alpha_1
Alpha[i,k,3]<-mu[i,3]+d[t[i,k],3]-d[t[i,1],3]   # model for linear predictor of alpha_2
  }
```

```
for (k in 2:na[i]) {                                    # LOOP THROUGH ARMS
    delta[i,k] ~dnorm(md[i,k],taud[i,k])                # trial-specific RE distributions
                                                        # for alpha_0 tx effects
    md[i,k]<-d[t[i,k],1]-d[t[i,1],1] +sw[i,k]           # mean of distributions (with
                                                        # multi-arm trial correction)
    w[i,k] <- (delta[i,k] - d[t[i,k],1] + d[t[i,1],1])  # adjustment for multi-arm RCTs
    sw[i,k] <- sum(w[i,1:k-1])/(k-1)                    # cumulative adjustment for multi-arm
                                                        # trials
    taud[i,k]  <- tau *2*(k-1)/k                        # precision of distributions (with
                                                        # multi-arm trial correction)
}
}

#priors
for (i in 1:ns) {                                       # LOOP THROUGH STUDIES
    mu[i,1:3] ~ dmnorm(mean[1:3],prec[,])               # vague priors for all trial baselines
}
d[1,1]<-0                     # alpha_0 treatment effect is zero for reference treatment
d[1,2]<-0                     # alpha_1 treatment effect is zero for reference treatment
d[1,3]<-0                     # alpha_2 treatment effect is zero for reference treatment
for (k in 2:nt){                                        # LOOP THROUGH TREATMENTS
    d[k,1:3] ~ dmnorm(mean[1:3],prec[,])                # vague priors for treatment effects
}
sd~dunif(0,2)                                           # vague prior for between-trial SD
tau<-1/(sd*sd)                                          # between-trial precision = (1/between-
                                                        # trial variance)
                                                        # *** PROGRAM ENDS
}
```

The dataset has three components and all need to be loaded for the model to run. First, a list specifying the transformation of time given the FP of choice (powers P1 and P2), the number of studies (ns; in the example ns = 8), the total number of event intervals across all arms of all trials (N; in the example N = 443) and the mean and precision of the required prior distributions need to be defined.

The second component are data in column format where the treatments in each study are defined (t [,1], t [,2], etc.) along with the number of treatment arms na [] per study, similar to the data set-up used in Chapter 4.

```
t[,1]  t[,2]  na[]    # Study label
1      2      2       # Facon 2007
1      2      2       # Hulin 2009
1      2      2       # Palumbo 2008
1      2      2       # Wijermans 2010
1      2      2       # Waage 2010
1      2      2       # Beksac 2010
1      3      2       # Intensive
1      4      2       # Mateos 2010/San Miguel 2008
END
```

Note: For three-arm trials an extra column t [, 3] needs to be added.

The third component of the data to load is in column format where for each study s [] and treatment arm a [], the number of events r [] out of number of patients at risk z [] for each time interval with length dt [] ending at time [] (in months) are defined. An extract of the third component of the data used in this example is presented as follows:

```
s[]   r[]    z[]    a[]    time[]  dt[]
1     12     196    1      2       2
1     9      184    1      4       2
1     5      175    1      6       2
1     6      170    1      8       2
:
:

1     1      11     1      66      2
1     1      10     1      72      6
1     5      125    2      2       2
1     1      120    2      4       2
1     2      119    2      6       2
1     2      117    2      8       2
1     1      114    2      10      2
:
:
```

```
1        0        10       2        64       2
1        1        10       2        66       2
1        0        9        2        72       6
2        5        116      1        2        2
2        5        111      1        4        2
:
:
8        4        54       2        45       3
8        2        23       2        48       3
END
```

Each row represents an event interval. The values in a [] indicate the treatment arm in each study that the data corresponds to, not the actual treatment compared in that arm. For example, a [] =1 corresponds to t [, 1] and a [] =2 corresponds to t [, 2]. Thus, the treatment arm number a [] restarts at 1 for each new study.

In Table 10.2 the model fit statistics for the different models are presented. There were no appreciable differences between the fixed and random effects versions of the different models. The second-order FP provided better fit to the data than the models where survival times were assumed to follow a Weibull or Gompertz distribution.

Out of all the models evaluated, the fixed effects second-order FP with $p_1 = 0$ and $p_2 = 1$ was considered most appropriate. In Table 10.3 the basic parameters of the relative treatment effects in terms of scale and shape of the log hazard function are presented for this model.

Based on the basic parameters in Table 10.3, functional parameter estimates in terms of scale and shape can be obtained for any treatment

Table 10.2 Model fit statistics for the different competing network meta-analysis models for the multiple myeloma example.

Model	Dbar	p_D	DIC
Fixed effects Weibull	1711.1	21.9	1733.0
Random effects Weibull	1707.9	25.1	1733.0
Fixed effects Gompertz	1694.1	21.9	1716.0
Random effects Gompertz	1690.0	25.0	1715.0
Fixed effects second-order FP with $p_1 = 0$ and $p_2 = 1$	1669.3	32.7	1702.0
Random effects second-order FP with $p_1 = 0$ and $p_2 = 1$	1665.4	36.6	1702.0
Fixed effects second-order FP with $p_1 = 1$ and $p_2 = 1$	1672.0	33.0	1705.0
Random effects second-order FP with $p_1 = 1$ and $p_2 = 1$	1668.1	35.9	1704.0

Table 10.3 Model parameter estimates representing multidimensional treatment effects of each intervention (MPT, CTDa, MPV) relative to the baseline treatment (MP) as obtained with fixed effects second-order fractional polynomial model with $p_1 = 0$ and $p_2 = 1$.

Treatment	$d_{0,1k}$	$d_{1,1k}$	$d_{2,1k}$
1 (MP)	Reference	Reference	Reference
2 (MPT)	0.330 (−0.246, 0.914)	−0.229 (−0.576, 0.120)	0.004 (−0.017, 0.025)
3 (CTDa)	0.403 (−0.799, 1.645)	−0.177 (−0.976, 0.601)	0.003 (−0.049, 0.055)
4 (MPV)	0.383 (−0.501, 1.278)	−0.638 (−1.232, −0.032)	0.045 (0.001, 0.090)

contrast in the network and the corresponding time-varying HRs according to equation (10.8). The following syntax needs to be added to the code to create this output.

Additional WinBUGS code to create time-varying hazard ratios based on treatment effect parameters obtained with second order fractional polynomial model

```
#Output
for (m in 1:maxt){     # create time points for output;
                       maxt reflects maximum time point
  time1[m] <- (equals(P1,0)*log(m) + (1-equals(P1,0))
  *pow(m,P1)     )
  time2[m] <- ((1-equals(P2,P1))*(equals(P2,0)*log(m)
  +(1-equals(P2,0))*pow(m,P2)) +
  equals(P2,P1)*(equals(P2,0)*log(m)*log(m)+
  (1-equals(P2,0))*pow(m,P2) *log(m)))
  }

#Hazard ratios over time for all possible contrasts
for (c in 1:(nt-1)){
  for (k in (c+1):nt){
    for (m in 1:maxt){
      log(HR[c,k,m]) <- (d[k,1]-d[c,1])+(d[k,2]-d[c,2])
      *time1[m]+(d[k,3]-d[c,3])*time2[m]
      }
    }
  }
```

In Figure 10.3 the corresponding HRs of MPV, CTDa and MPT relative to MP over time are presented. Between about 7 and 34 months of follow-up,

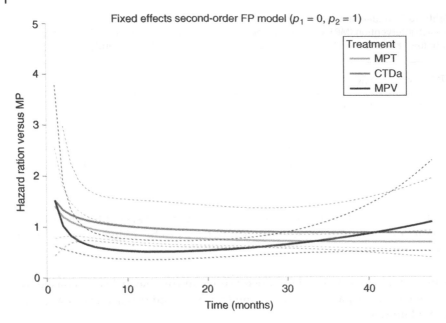

Figure 10.3 Relative treatment effect estimates of each intervention (MPT, CTDa, MPV) versus reference treatment (MP) expressed as hazard ratios as obtained with fixed effects second-order FP network meta-analysis model with $p_1 = 0$ and $p_2 = 1$. 95% credible intervals indicated by broken lines.

95% credible intervals of the HRs for MPV relative to MP (as represented with the dashed black lines) exclude 1 (no effect) for all four models, indicating that MPV is more efficacious than MP. MPT is more efficacious than MP after about 15 months. The 95% credible intervals of the HRs with CTDa relative to MP do include 1, consistent with no evidence that CDTa differs from MP.

In Figure 10.4 the HRs with MPV relative to the other interventions are provided. There are no meaningful differences between MPV and MPT; the 95% credible intervals for the estimates of these indirect comparisons include 1, with the exception of the HRs obtained with the second-order FP model with $p_1 = 0$ and $p_2 = 1$ between about 10 and 20 months.

In Figure 10.5 the modelled OS curves for each intervention are provided, which are based on the time-varying HRs of each intervention relative to MP as obtained with the different network meta-analysis models, and subsequently applied to a parametric OS reference curve with MP, obtained from the MPV versus MP study. The following syntax needs to be added to the code to create this output.

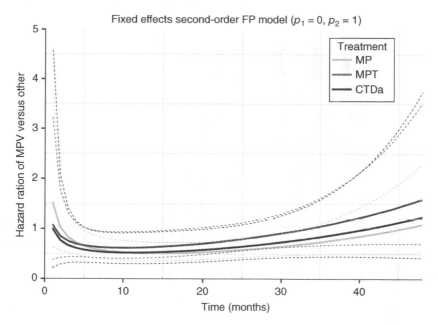

Figure 10.4 Relative treatment effect estimates of MPV versus MPT, CTDa and MP expressed as hazard ratios as obtained with second-order FP network meta-analysis model with $p_1=0$ and $p_2=1$. 95% credible intervals indicated by broken lines.

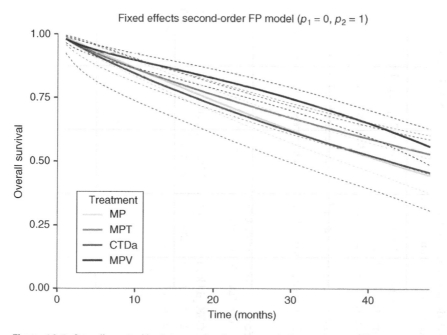

Figure 10.5 Overall survival by intervention based on relative treatment effect estimates of each intervention (MPT, CTDa, MPV) versus reference treatment (MP) as obtained with second-order FP network meta-analysis model with $p_1=0$ and $p_2=1$ and applied to OS curve with MP from study 8. 95% credible intervals indicated by broken lines.

Additional WinBUGS code to create survival curves based treatment effect parameters obtained with second order fractional polynomial model

```
# Provide estimates of survival probabilities over time by treatment
for (k in 1:nt) {
    alpha0[k]<-mu[study of interest,1]+d[k,1]       # alpha_0 by treatment using baseline from
                                                      study of interest
    alpha1[k]<-mu[study of interest,2]+d[k,2]       # alpha_1 by treatment using baseline from
                                                      study of interest
    alpha2[k]<-mu[study of interest,3]+d[k,3]       # alpha_2 by treatment using baseline from
                                                      study of interest

for (m in 1:maxt) {
    log(HAZARD[k,m])<-alpha0[k]+alpha1[k]*time1[m]+alpha2[k]*time2[m]   #hazard over time
                                                                         by treatment

    CUM_H[k,m]<-sum(HAZARD[k,1:m])       # cumulative hazard over time by treatment
    T[k,m]<-1-exp(-CUM_H[k,m])           # mortality over time by treatment
    S[k,m]<-1-T[k,m]                     # survival over time by treatment
                }
        }
```

10.11 Network Meta-Analysis of Survival Outcomes for Cost-Effectiveness Evaluations

For cost-effectiveness evaluations of competing interventions that aim to improve survival, differences in expected survival or expected QALYs between the competing interventions are of interest. When the evidence base consists of multiple trials, it is common practice to assume a certain parametric survival function for the baseline intervention and apply the treatment-specific constant HRs estimated from a (network) meta-analysis to obtain corresponding survival functions for the other interventions of interest, enabling comparisons of expected survival and QALYs. The PH assumption implies that only the scale of the corresponding parametric hazard functions is affected by treatment, and accordingly all the competing interventions have the same shape. Since the tail of the survival function has a great impact on expected survival, the PH assumption may lead to bias in differences in expected survival and QALYs.

A network meta-analysis model for survival outcomes using multidimensional relative treatment effects in combination with a multidimensional (meta-analysis) model for the outcome with the reference treatment will provide 'pooled' survival curves for the competing interventions of interest and can form the 'backbone' of a cost-effectiveness model (see Chapter 5). This approach with time-varying relative treatment and baseline effects allows survival functions for all the interventions to be more closely fitted to the available data.

In the case of a relatively short and censored follow-up, there is considerable uncertainty as to the outcomes with the reference treatment and relative treatment effects for the time period beyond the available evidence. Frequently, one of the following approaches is taken for the relative treatment effects:

1) The assumption is made that beyond the available data there are no differences in effect between the compared interventions and the HR is set equal to 1.
2) The relative treatment effect estimates are extrapolated.

When constant HRs were estimated based on the available data, an extrapolation strategy implies that these remain constant at the same level for the time period beyond the available evidence. In the context of time-varying HRs estimated with the multidimensional network meta-analysis models, extrapolation implies obtaining HRs at time points beyond the available follow-up according to the estimated scale- and shape-related treatment effect parameters (see equation (10.8)).

With limited follow-up data, favouring a constant HR network meta-analysis model over a multidimensional model on the basis that we cannot obtain stable parameter estimates with the latter effectively replaces parameter uncertainty with structural uncertainty. Structural uncertainty is typically ignored in the uncertainty of the cost-effectiveness estimates. However, even with network meta-analysis models with a multidimensional treatment effect, there is still a danger of understating the uncertainty in extrapolating the curves in case of censored follow-up because the choice of the models is still based on model fit criteria. We would therefore recommend searching for external evidence (e.g. long-term registries or cohorts) together with clinical input to put some constraints on the long-term predictions from the multidimensional models in order to obtain mean survival benefits that are credible (Guyot et al., 2016).

10.12 Summary and Further Reading

Survival times can be summarised with the density function, survival function and hazard function. Network meta-analysis of reported proportions alive at a certain time point of interest, median survival and reported HR can be performed using the core network meta-analysis model with an appropriate likelihood and link function as presented in Chapter 4. If IPD or recreated data from published Kaplan–Meier curves are available, network meta-analyses using all the information over the available follow-up can be performed. Network meta-analysis models with multidimensional treatment effects do not rely on the strong assumption of constant treatment effects over time and may provide estimates that are a more accurate reflection of the available evidence. In combination with a baseline survival function for a reference treatment, these multidimensional network meta-analysis models embedded in parametric survival functions can form the basis for cost-effectiveness models.

For general texts on survival analysis, we refer to relevant chapters in Collett (2003) and Cox and Oakes (1984). An excellent introduction to the topic of survival analysis is provided by Kleinbaum and Klein (2012).

The benefits of extracting data from published survival curves for network meta-analysis of time-to-event data have been emphasised in this chapter. In addition to the algorithm by Guyot et al. (2012) and the approximate method by Jansen and Cope (2012), other approaches have been proposed in the literature (Dear, 1994; Parmar et al., 1998; Arends et al., 2008; Fiocco et al., 2009; Hoyle and Henley, 2011). All these approaches use the Kaplan–Meier relationships to reconstruct the data used to produce the curves, but the methods have been refined over time; in particular the development of digitising software has meant that very close approximations to the IPD are now possible using the method of Guyot et al. (2012).

In this chapter we only touched briefly upon the concept of extrapolation of censored survival data, which is needed for cost-effectiveness analysis to obtain estimates of expected QALYs and costs. Relevant papers on the topic of extrapolation of time-to-event data have been provided by Latimer (2013), Grieve et al. (2013), Demiris and Sharples (2006) and Demiris et al. (2015).

In clinical trials of cancer treatments, switching of the control group to the new intervention group is frequently allowed after disease progression. As such, an intention-to-treat analysis results in biased estimates of the relative treatment effects for OS. Different methods, based on IPD, have been proposed to adjust the relative treatment effect in the presence of treatment switching. Latimer and Abrams (2014) provide an excellent overview of the different methods including rank-preserving structural failure time models, the iterative parameter estimation algorithm and the inverse probability of censoring weighting method and explain their assumptions and limitations (Latimer and Abrams, 2014). When Kaplan–Meier curves adjusted for treatment switching are available for studies of interest, these can obviously be used in the network meta-analysis. However, given the limitations of the adjustment methods, a network meta-analysis based on the unadjusted Kaplan–Meier curves is recommended as well.

We conclude with some brief thoughts about future research topics of interest. Very commonly studies report data on both PFS and OS as both are important for decision making. Separate meta-analyses of OS and of PFS data ignore the relationship between these outcomes. Early work to develop a method for the joint meta-analysis of OS and PFS that is based on a tri-state (pre-progression, progression and death) transition model where time-varying hazard rates and relative treatment effects are modelled with known parametric survival functions or FP suggests that this would be a very powerful tool for evidence synthesis and cost-effectiveness evaluations (Jansen and Trikalinos, 2013). Such a model provides a coherent framework to evaluate whether transition rates between pre-progression and progression, progression and death and pre-progression and death are time varying or constant over time and which of these are affected by treatment. Given the structure of the model and therefore the simultaneous analysis of PFS and OS, the studies with only PFS data or only OS data still contribute indirectly to the other endpoint as well. It also facilitates the inclusion of external evidence for post-progression in a straightforward manner.

It is standard practice for papers to provide Kaplan–Meier curves for PFS and OS for the overall study population of a trial along with corresponding constant HRs, while forest plots often provide HRs for relevant subgroups (according to predefined levels of effect modifiers) without subgroup specific Kaplan–Meier curves. With the aim of performing a network meta-analysis for a specific population ('the target population'), the relevant evidence base may consist of studies where the overall study population 'matches' this target population for some interventions, but for other relevant interventions, the available studies are limited to those where a subgroup of the overall study population 'matches' the target population. Rather than ignoring these latter studies and thereby potentially limiting the feasible indirect comparison between relevant competing interventions of interest, we can in principle use the reported Kaplan–Meier curves of the overall study population for these studies to estimate treatment effects in terms of scale and shape and use the difference between the reported constant HR for the study-specific subgroup and the reported constant HR for the overall study population to adjust (or shift) the treatment effects in terms of scale and obtain a time-varying HR function representative for the subgroup of that trial (and therefore the target population of interest). With such an approach we rely on the same assumption as a meta-regression analysis where the patient characteristic only affects the treatment effects in terms of scale, but not shape. For a proper implementation we need to develop synthesis models that capture both the overall population and relevant subgroup results reported per study.

Other topics of interest for future research include advancing the methods for evaluation of consistency with multidimensional treatment effects (see Section 10.6.2) and efficient approaches for meta-regression analysis with covariates impacting treatment effects in terms of scale, shape or both (see Section 10.6.3).

10.13   Exercises

10.1   The Weibull and Gompertz model in Section 10.10 assumes treatment effects in terms of scale and shape. Modify the random effects network meta-analysis Weibull model (*Ch10_RE_1st_order_model.odc*) with treatment effects on both scale and shape to obtain a random effects PH model assuming Weibull distributed survival times where treatment effects impact the scale only.

10.2   The treatment effects regarding the (second) shape parameters may be unstable. How can we adapt the model to obtain more stable estimates?

10.3   What are the implications of removing non-significant relative treatment effects regarding shape parameters from the multidimensional network meta-analysis model?

11

*Multiple Outcomes

11.1 Introduction

In analyses of which treatment is most effective or most cost-effective, there are, more often than not, several outcomes to consider. Some of these evidence structures, those relating to ordered categories and competing risks, have already been touched on in Chapter 4. In this chapter we turn to structures where the relationships between the outcomes are more complex. These might be multiple observations taken at different follow-up times, different outcomes measured at the same time or simply different endpoints such as stroke, bleed and death. Almost invariably we find that while many trials report more than one outcome, each trial reports a different set of outcomes. Evidence structures are therefore always incomplete and sometimes very sparse.

Separate analysis of each outcome is certainly an option and may serve as a useful preliminary to some of the more sophisticated analyses described in this chapter. But a combined analysis of multiple outcomes is likely to be preferable for several reasons. Firstly, the outcomes are likely to be correlated, both within and between trials, so that separate analyses would involve a degree of double counting. In addition, if several outcomes are used in a cost-effectiveness analysis or in a multi-criterion decision analysis, it is essential that correlations are propagated into the decision model.

Secondly, it may produce a more richly connected network that includes comparisons between treatments that would not otherwise have been made. Inevitably, this requires assumptions being made about the *relationships* between treatment effects on different outcomes. This, in turn, leads to a need to ensure that the relationships reflect clinical opinion and, where sufficient data exist, that they are supported by the evidence.

In many of the examples we develop later in this chapter, these structural relationships between outcomes allow us to incorporate all the evidence within a coherent, unifying, mathematical framework. This not only leads to more

Network Meta-Analysis for Decision-Making, First Edition. Sofia Dias, A. E. Ades, Nicky J. Welton, Jeroen P. Jansen and Alexander J. Sutton.
© 2018 John Wiley & Sons Ltd. Published 2018 by John Wiley & Sons Ltd.
Companion website: www.wiley.com/go/dias/networkmeta-analysis

robust conclusions, based on a larger body of evidence, but also rules out the arbitrary selection of outcomes and the inefficient discarding of what, in some cases, may be the majority of the data.

In Section 11.2 we begin by reviewing multivariate normal (MVN) random effects (RE) meta-analysis, an increasingly popular method for multi-outcome synthesis (Nam et al., 2003; Riley et al., 2007a; Jackson et al., 2011; Bujkiewicz et al., 2013; Mavridis and Salanti, 2013; Wei and Higgins 2013b). We then briefly reconsider (Section 11.3) some of the multivariate likelihood structures already considered in Chapter 4 and suggest some simple extensions of univariate methods that may be preferable to more sophisticated methods in many circumstances. The main material of the chapter is presented in Sections 11.4–11.8. Each of these sections is a case study looking at a different multi-outcome evidence structure, exemplifying different clinical and logical relationships between outcomes.

11.2 Multivariate Random Effects Meta-Analysis

Multivariate RE meta-analysis allows for correlations between treatment effects at two levels: the within-trial or likelihood level and the between-trial level. Setting out the MVNRE model in vector notation clarifies its relation to standard univariate meta-analysis. Here we consider the two-treatment case for a set of continuous treatment differences $\mathbf{Y}$:

$$\mathbf{Y}_i \sim MVN\left(\boldsymbol{\delta}_i, \boldsymbol{\Sigma}_i^W\right)$$
$$\boldsymbol{\delta}_i \sim MVN\left(\mathbf{d}, \boldsymbol{\Sigma}^B\right) \tag{11.1}$$

At the first stage the observed treatment differences in trial i are distributed with the mean $\boldsymbol{\delta}_i$ and a *within-trial* variance–covariance (VCV) matrix $\boldsymbol{\Sigma}_i^W$ determined by the variances of the observed data and the within-trial correlations. At the second stage, the trial-specific treatment effects $\boldsymbol{\delta}_i$ have RE mean $\mathbf{d}$ and a between-trial VCV matrix $\boldsymbol{\Sigma}^B$. The between-trial correlations are between the trial-specific *mean* treatment effects, so the model captures the idea that trials in which treatment effects on one outcome tend to be large may have large (positive correlation), or small (negative correlation), treatment effects on another outcome. The model can of course be extended to a network of treatments (Schmid et al., 2014; Efthimiou et al., 2014) and can be applied to continuous data or to binomial or Poisson data if normal approximations to the likelihood are employed (Trikalinos and Olkin, 2008, 2012).

All multi-outcome models must start by considering the correlations between mean treatment effects at the first (likelihood) stage. A valuable paper by Wei and Higgins (2013a) shows how these can be derived from the joint distribution of variables within trials. Unfortunately, trial investigators seldom report correlations between variables, which creates a problem for any multi-outcome analysis, not just MVNRE models. This can usually be remediated by the use of external data on correlations accompanied by sensitivity analyses (Lu et al., 2014; Ades et al., 2015; Jansen et al., 2015). In some cases correlations are also partly the result of the structural relationship between variables. For example, time to tumour progression and time to death are necessarily correlated within trials because tumour progression cannot follow death, and the closer in time the two outcomes are, the higher the degree of positive within-trial correlation between them.

Experience in the use of MVNRE models is increasing rapidly and it is becoming clear that the second stage formulation has a number of limitations. These derive from their 'weakness' as models: the only claim they make about the relationship between the mean treatment effects on different outcomes is that there is a correlation somewhere between -1 and 1. At the same time, the number of parameters requiring estimation in an L-outcome dataset is L variances and $L(L-1)/2$ correlations. The number of correlations therefore increases rapidly to near-unmanageable levels: 3 correlations with 3 outcomes, 15 with 6 outcomes, 45 with 9 and so on. This, of course, applies to a two-treatment synthesis: for a S-treatment synthesis, larger matrices and further parameters are required, although a range of simplifications are possible (Schmid et al., 2014; Efthimiou et al., 2014). Bearing in mind the extremely sparse data usually available, there is often an almost complete lack of data to identify these parameters. One should also mention the technical difficulty in implementing the constraints required to ensure the matrix Σ^B is a feasible VCV matrix, namely, that it is positive and semi-definite. Wei and Higgins (2013b) illustrate the main options.

Besides this, investigation of MVNRE models has demonstrated that the extent to which they 'borrow strength' across outcomes is often very modest (Jackson et al., 2011), and under some circumstances it appears they may actually 'borrow weakness' (Bujkiewicz et al., 2013). In addition, there is no borrowing of strength at all in datasets where every trial reports on every outcome (Riley et al., 2007a). This is simply because the existence of a correlation between outcomes has no bearing at all on the marginal distribution of the mean treatment effect, only on their joint distribution. But the joint distribution often has no role in decision-making. For example, in the form of HTA practised by NICE, the clinical benefit of treatment may be represented by a single, preference-based generic measure of health-related quality of life (HRQoL), such as the EuroQol 5D (EQ-5D) (Rabin and Charro, 2001). Only the marginal distribution of the treatment effect on EQ-5D is, therefore, relevant, and the benefit of MVNRE models then falls entirely on the extent to which the EQ-5D outcome can borrow strength from other outcomes.

Before closing this introduction, it is worth mentioning a class of models that are very closely related to MVNRE models. These are the regression models now routinely used to study surrogate outcomes, originated by Daniels and Hughes (1997). The MVN structure at the between-trial level, assuming just two outcome variables and two treatments, is replaced by a linear regression:

$$\mathbf{Y}_i \sim MVN\left(\delta_i, \Sigma_i^W\right)$$
$$\delta_{2i} \sim Normal\left(\alpha + \beta\delta_{1i}, \sigma_2^2\right) \tag{11.2}$$

Here, the relative treatment effect in trial i on the second 'clinical' outcome, for example, time to AIDS or death, is a linear function of the true relative treatment effect on the first 'surrogate' outcome, CD4 cell count, with intercept α and slope β, and with σ_2^2 representing error variance in the clinical outcome around the regression line. The regression model in equation (11.2) is slightly weaker than the MVN model in equation (11.1) because it makes no assumptions about the between-trial distribution of treatment effects δ_{1i}. However, if investigators were contemplating a univariate meta-analysis of CD4 cell count outcome, there is little doubt that they would make the routinely accepted assumption that treatment effects are normally distributed on an appropriate scale: $\delta_{1i} \sim \text{Normal}(d_1, \sigma_1^2)$. With this almost trivial additional assumption, equation (11.2) is simply a re-parameterisation of equation (11.1). Experience on borrowing strength across outcomes in MVNRE models therefore applies equally to surrogacy models. This leads us to question how effective information on the surrogate outcome will be in adding strength to inferences on the clinical outcome, which will be the focus of decision models.

Finally, it is important to note that in some of the most commonly occurring situations, for example, when the multiple outcomes are patient scores on different, but very similar, test instruments or from the same test instruments on repeat occasions, MVNRE models have properties that make them less than ideal in a decision-making context. For example, depression may be measured by the Beck Depression Inventory, the Hamilton Depression Scale, the Montgomery–Asberg Depression Rating Scale or many other similar instruments. Psychometric analysis, as well as expert clinical opinion, tells us that these instruments are testing essentially the same underlying construct. As a result, we expect that a treatment that is best on one scale will be best on all the others as well. However, MVNRE models do not have this property: it would be quite possible – particularly when data are sparse – to arrive at different treatment recommendations from each scale. The assumption of an explicit correlation between outcomes does nothing to prevent this. In these situations MVNRE models fail to provide a coherent basis for treatment recommendations. This is a serious shortcoming, and we propose some alternative models designed for a decision-making context in Section 11.5.

11.3 Multinomial Likelihoods and Extensions of Univariate Methods

Some multivariate methods have already been introduced in Chapter 4 and applied to ordered category data and competing risks (see Section 4.5). In these examples multinomial likelihoods were used to capture the negative correlations between outcomes at the within-trial level. It is also possible to construct approximate MVN likelihoods for this kind of data, based on contrasts between treatment arms rather than arm-level data (Trikalinos and Olkin, 2008; Efthimiou et al., 2014). In the case of ordered category data, we suggested a model in which the treatment effects were the same at every cut-point in order to avoid the possibility of arriving at different conclusions on relative efficacy from different outcomes (see Exercise 4.4). Of course, such assumptions must be tested empirically as far as this is possible and must be supported by informed clinical opinion.

The astute reader will have noticed that while the within-trial correlations were expressed by a multinomial likelihood in Chapter 4, there was no explicit model for the correlations between outcomes at the between-trial level. In the probit model for ordered categories (Section 4.5.1), the between-trial correlations are expressed implicitly through the cut-point position parameters z_c, which can either be fixed or vary randomly between trials. This allows considerable flexibility at the between-trial level, without the complexity of specifying an explicit between-trial correlation matrix. For the competing risk models (Section 4.5.2), between-trial variances were specified, but not between-trial correlations, either implicitly or explicitly. Given that between-trial correlations will be expressed in the joint posterior distribution, and because there are so few data to inform these correlations, it is not clear that there would be benefit in introducing further between-trial structure unless strongly informative prior distributions were being considered.

Incomplete data are a common feature of competing risk structures with many trials reporting collapsed categories. For example, while some trials on statins report fatal stroke, fatal cardiovascular disease and fatal non-cardiovascular disease, others might report all fatal cardiovascular disease as a single category. Multinomial likelihood models for incomplete data have been discussed by Schmid et al. (2014).

Multinomial likelihoods can also be expressed as a series of conditional binomials. This allows one to use univariate approaches, for example, with a logit link, to analyse multivariate data while still taking account of correlations at the within-trial level. This is a very useful technique for dealing with response and dropout data, because it is natural to conceptualise these data as first dropout and then response conditional on not having dropped out (National Clinical Guideline Centre, 2012; National Collaborating Centre for Mental Health, 2014).

Once again, the between-trial level can be ignored, unless there is a specific requirement to represent known relationships between these outcomes. The conditional binomial approach has the advantage of simplicity while at the same time representing the outcomes in a way that reflects the logical relationship between them. Further, these relationships would need to be captured in any cost-effectiveness analysis. MVNRE models can also be applied to data on dropout and response (Efthimiou et al., 2014), but leaving aside the complexity this introduces, there are potential risks in ignoring the underlying structure of the data. Specifically, if response is modelled rather than response conditional on not dropping out, marked differences between treatments in dropout may introduce inconsistent treatment effects on the response outcome.

11.4 Chains of Evidence

'Chain of evidence' structures involve a temporal relationship between outcome variables within the natural history of a disease. Suppose there are three states temporally ordered 1, 2 and 3 and that there are three sets of trials reporting treatment effects on (a) the $1 \rightarrow 2$ transition, (b) the $2 \rightarrow 3$ transition and (c) the $1 \rightarrow 3$ transition. The essential property here is that one should be able to predict results on the $1 \rightarrow 3$ transition by combining the evidence on the other two elements. If this can be done, it follows that the observed $1 \rightarrow 3$ evidence can be combined with the evidence based on the other two sources of evidence. If three separate analyses are conducted, the results will not be internally consistent. As always, the 'direct' evidence on $1 \rightarrow 3$ could be inconsistent with what is predicted from the other two sources, and this should be investigated whenever possible. Chain of evidence structures can themselves take different forms, depending on the types of data that are reported and the logical relationship between the outcomes. We look at two such structures in this section.

11.4.1 A Decision Tree Structure: Coronary Patency

Our first example comes from the literature on the confidence profile method (CPM) (Eddy, 1989; Shachter et al., 1990; Eddy et al., 1992), the original inspiration for much of the work in this book. Although now somewhat dated, it is conveniently simple and illustrates the main issues. The CPM data (Table 11.1) comprises three kinds of evidence linking acute myocardial infarction (AMI), coronary patency (CP) and 1-year survival (1-YS) (Figure 11.1):

1) A three-trial network of evidence on the effect of three treatments acting on CP following AMI: tissue type A plasminogen (t-PA), intravenous streptokinase (IVSK) and conventional care (CC). This is the AMI $\rightarrow$ CP evidence.

Table 11.1 Coronary patency example (Ades 2003). Reproduced with permission of John Wiley & sons.

Study	t-PA	IVSK	CC
Probability of coronary patency given treatments			
TIMI	78/118	44/122	
Collen	25/33		1/14
Kennedy		93/134	14/116
Probability of survival conditional on patency status			
Kennedy: patency		88/93	14/14
Kennedy: no patency		35/41	85/102
Probability of survival conditional on treatment			
20 trials		2260/2672	2111/2612

Three evidence sources for a chain of evidence analysis. Each cell shows the numerators/denominators.

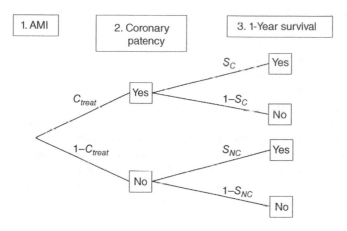

Figure 11.1 Decision tree for coronary patency example (Ades, 2003). Reproduced with permission of John Wiley & Sons.

2) The CP → 1-YS evidence comes from a single trial looking at 1-YS in patients who had achieved patency and patients who had not achieved patency. The trial is effectively treated as a pair of cohort studies, one for each arm.
3) AMI → 1-YS information from trials comparing CC and IVSK with 1-YS as an outcome.

In the original CPM work, 20 trials were 'pooled' by simply adding the numerators (numbers of survivors) to form a single overall numerator and denominators. This is not a procedure that should ever be recommended, but we have used the same data as in the original CPM work in the interests of simplicity.

The premise of the analysis (Ades, 2003) is that treatment acts only on the probability of CP and that survival conditional on patency status is independent of treatment. Based on the decision tree structure of Figure 11.1, we can derive the probability of 1-YS conditional on treatment A, S_A, from the probability of CP given treatment A, C_A, and the probability of survival conditional on achieving patency, s_C, or not, s_{NC}, assumed independent of treatment. Thus the probability of 1-YS on treatment A is

$$S_A = C_A s_C + (1 - C_A) s_{NC} \tag{11.3}$$

The term C_A, the probability of CP given a treatment A, is estimated by the network meta-analysis of the TIMI, Collen and Kennedy trials, while two estimates of the parameters s_C and s_{NC} (one from each arm) are provided by the Kennedy trial. The 20 trials data thus estimates a function of the parameters estimated by the two other kinds of data for patients treated with IVSK. The synthesis of all three data types, subject to the constraint represented by equation (11.3), is therefore a single, unified and coherent account of all the data.

Because the network meta-analysis combined with the conditional survival data can be used to predict the results of the 1-YS trials, given that we have direct observations on that parameter, there is a possibility of inconsistency in the data, which can and should be investigated, for example, with cross-validation (Ades, 2003) (see also Chapter 3).

Note, however, that equation (11.3) sets out the relationship between *absolute* probabilities of survival and patency, not the relationship between *relative* treatment effects. In a previous model for these data (Ades, 2003), the network meta-analysis of the AMI → CP evidence was a standard logistic model. As a result it was necessary to put a model on the control arms, which, as discussed in Chapter 5, is not the best practice. However, if treatment effects are expressed on the risk difference scale, rather than a logit scale, this can be avoided (see Exercise 11.1).

11.4.2 Chain of Evidence with Relative Risks: Neonatal Early Onset Group B Strep

Our second example of a chain of evidence differs in that individuals are not at risk at stage L unless they reached a particular outcome on stage $L-1$ (Figure 11.2). The three stages are maternal carriage (MatC) of group B streptococcus (GBS), neonatal carriage (NC) and neonatal early onset (EO) GBS disease. In this example, neonates cannot be contaminated by GBS unless the mother is carrying GBS at the time of delivery, and they cannot acquire neonatal EOGBS unless they are contaminated with GBS (Colbourn et al., 2007a, 2007b).

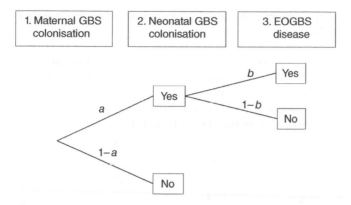

Figure 11.2 Decision tree for early onset group B streptococcus example (Colbourn et al., 2007a).

This apparently minor difference induces a model structure that is quite different from the CP example.

There have been a number of trials of intravenous antibacterial prophylaxis (IVAP) during delivery compared with placebo. In these trials, only women carrying GBS are included. Some trials report the probability of neonatal contamination given maternal carriage, the MatC → NC transition in Figure 11.2. Some trials also report outcomes of the NC → EOGBS transition, and one trial reports *only* the MatC → EOGBS transition, the probability of EOGBS conditional on maternal carriage. Note that studies that report both MatC → NC and NC → EOGBS are also reporting MatC → EOGBS, but we only record the latter results in Table 11.2 if results at the lower two levels are unavailable. This allows us to examine the locus of the treatment effect, as well as obtaining an estimate of the main parameter of interest, which is the overall (MatC → EOGBS) effect of IVAP on EOGBS.

To analyse this further, let us expand the notation in Figure 11.1, so that subscript T denotes the IVAP treatment arm and subscript C the placebo control arm. Then, the treatment effect for acquisition of EOGBS given maternal carriage expressed as a relative risk (RR) is

$$RR_{MatC \to EOGBS} = \frac{a_T b_T}{a_C b_C} = \left(\frac{a_T}{a_C}\right)\left(\frac{b_T}{b_C}\right) = RR_{MatC \to NC} RR_{NC \to EOGBS} \qquad (11.4)$$

In other words, the RR for the overall MatC → EOGBS transition is the product of the RRs for the two component transitions. It can be readily confirmed that an equivalent relationship does not exist for odds ratios (ORs). The implication is that on a log RR scale, consistency equations on any two of the three transitions imply consistency equations on the third, allowing a combined synthesis to proceed, while still conforming to the basic principles of

Table 11.2 Trials on intravenous antibiotic prophylaxis to prevent neonatal early onset group B streptococcal (EOGBS) infection.

Study	IVAP		Placebo	
	r	N	r	N
Newborn colonisation conditional on maternal colonisation $(1 \rightarrow 2)$				
Boyer (a)	2	69	46	82
Boyer (b)	1	43	13	37
Boyer (c)	8	85	40	79
Matorras	2	54	24	56
Easmon	0	38	17	49
Yow	0	34	14	24
EOGBS conditional on newborn colonisation $(2 \rightarrow 3)$				
Boyer (a)	0	2	4	46
Boyer (b)	0	1	1	13
Boyer (c)	0	8	5	40
Matorras	0	2	3	24
EOGBS conditional on maternal colonisation $(1 \rightarrow 3)$				
Tuppurainen	1	88	4	111
Morales	0	135	2	128

Adapted from Colbourn et al. (2007a).

network meta-analysis set in Chapter 2, and the preference for keeping treatment effect and absolute effects separate (Chapter 5). See Exercise 11.2.

Chains of evidence with the kind of decision tree structure exemplified in Figure 11.2 are perhaps quite common. Other examples could be found in research on reproduction, where the chain of outcomes might be conception, delivery and live birth, or studies of induction of labour, where vaginal delivery within 24 h could be one outcome, followed by later vaginal delivery or caesarean section.

11.5   Follow-Up to Multiple Time Points: Gastro-Esophageal Reflux Disease

Gastro-esophageal reflux disease can be healed by a number of pharmaceutical treatments, and trials tend to be run over a 12-week period (Goeree et al., 1999; Lu et al., 2007). The illustrative dataset (Figure 11.3) consists of a six-treatment

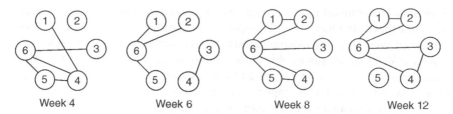

Figure 11.3 Evidence structure in the gastro-esophageal reflux disease example (Lu et al., 2007). Reproduced with permission of John Wiley & Sons.

network of trials in which patient status, healed or not healed, is reported at one or more follow-up times drawn from the set: 4, 6, 8 or 12 weeks. Each trial may report the proportion healed at one or more of these follow-up times. The fact that there are a discrete number of follow-up times simplifies the data structure and coding, but the underlying model we use here can be applied to a set of trials with irregular follow-up times (see succeeding text).

Note that if the analysis is undertaken separately at each follow-up time, then none of the networks would be connected. This is one of many instances in which a combined analysis of multiple outcomes is the only way to achieve a connected network.

The first step in the analysis is to transform the data. The data as published appear as the total number healed at the end of each reporting period; this should be recast as a series of independent conditional binomial observations, giving the number healed *during* each interval in relation to the number at risk at the start of the interval (Table 11.3). We can conceptualise the general model for this type of data as a two-state transition model with transitions only in one

Table 11.3 Gastro-esophageal reflux (Lu et al., 2007).

Study	Treatment	Start time	Duration	Total patients in arm	Healed at end of period	Healed during period	At risk at start
1	3	0	4	20	5	5	20
		4	4	20	7	2	15
	6	0	4	21	8	8	21
		4	4	21	12	4	13
2	4	0	6	65	21	21	65
		6	2	65	24	3	44
...		...			...		

Reproduced with permission of John Wiley & Sons.
Fragment of data to illustrate conversion of data to independent binomial observations. Time in weeks.

direction. The simplest model is to assume a parametric form for time to healing. An exponential model would imply a constant healing rate over time for each treatment. This seems unnecessarily restrictive as one can almost always expect a 'deletion of susceptibles' phenomenon, in which the most readily healed patients are healed earlier, and the incidence falls progressively over time. A Weibull model with shape parameter greater than 1 would capture this.

A piecewise constant hazard within each of the four time periods was chosen as it confers flexibility with relatively simple coding (Lu et al., 2007). A number of models were examined and $\bar{D}$ and DIC statistics were used to help select the final model. A random walk process over time was assumed: this captures the idea that the log hazard ratios δ_{iku} for treatment k in trial i in each time period u are likely to be similar to the log hazard ratio on the previous period. This is achieved by putting a standard RE model on the first time period and having subsequent periods controlled by the random walk process:

$$\delta_{ik1} \sim \text{Normal}\left(d_{t_{i1},t_{ik}}, \sigma^2\right)$$
$$\delta_{iku} \sim \text{Normal}\left(\delta_{ik,u-1}, \sigma_{RW}^2\right), \quad u = 2, 3, 4$$

A vague prior distribution was put on the random walk variance parameter σ_{RW}^2, which allows the data to determine the level of smoothing. In effect, the data on the treatment effects in the subsequent time periods propagates back through the random walk model to contribute information to the overall treatment effect hyper-parameters d_{1k} and σ^2.

In parallel with the model for the treatment effect, the model for the control treatment on each trial has vague unrelated prior distributions for each trial for the first time period and a second random walk process for the subsequent time periods. The model thus keeps to the principle that trial control arms are free to vary independent of treatment effects and independent of other trials' control arms, while treatment effects are subject to consistency constraints. This same model has also been applied to trial data on drug-eluting stents (Stettler et al., 2007, 2008).

The piecewise constant hazard model can be readily applied in situations where the reported follow-up times do not all belong to a small set and where instead observations are reported at essentially random times. In this case it is still open to investigators to set up a small number – perhaps three or four – of follow-up periods, for example, month 1, months 2–6, 7–12, 13 or more. To incorporate a trial with two observations at, say, 3 months and 12 months, it would be necessary to calculate how many months at risk were contributed by each of the four time periods to each of the two observations.

This kind of approach solves a number of problems created when trials report at a variety of follow-up periods. First, it frees investigators from unrealistic assumptions that hazards remain constant over long periods. Second, one often sees models based on ORs or RRs that assume that the OR or RR at 2 months is the same as the OR or RR at 6 or 12 months. These assumptions can only be true under very specific and also very implausible circumstances: in many situations ORs and RRs will tend towards unity as the time interval increases, because the number of susceptibles diminishes in both arms. Piecewise constant hazard models offer a flexible alternative.

It is instructive to compare the aforementioned approach with MVNRE, which was applied by Trikalinos and Olkin (2012) to a very similar dataset with binomial outcomes at four time points. In contrast to our reconstruction of the data in terms of independent binomial observations, the MVNRE approach is based on the between-treatment contrasts in the proportion surviving at each time point: an approximate within-trial covariance matrix can be calculated from the data. At the between-trial level, five correlation matrices were compared. The choice of between-trial correlation structure had little impact on estimates, and the authors noted that the benefits of the non-zero between-trial correlations between outcomes were 'modest at best'. In the model we have proposed here, the between-trial correlation between outcomes is again implicit, carried by the random walk smoothing variance, which is readily estimated from the data. A high variance indicates a low between-trial correlation.

More generally, data of this type can be handled with survival analysis; the main modelling issues are first the choice of survival time distribution and second the parameters of the survival time distribution that should carry the treatment effect (see Chapter 10).

11.6 Multiple Outcomes Reported in Different Ways: Influenza

This example (Welton et al., 2008b; Burch et al., 2010) shares many of the features of a chain of evidence problem, in that it revolves around clinical states that have a fixed temporal order within a natural history model. In this case, although there is an underlying three-state Markov model with only forward transitions (Figure 11.4a), evidence is only available on the transition $1 \rightarrow 2$ (start to end of fever) and on the implied transition $1 \rightarrow 3$ (start to end of symptoms) (Figure 11.4b). In this situation the ideal would be to use this information to estimate the rate parameters of the underlying Markov model, and this would be feasible if one could write down the distribution of times from $1 \rightarrow 3$ as a function of the distributions of times $1 \rightarrow 2$ and $2 \rightarrow 3$. This can be done, but only for a very limited range of distributions, such as the exponential. Because it was established that the data were not in fact compatible with these distributions, an alternative approach was adopted, in which the $1 \rightarrow 2$ transitions and the $1 \rightarrow 3$ transitions were modelled as Weibull distributions, in which the shape and scale parameters attaching to the two outcomes were related, and with the treatment effect carried, as is conventional, on the scale parameter (Welton et al., 2008b).

The dataset posed further difficulties. Some trials reported time to end of fever and some time to end of symptoms (which cannot precede the former), while other trials reported both (Table 11.4). In addition some trials reported median times and their standard errors, and others mean times and their standard errors. A few trials also reported the number of patients with symptoms at 21 or 28 days' follow-up, which together with median times provide information on the shape as well as scale of the distribution. It is evident that separate analyses for each outcome would make it impossible to form a connected network of comparisons. Synthesis can be achieved, however, by noting that both the mean and the median of a Weibull distribution can be written in terms of the shape α_{Cik} and scale β_{Cik} parameters, where $C = \{F,S\}$ is an index

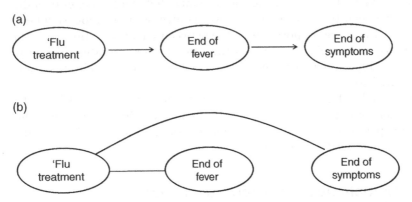

Figure 11.4 Influenza example (Welton et al., 2008b). (a) The underlying Markov model. (b) Structure of evidence and relation to the model. Reproduced with permission of John Wiley & Sons.

Table 11.4 Influenza treatment (Welton et al., 2008b).

Treatment	Number of trials	Time to end fever	Time to end symptoms	% Symptom-free (days)
Amantadine	6	Mean	—	
Oseltamivir	8	Median	Median	21 days
Zanamivir	5	—	Median	28 days

Placebo-controlled trials evidence structure.

for fever or symptom outcomes, i indexes the trial and k the treatment. We can therefore assume these data have a normal likelihood, with expectations that are functions of shape and scale (Welton et al., 2008b):

$$Y_{Cik}^{Median} \sim \text{Normal}\left(\beta_{Cik}\ln(2)^{1/\alpha_{Cik}}, Var_{Cik}^{Median}\right)$$

$$Y_{Cik}^{Mean} \sim \text{Normal}\left(\beta_{Cik}\Gamma\left(1+\frac{1}{\alpha_{Cik}}\right), Var_{Cik}^{Mean}\right) \quad\quad (11.5)$$

$$r_{Cik}^{21} \sim \text{Binomial}\left(p_{Cik}^{21}, n_{Cik}^{21}\right), \; p_{Cik}^{21} = E\left[\Pr\left(T_{Cik}<21\right)|Y_{Cik}^{Median}\right]$$

The third line in equation (11.5) is for the binomial data on the proportion of patients who are symptom-free at 21 days, conditional on the median. The conditioning is because these proportions are not independent of the median (we know 50% are symptom-free at the median time). The statistical synthesis of evidence on different functions of the same underlying parameters is a typical example of multi-parameter evidence synthesis (Ades and Sutton, 2006).

11.7 Simultaneous Mapping and Synthesis

The term 'mapping' is often used in HTA when there is a need to assess a treatment effect in terms of quality-adjusted life years, but this has not been measured in trials. Instead, investigators estimate the treatment effect on a generic HRQoL instrument, such as EQ-5D, by multiplying a treatment difference on a disease-specific measurement scale by a 'mapping coefficient' (Brazier et al., 2010; Longworth and Rowen, 2011, 2013). For example, in the absence of trial evidence measuring a treatment effect of biologic therapies in ankylosing spondylitis on EQ-5D, one might use the treatment effect δ_{BASDAI} on the Bath Ankylosing Spondylitis Daily Activity Index (BASDAI) and multiply this by a mapping coefficient $\beta_{BASDAI \rightarrow EQ-5D}$:

$$\delta_{EQ-5D} = \beta_{BASDAI \rightarrow EQ-5D} \cdot \delta_{BASDAI}$$

Typically the mapping coefficients are based on ordinary least squares (OLS) regression analyses of cohort studies in which both EQ-5D and the disease-specific measure have been measured. However, it is widely recognised that the OLS estimate is inappropriate and systematically underestimates the QALY benefit of treatments, unless adjusted for error in covariates (Lu et al., 2013; Fayers and Hays, 2014). Indeed, there is an entire literature on 'test equating' and 'test linking' methods that are intended precisely to capture the essentially symmetrical functional relationships between multiple outcomes measured with error (Kolen and Brennan, 1994; Dorans et al., 2007).

In what follows the mapping coefficients linking several test instruments will be estimated from the RCT evidence itself. The method will simultaneously synthesise both the mapping coefficients and the treatment effect information within and between trials. Unlike estimates based on OLS regression, these estimates will be invertible and transitive:

$$\beta_{X \to Y} = \frac{1}{\beta_{Y \to X}} \quad \text{and} \quad \beta_{X \to Z} = \beta_{X \to Y} \beta_{Y \to Z} \tag{11.6}$$

which are necessary conditions for coherent mappings between scales (Lu et al., 2013). However, the method can only be applied to connected networks of outcomes. Figure 11.5 shows two such networks, the first from a study of 6 outcomes in 8 trials on biologic therapy in ankylosing spondylitis (Lu et al., 2014) and the second, 9 outcomes in 22 trials on selective serotonin reuptake inhibitors for social anxiety (Ades et al., 2015). The ankylosing spondylitis dataset (Table 11.5) illustrates a typical data structure suitable for this form of analysis.

In the social anxiety dataset, with $L = 9$ outcomes, there are a total of $L(L-1)/2 = 45$ mappings between outcomes to be estimated. However, because of the properties inherent in equation (11.6), there are only $L - 1 = 8$ independent mappings, which can be considered as basic parameters (Eddy et al., 1992). The remaining 37 are functions of those eight, which can be inferred from equation (11.6). It is interesting to note the parallel here with network meta-analysis where a similar transitivity assumption derives from an underlying assumption of exchangeability of treatment effects (Lu et al., 2011), allowing us to reduce the dimensionality of the problem in exactly the same way. The network meta-analysis exchangeability assumption is equivalent to assuming that treatments are missing at random (MAR) with respect to relative efficacy. The mapping models make the same assumption about outcomes.

Two forms of the model have been discussed (Lu et al., 2014; Ades et al., 2015). In the fixed mapping models, the mapping ratios are constant across trials. A random mapping model has also been implemented, which allows the mapping coefficients to vary from trial to trial around a mean value. In work carried out so far, the random mapping model fits the data considerably better than a fixed mapping model (Lu et al., 2014; Ades et al., 2015), but the extent of between-trial variation in mappings is limited, with a coefficient of variation around 15%. The fact that such variation exists is not surprising: we need only to admit that the measurement scales are not exactly linearly related or equivalently that some are more sensitive in some parts of the underling severity continuum than others. Under these circumstances, the presence of between-trial heterogeneity in severity will automatically generate a degree of variation in the mapping ratios.

The key assumption behind these models is that the treatment effects on the different outcomes are proportional, or nearly proportional, to each other. Further, if there is no (true) effect on one outcome, there is no effect on any. These models can therefore be seen as approximately special cases of the

(a)

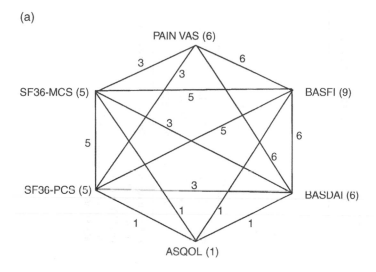

(b)

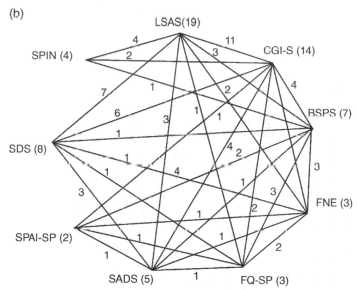

Figure 11.5 Simultaneous mapping and synthesis: (a) connected network of six outcomes in trials of biologic therapy in ankylosing spondylitis (Lu et al., 2014). ASQOL, ankylosing spondylitis quality of life; BASDAI, Bath Ankylosing Spondylitis Disease Activity Index; BASFI, bath ankylosing spondylitis functional index; PAIN VAS, pain visual analog scale; SF36-PCS/ MCS, short form 36 physical/mental summary. (b) A connected network of nine outcomes in treatments for social anxiety (Ades et al., 2015). BSPS, brief social phobia scale; CGI-S, clinical global impression – severity; FNE, fear of negative evaluation; FQ-SP, fear quotient – social phobia; LSAS, Liebowitz social anxiety scale; SADS, social avoidance and distress scale; SDS, Sheehan disability scale; SPAI-SP, social phobia and anxiety inventory – social phobia; SPIN, social phobia inventory. Reproduced from John wiley & sons.

Table 11.5 Ankylosing spondylitis (Lu et al., 2014).

Trial	Treatment	Weeks	PAIN VAS	BASFI	BASDAI	SF-36 PCS	SF-36 MCS	ASQOL
1. Gorman 2002	ETA	16	−4.15 (0.803)	−2.2 (0.772)				
2. Brandt 2003	ETA	6		−1.7 (0.80)	−2.2 (0.553)			
3. Davis 2003	ETA	24	−1.83 (0.410)	−1.41 (0.285)	−1.91 (0.258)			
4. Van der Heijde 2006	ADA	12	−1.956 (0.350)	−1.414 (0.267)	−1.8 (0.283)			
5. Braun 2002	INF	12		−2.0 (0.561)	−2.6 (0.433)	8.3 (5.74)	3.85 (4.33)	
6. Van der Heijde 2009	ADA	24		−1.5 (0.236)	−1.8 (0.286)	5.2 (1.026)	1.6 (1.162)	−2.4 (0.488)
7. Van der Heijde 2005	INF	24	−2.6 (0.333)	−1.7 (0.230)	−2.5 (0.274)	9.4 (0.957)	0.7 (1.045)	
8. Inman 2008	GOL 50	12	−2.7 (0.412)	−1.5 (0.264)		4.9 (1.164)	1.4 (1.023)	
9. Inman 2008	GOL 100	12	−2.8 (0.424)	−1.6 (0.274)		6.0 (1.069)	3.6 (1.252)	

Treatment effects on change from baseline, relative to placebo, and their standard errors in eight trials of TNF-α inhibitors.
ADA, adalimumab 40 mg; ETA, etanercept 25 mg; GOL 50, golimumab 50 mg; GOL 100, golimumab 100 mg; INF, infliximab 5 mg.
ASQOL 0–18 scale; BASDAI 0–10 scale; BASFI 0–10 scale; PAIN VAS 0–10 scale; SF-36 PCS 0–50 scale; SF-36 PCS (0–50).

general MVNRE model, or of its regression-based equivalent, in which at the between-trial level there is a zero intercept and no additional error about the regression line.

One of the properties of these models is that once the synthesis is conducted, investigators are able to report the treatment effect on *any* of the measurement instruments in the network, as illustrated in Table 11.6. In this sense they represent an alternative to standardisation of treatment effects by dividing them by the sample standard deviation, which is a very commonly used strategy for combining information from trials that use different, but similar, test instruments (Cohen, 1969; Hedges, 1981; Higgins and Green, 2008). In spite of its popularity, particularly in education and social sciences, standardisation has been criticised (Greenland et al., 1986, 1991), and it has even been described as 'useless for meta-analysis' (Rothman et al., 2012) because of the instability resulting from dividing treatment effects by a sample statistic and particularly the distortion resulting from variation in the true standard deviation in the trial populations.

Although the mapping model makes stronger assumptions than MVNRE – that is, a treatment that is most effective in one outcome is most effective on all outcomes – these are assumptions that clinicians regard as reasonable in the two examples discussed previously. It is interesting to note that standardisation makes even stronger and in fact entirely implausible assumptions, namely, that

Table 11.6 Social anxiety (Ades et al., 2015).

Test Instrument	Mean treatment effect		Between-trial standard deviation	
	Mean	95% CrI	Median	95% CrI
Liebowitz social anxiety scale	−11.7	(−9.8, −13.7)	3.2	(1.64, 5.21)
Clinical global impression–severity	−0.523	(−0.42, −0.63)	0.14	(0.07, 0.24)
Brief social phobia scale	−4.95	(−3.55, −6.59)	1.34	(0.65, 2.32)
Fear of negative evaluation	−2.57	(−1.66, −3.72)	0.69	(0.32, 1.24)
Fear questionnaire – social phobia	−4.16	(−2.84, −5.92)	1.11	(0.54, 2.03)
Social avoidance and distress scale	−2.66	(−1.87, −3.57)	0.72	(0.35, 1.25)
Social phobia and anxiety inventory – social phobia	−19.9	(−12.7, −30.1)	5.26	(2.46, 10.1)
Sheehan disability scale	−4.44	(−1.88, −3.08)	0.66	(0.34, 1.11)
Social phobia inventory	−5.66	(−3.84, −7.86)	1.52	(0.72, 2.73)

Reproduced from John wiley & sons.
Posterior mean and 95% CrI of the mean treatment effects. Posterior median of the between-trial standard deviation and 95% CrI for each of the nine test instruments from a random mapping coefficients model.

mappings between standardised scores are all 1 and that all the test instruments therefore must have exactly the same validity and measurement error. In spite of this, treatment effects estimated by the mapping models may be estimated with greater relative precision than under equivalent standardisation models, and they can be applied to a far wider range of datasets (Ades et al., 2015), including to cases where trials report a mixture of change scores and scores at follow-up.

11.8   Related Outcomes Reported in Different Ways: Advanced Breast Cancer

In this example (National Institute for Health and Clinical Excellence, 2009a; Welton et al., 2010), trials reported up to three outcomes: proportion responding to treatment, time to tumour progression and overall survival. These outcomes are clearly related, and this should be captured in a single coherent model. In this case, however, the three trials comparing the four treatments of interest contained only limited information that could inform the relationship between the outcomes (Table 11.7), so two further trials involving treatments that were not part of the decision problem were included in the analysis to provide more information (Figure 11.6).

In order to drive the economic analysis (National Institute for Health and Clinical Excellence, 2009a), the model is required to provide estimates of the category breakdown (Responder, Stable Disease, Progressive Disease, and Not Assessable) in each trial by treatment and the distribution of both time to tumour progression and overall survival in each category. In other words, estimates are needed for every cell in Table 11.7.

Table 11.7 Advanced breast cancer data structure: overall proportion in each category; median time to tumour progression, by category; median overall survival, by category.

Trial	Overall response, proportion in each category				Median time to progression, by patient category					Median overall survival, by patient category				
	Res	StDis	PrDis	NA	Res	StDis	PrDis	NA	Total	Res	StDis	PrDis	NA	Total
1	x	x	x	x	x				LHR					LHR
2	x	x	x	x	x				x					x
3	x	x	x	x					x					x
4	x	x			x				x					x
5	x				x				x					x

'LHR' indicates that the log hazard ratio was given rather than median time.
NA, not assessable; PrDis, progressive disease; Res, responder; StDis, stable disease.

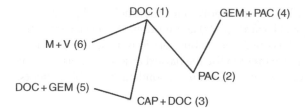

Figure 11.6 Advanced breast cancer: network of evidence (Welton et al., 2010). CAP, capecitabine; DOC, docetaxel; GEM, gemcitabine; M + V, mitomycin and vinblastine; PAC, paclitaxel. The numbers in brackets reflect the treatment ordering used in modelling. Reproduced with permission of John Wiley & Sons.

We can begin by considering how this would be achieved if there was a complete set of data from a large number of trials. Firstly, the proportions in each category would be modelled by a multinomial logistic regression, with separate baseline and relative treatment effects on a log odds scale for the probabilities of response, stable disease and progressive disease. The times to progression and mortality could be modelled as Weibull distributions with shape and scale parameters for each outcome related via a MVNRE model.

The extreme sparseness of the data, however, forced investigators to adopt a model with relatively few parameters. Here are some of the major features of the model adopted:

1) All treatment effects models were fixed effects as there is only one trial on each link in the network (Figure 11.6).
2) The treatment effects on progression and overall survival were considered to be identical.
3) To model the proportion of patients in each category, it was necessary to assume a RE model for the proportion non-responders who were stable and the proportion of non-stable non-responders with progressive disease.
4) Because there was only a single summary estimate on time to progression and overall survival, it was not possible to estimate the Weibull shape parameter. Experiments with setting the shape to fixed value showed that a shape value of 1 gave the best fit, so exponential distributions were adopted.
5) Median time to outcome informs the function $\log(2)/\lambda$ of the rate parameter λ for an exponential distribution. Data on median time to outcome in all patients can be incorporated by characterising this as a weighted average of the median survival in the subcategories. 'Collapsed category' data occurs commonly in synthesis problems (Ades and Sutton, 2006). However, the median of a weighted sum of exponentials is not itself an exponential, and it is necessary to obtain an expression for this function (Welton et al., 2010).
6) It is necessary to put a model on control arms for some of the failure time outcomes.

7) In some trials a log hazard ratio was reported for the whole trial population. This assumes a proportional hazards (PHs) assumption. However, even if the PH assumption is true in the constituent subpopulations, it cannot be true in a pooled dataset. As an approximation the log hazard ratio was interpreted as if it represented a common treatment effect in responders and stable patients.

11.9 Repeat Observations for Continuous Outcomes: Fractional Polynomials

Fractional polynomials (FPs) are a family of flexible parametric functions that can describe a wide range of shapes (Royston and Sauerbrei, 2008):

$$f(Y) = \begin{cases} \alpha_0 + \displaystyle\sum_{m=1}^{M} \alpha_m u^{p_m} & \text{if } p_1 \neq \cdots \neq p_M \\ \alpha_0 + \alpha_1 u^{p_1} + \displaystyle\sum_{m=2}^{M} \alpha_m u^{p_1} \left(\ln(u) \right)^{m-1} & \text{if } M > 1, \ p_1 = \cdots = p_M \end{cases}$$

The powers are usually chosen from the set $p = \{-2, -1, -0.5, 0, 0.5, 1, 2\}$ with $u^0 = \ln(u)$. Typically an FP of the second order ($M = 2$; two time-related parameters) is considered sufficiently flexible to capture a wide range of functional forms. FPs can be incorporated in the network meta-analysis framework to model treatment effects over time (Jansen et al., 2015). The between-trial model is parameterised in a way that parallels the standard model for a single outcome: a baseline FP and separate FPs for the treatment effects. Note that the consistency equations apply to each of the FP coefficients. The RE model based on a second-order FP can be described as

$$\theta_{iku} = \begin{cases} \alpha_{0,ik} + \alpha_{1,ik} u^{p_1} + \alpha_{2,ik} u^{p_2} & \text{if } p_1 \neq p_2 \\ \alpha_{0,ik} + \alpha_{1,ik} u^{p} + \alpha_{2,ik} u^{p} \ln(u) & \text{if } p_1 = p_2 = p \end{cases}$$

$$\begin{pmatrix} \alpha_{0,ik} \\ \alpha_{1,ik} \\ \alpha_{2,ik} \end{pmatrix} = \begin{pmatrix} \mu_{0,i} \\ \mu_{1,i} \\ \mu_{2,i} \end{pmatrix} + \begin{pmatrix} \delta_{0,t_{i_1}t_{ik}} \\ \delta_{1,t_{i_1}t_{ik}} \\ \delta_{2,t_{i_1}t_{ik}} \end{pmatrix}$$

$$\begin{pmatrix} \delta_{0,t_{i_1}t_{ik}} \\ \delta_{1,t_{i_1}t_{ik}} \\ \delta_{2,t_{i_1}t_{ik}} \end{pmatrix} \sim \text{Normal} \left(\begin{pmatrix} d_{0,1t_{ik}} \\ d_{1,1t_{ik}} \\ d_{2,1t_{ik}} \end{pmatrix} - \begin{pmatrix} d_{0,1t_{i1}} \\ d_{1,1t_{i1}} \\ d_{2,1t_{i1}} \end{pmatrix}, \Sigma \right) \tag{11.7}$$

$$\Sigma = \begin{pmatrix} \sigma_0^2 & \sigma_0\sigma_1\rho_{01} & \sigma_0\sigma_2\rho_{02} \\ \sigma_0\sigma_1\rho_{01} & \sigma_1^2 & \sigma_1\sigma_2\rho_{12} \\ \sigma_0\sigma_2\rho_{02} & \sigma_1\sigma_2\rho_{12} & \sigma_2^2 \end{pmatrix}$$

with i and k indexing trials and treatments and θ_{iku} representing the true value of the outcome at follow-up time u. In this example (Jansen et al., 2015), FP functions are used to describe the functional form of the treatment effect over time in a network meta-analysis of six hyaluronan-based viscosupplements for osteoarthritis of the knee fitted to pain outcomes reported at between 2 and 11 time points in 16 trials. Since the outcome was reported as change from baseline, α_0 was removed from the model. Based on $\bar{D}$ and *DIC* criteria, a model with powers $p_1 = 0.5$ *and* $p_2 = 1$ and only a between-study heterogeneity parameter for treatment effect d_1 was considered appropriate. At the within-study level, the correlations between observations at different time points need to be taken into account. Correlations are likely to depend on time between observations, and with so much variation between trials in the number and timing of repeat observations, this can be complex to programme. As an alternative a range of simple univariate models were run with extensive sensitivity analyses.

Fitted curves from these models are presented in Jansen et al. (2015). Note that the FP framework not only analyses studies with different follow-up periods and outcomes measured at different time points but also allows for non-linear trends in treatment effects and changes in relative efficacy of treatments over time that could lead to treatment effects curves relative to placebo crossing. See Chapter 10 for more details.

11.10 Synthesis for Markov Models

Although not very common, situations where several trials report at a level of detail that allows the construction of a Markov transition model present some particularly interesting problems. Trials of treatments for asthma patients have been reported in terms of the week-to-week transitions observed over a 52-week period (Price and Briggs, 2002). A four-state model was used: successfully treated week (STW), unsuccessfully treated week (UTW), exacerbation (X) and, an absorbing state, treatment failure (TF) (Figure 11.7a). A nine-parameter Markov state transition model was constructed, giving a nominal 18 parameters in all, although the presence of zero cells meant that only 14 could be estimated from data. A typical set of data is shown in Table 11.8.

Five trials providing evidence on four asthma treatments reported results in the same format (Price et al., 2011). The analysis began by replacing the Markov transition probability model by a Markov rate model (Figure 11.7b) in which transitions were only possible between adjacent states (the result of a model selection process). Transitions from one state to another and then on to the next within one week are still possible within this framework. Because the likelihood is unaffected by this structural change, as it is based on a set of multinomial distributions in both cases, it was possible to show that the rate

(a)

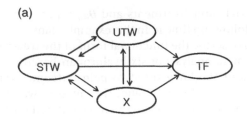

(b)

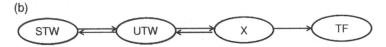

Figure 11.7 (a) Markov transition probability model, (b) Markov rate model. STW, successfully treated weeks; TF, treatment failure; UTW, unsuccessfully treated weeks; X, exacerbation (Price et al., 2011). Reproduced with permission of John Wiley & Sons.

Table 11.8 Data to inform the Markov transition probability or rate models of Figure 11.4 for a single arm.

		Time period T			
		STW	UTW	X	F
Time period $T-1$	STW	310	70	2	0
	UTW	96	1683	11	13
	X	1	11	1	0
	F	0	0	0	94

From Price et al. (2011).
Number of transitions observed over a 52-week period in a single trial arm.
F, treatment failure (absorbing state); STW, successfully treated week;
UTW, unsuccessfully treated week; X, exacerbation.

model with six parameters fitted as well as a probability model with a nominal nine parameters.

To estimate transition rates from data on transition probabilities, it is necessary to be able to specify the relationship between them. In general these are given by the Kolmogorov forward equations (Grimmett and Stirzaker, 1992):

$$\frac{d}{du}\mathbf{P}(\mathbf{u}) = \mathbf{P}(\mathbf{u})\mathbf{G}$$

which relate the transition probability matrix $\mathbf{P}(\mathbf{u})$ with cycle length u and transition probability elements p_{ij} from state i to state j to rate matrix $\mathbf{G}$ with rate elements λ_{ij}. The rate model in Figure 11.7b has the structure

$$\mathbf{G} = \begin{pmatrix} -\lambda_{12} & \lambda_{12} & 0 & 0 \\ \lambda_{21} & -(\lambda_{21}+\lambda_{23}+\lambda_{24}) & \lambda_{23} & \lambda_{24} \\ 0 & \lambda_{32} & -(\lambda_{32}+\lambda_{34}) & \lambda_{34} \end{pmatrix}$$

This matrix can be written out as a series of ordinary first-order differential equations, whose solutions have the form

$$\mathbf{P}(\mathbf{u}) = e^{u\mathbf{G}} = \sum_{n=0}^{\infty} \frac{u^n}{n!}\mathbf{G}^n$$

For simpler structures this can be solved in closed form, or solutions can be found by analytic packages such as MAPLE. In this case solutions were not found, so the WinBUGS add-on WBDiff (Lunn, 2004), which solves the differential equations on every MCMC cycle using the Runge–Kutta method, was used.

The objective of this analysis was to identify the transition on which the treatment effect was operating. Looking at each transition in turn, a network meta-analysis model with baseline and relative treatment effect subject to consistency equations was applied to the rate parameters on the log scale. Again using $\bar{D}$ and *DIC* criteria for goodness of fit and model selection, it was shown that the locus of the treatment effect was on the backward transition from UTW to STW. This makes clinical sense: when asthma is under control, patients do not use medication, but during an attack they take medication, and the best treatment is the one that restores them to normal functioning fastest.

With the treatment effect located on the UTW to STW transition, transition hazards between other states may remain the same for all five treatments. Thus, this model has far fewer parameters than models with a treatment effect on every transition. As a result, when the network meta-analysis model was embedded in a cost-effectiveness analysis, there was considerable impact on decision uncertainty (Price et al., 2007).

11.11 Summary and Further Reading

Each of the aforementioned examples illustrates how a particular evidence structure can be modelled in a way that captures functional or logical relationships between outcomes, and we would refer readers to the original papers for a more detailed account and for sample WinBUGS code. These examples, of course, come nowhere near exhausting the many evidence structures and models in which multiple outcomes can occur. WinBUGS and other similar software provide investigators with enormous flexibility.

Here, we give some further examples of multiple outcome evidence structures that readers can consult:

1) Linear functional relationships between treatment effect parameters, similar to the models for mapping and advanced breast cancer, have been used elsewhere. Hazard ratios for time to tumour progression and for overall survival have been related in this way in the development of clinical guidelines for colorectal cancer (National Institute for Health and Clinical Excellence, 2011). A similar approach was used to model recovery and response outcomes in treatments for social anxiety (National Collaborating Centre for Mental Health, 2013; Mayo-Wilson et al., 2014).
2) A particularly simple approach to multi-outcome synthesis can be used for repeat measures over time, if it can be assumed that the treatment effect is constant over time. This is to use a univariate model, but adjust the variance of the observations to take account of the correlation structure (National Institute for Health and Care Excellence, 2013b).
3) In a study of the effect of smoking cessation interventions on the probability of relapse after quitting smoking, data were available from repeat observations of the proportion surviving to different times (Madan et al., 2014b). This enabled investigators to estimate an underlying survival distribution.

It is worth noting, in conclusion, that in many of the examples given in this chapter, particularly those involving events in time (Sections 11.6, Influenza, and 11.8, Advanced breast cancer), much simpler analyses would be possible if IPD had been available. Much of the complexity of the problem derives from the use of summary measures for survival time problems rather than life-table data, which are the sufficient statistics in this context. It is therefore relevant to note that there are methods for reconstructing life-table data from Kaplan–Meier curves (Guyot et al., 2012), which could be deployed in these cases. Further complication is caused by the tendency to report the $1 \rightarrow 3$ (overall survival) rather than $2 \rightarrow 3$ (survival conditional on tumour progression).

But in other cases, particularly those involving continuous measures (social anxiety, ankylosing spondylitis), aggregated mean and variance data, with external data on correlations, are sufficient as these are the sufficient statistics for such data. IPD would not have much impact on the analysis, except to provide better estimates of within-trial correlations and to check the implied linearity assumptions. For analysis of covariate effects, of course, IPD are always to be preferred (Chapter 8).

11.12 Exercises

11.1 *In the chain of evidence coronary patency example (Section 11.4.1), show that adopting a risk difference scale for the relative treatment effect on the coronary patency outcome implies a risk difference scale for relative effects on the survival outcome.

Convert the data in Table 11.1 into risk differences and their normal approximation variances and fit a chain of evidence model. Note that 0.5 should be added when calculating the variance of the risk difference when all individuals had the event (Higgins and Green, 2008). Investigate how well this model fits the three kinds of data using the appropriate residual deviance formula (Table 4.1).

Use cross-validation to assess the prediction of the treatment effect on the $1 \rightarrow 3$ evidence from the $1 \rightarrow 2$ and $2 \rightarrow 3$ evidence (see Chapter 3).

11.2 *For the EOGBS example (Section 11.4.2), fit a relative risk synthesis model to the data in Table 11.2 (given in file *Ch11_eogbsData.txt*) based on the relationship in equation (11.4). Use an arm-based likelihood model with log relative risks modelled according to Warn's method (Warn et al., 2002) (see Chapter 2).

Use cross-validation to assess the prediction of the treatment effect on the $1 \rightarrow 3$ evidence from the $1 \rightarrow 2$ and $2 \rightarrow 3$ evidence.

Note: use a fixed effects model for the $2 \rightarrow 3$ data (neonatal EOGBS conditional on neonatal carriage) with a weakly informative prior distribution on the treatment effect, say, Normal(0,0.1): this is needed to stabilise the treatment effect in the cross-validation exercise, in the light of the zero cells. It is not needed in the full synthesis. A random effects model is needed for the $1 \rightarrow 2$ data (neonatal carriage conditional on maternal carriage): we suggest a truncated gamma prior distribution on the precision, `tau ~ dgamma(.2,.2)I(.05)`, to stabilise the between-study variance to a reasonable level in the presence of sparse and heterogeneous data.

12

Validity of Network Meta-Analysis

12.1   Introduction

Doubts have been expressed about the validity of network meta-analysis from the outset, and there has been a steadily growing literature around this that deserves to be evaluated and reviewed. Certainly, any statistical method has its limitations. In the case of network meta-analysis, the concerns have revolved mainly around its assumptions. There have been two main lines of empirical research.

Firstly, there is a series of reviews of applied literature that set out to document whether the assumptions of network meta-analysis have been checked or even mentioned. Unfortunately, there has been little consensus on what the assumptions actually are, and in a number of papers, the assumptions are stated in a form that is subtly different from how they have been stated in this book. For this reason we begin the chapter by reviewing the terminology surrounding network meta-analysis assumptions (Section 12.2) in the hope of dispelling some of the confusion.

A persistent assumption running throughout the clinical and methodological literature has been that direct evidence is superior to indirect evidence, although there seems to be no formal analysis that supports this claim. In Section 12.3 we present some 'thought experiments' that aim to clarify in quantitative terms the threats that unrecognised effect modifiers pose to valid inference in evidence synthesis. We ask whether indirect comparisons are more vulnerable, less vulnerable or just equally vulnerable to bias originating from unrecognised effect modifiers. The thought experiments also provide qualitative and quantitative insight into the threats to validity of inference in both pairwise and network meta-analyses.

A second line of empirical work in the literature has focussed on the validity of the 'consistency assumption'. This consists of meta-epidemiological studies that set out to check whether, in large collections of evidence

Network Meta-Analysis for Decision-Making, First Edition. Sofia Dias, A. E. Ades, Nicky J. Welton, Jeroen P. Jansen and Alexander J. Sutton.
Companion website: www.wiley.com/go/dias/networkmeta-analysis

networks, the empirical data accords with this key assumption. We review this area in Section 12.4.

It will be evident from Chapter 7 on inconsistency that it is generally not possible to confirm that the evidence in any specific network is 'consistent'. However carefully one checks for inconsistency, there is seldom sufficient data to reject the null hypothesis of consistency with any confidence. In networks without loops, the key assumptions cannot be tested at all. This has given rise to a legitimate concern with the validity of estimates from network meta-analysis, leading to attempts to assess the 'quality' of evidence that it generates, given the quality of input evidence. In Section 12.5 we suggest that attention should be shifted away from 'quality' and towards the robustness of the treatment recommendation to the quantitative assumptions about the evidence inputs.

12.2   What Are the Assumptions of Network Meta-Analysis?

12.2.1   Exchangeability

The assumptions of network meta-analysis can be stated in a variety of ways. One much-discussed question is whether or not network meta-analysis makes 'additional' assumptions over and above what is normally assumed in pairwise meta-analysis. This has taken on some importance because investigators want to know whether additional checks need to be made to ensure validity when conducting a network meta-analysis that were not necessary for a pairwise analysis. Specifically, does the guidance on conduct of 'high quality' pairwise meta-analysis, such as that contained in the Cochrane Handbook (Higgins and Green, 2008), need to be extended for a network meta-analysis? There is a premise here, of course, that the 'standard' methods for pairwise meta-analysis are sufficient to ensure *its* validity: as we shall see, this is open to question.

The issue has become further clouded because many of the assumptions of network meta-analysis can be derived from others, and it can become difficult to tell whether or not a particular assumption is 'additional' to assumptions that are already made for pairwise meta-analysis. We shall try to unpick which assumptions are derived from others, which are re-statements of the same assumptions in other terms and which, if any, are additional.

Random effects pairwise meta-analysis makes an assumption of *exchangeability* of trial-specific treatment effects (see Chapter 2). (Fixed effects models are a special case where there is no between-trial variation because every study estimates the same effect.) Note that exchangeability is a relationship between parameters in a hierarchical model. In a pairwise meta-analysis, predictive cross-validation (DuMouchel, 1996) is one way of demonstrating that the exchangeability assumption has *not* been met (see Chapter 3). Comparison of a random effects model with an unrelated mean effects model using DIC and $\bar{D}$ (Chapter 7) might also indicate a lack of exchangeability if the random effects model failed to demonstrate shrinkage. But it is probably impossible to verify that the exchangeability requirement *has* been met.

Moving to network meta-analysis, the only assumption required is, again, exchangeability of the true trial-specific effects. In the context of a network meta-analysis, it needs to be understood that this applies to the entire ensemble of trials, including treatment effect parameters that may have never been observed: for example, the $\delta_{i,BC}$ effects in AD trials are to be considered exchangeable with those in BC trials. One might imagine that the entire set of M trials had each included all S treatments and that subsequently some of the treatment arms went 'missing at random' (MAR). This is most simply interpreted as meaning that the missingness is without regard for the presence of effect modifiers. Note that there is no requirement that each treatment is missing with an equal probability. Nor is it necessary that missingness be unrelated to absolute efficacy of each treatment. The only requirement is that missingness is unrelated to *relative* efficacy: it is, after all, the trial-specific relative treatment effects that are assumed to be exchangeable. Arms are MAR, conditional on the original choice of trial design. Trialists are free to choose which treatments to include, without introducing bias.

It is easily shown that the *consistency* assumptions follow mathematically from exchangeability (Lu et al., 2011) (see Exercise 12.1). Consistency is a relationship between the true treatment effects or, in a random effects model, between the means of the random effects distributions. Technically, then, this completes everything that needs to be said about the assumptions of network meta-analysis: exchangeability is the only assumption and exchangeability implies consistency.

12.2.2 Other Terminologies and Their Relation to Exchangeability

However, other investigators have expressed different views, using different terminologies, not always in the same way. In a key paper, Song et al. (2009) refer to three separate assumptions: homogeneity, similarity and consistency. Homogeneity is said to be the standard assumption in pairwise meta-analysis. In a network meta-analysis, in this formulation, homogeneity should be fulfilled separately for each pairwise contrast. Similarity and consistency are seen as additional assumptions required for network meta-analysis. It appears that the term homogeneity is being used to stand for exchangeability, perhaps implying an additional requirement that the degree of variation is not too large. While we certainly agree that the extent of heterogeneity is a major issue, and in fact the most important one (see succeeding text), we do not believe that the assumptions required for valid evidence synthesis include specific limits on the extent of between-trial variation. Exchangeability implies variation, albeit variation with very specific properties (see Chapter 2), but it implies nothing about the *degree* of variation.

One of the difficulties in this literature is that definitions are somewhat informal and lacking in mathematical development. Exchangeability is clearly a property of the true trial-specific relative treatment effect parameters, $\delta_{i,XY}$, and consistency is a property of pooled mean relative treatment effects. The view that there are three assumptions, homogeneity, similarity and consistency, and that they are all separate assumptions that must be met, has been repeated many times, but the terms are not used in precisely the same way. Similarity has been described as a property of 'trials' or of 'moderators of relative treatment effects' (Song et al., 2009). Elsewhere the consistency assumption is said to refer to 'evidence' (Edwards et al., 2009). But other authors using the same homogeneity, similarity and consistency terminology are clear that these are properties of parameters and also note that they are very closely related (Donegan et al., 2011).

Cipriani et al. (2013) explain that 'the main assumption (…) is that there are no important differences between the trials making different comparisons, other than the treatments being compared'. This is an elegant and simple way of stating the exchangeability assumption and an endorsement that it is the only assumption required.

Another term that has been introduced is *transitivity* (Baker and Kramer, 2002; Salanti, 2012; Cipriani et al., 2013), which is, like consistency, a relationship between parameters, and it is usually accepted that the terms are equivalent (Salanti, 2012). Some experts prefer it because it reminds us that the mathematical relationships between parameters that need to be true for valid inference must also be in place even when no evidence loops are involved, for example, with indirect comparisons. Both transitivity and consistency are relationships between expectations of parameters in a loop; both are implied by exchangeability, but neither implies exchangeability (see Exercise 12.2). However, once again, there is a lack of formal development and an unclear and inconsistent use of terms. In some papers it is implied that consistency can hold when transitivity does not (Salanti, 2012). Elsewhere transitivity is supposed to be another way of referring to similarity of study characteristics (Puhan et al., 2014). Yet another usage has transitivity as a property required of estimates rather than a property of parameters (Baker and Kramer, 2002).

A series of empirical reviews of the methodological quality of applied network meta-analysis (Edwards et al., 2009; Song et al., 2009; Donegan et al., 2011) have tried to document the extent to which researchers have checked or even shown awareness of particular assumptions. This is a difficult exercise to carry out and interpret in the absence of clarity on what the assumptions actually are. We expect we are not the only authors who have submitted a network meta-analysis with a thorough analysis of consistency only to have a reviewer complain that we have not examined, or even mentioned, transitivity (when in fact this is just another name for consistency). All this gives weight to calls to regularise the way network meta-analyses are reported (Bafeta et al., 2014; Hutton et al., 2014).

In view of the overlap and redundancy in these conflicting formulations and the inconsistent use of terminology, we prefer to couch the assumptions of network meta-analysis in terms of exchangeability, a long-standing concept whose properties are well studied (Anon, 2015).

However, none of this addresses the legitimate concerns of investigators wishing to ensure that their estimates are valid. Although it is easy to criticise the literature on the assumptions of network meta-analysis for spreading terminological confusion, it is a body of literature that correctly identifies the issues that must be addressed if valid inferences are to be drawn from network meta-analysis. In Section 12.3, we explore this in a slightly more formal way through a series of thought experiments, which have two purposes. The first is to determine whether direct evidence is more likely to lead to valid inference than indirect evidence. The second is to identify the main factors that determine whether inference in evidence synthesis is likely to be valid or not and to gain a more quantitative understanding of their impact.

12.3 Direct and Indirect Comparisons: Some Thought Experiments

Many investigators have assumed that 'direct' evidence is superior to 'indirect' (Song et al., 2009; Donegan et al., 2011). Cranney et al. (2002) state

> ...an apparently more effective treatment may have been tested in a more responsive population. ... Conclusions about the relative effectiveness of therapies must await results of head-to-head comparisons.

But if this is interpreted as a concern about the distribution of an unrecognised effect modifier, how can we be sure that any additional direct evidence will be any better? If the effect modifier is unrecognised, there will be no way of knowing whether it is present in the proposed head-to-head comparison or not.

Another claim is that indirect comparisons are 'observational studies' (Higgins and Green, 2008):

> Indirect comparisons are not randomized comparisons.... They are essentially observational findings across trials, and may suffer the biases of observational studies, for example due to confounding.... unless there are design flaws in the head-to-head trials, the two approaches should be considered separately and the direct comparisons should take precedence as a basis for forming conclusions.

But other authorities have regarded even pairwise meta-analysis as 'observational' (Victor, 1995; Egger et al., 1997).

Besides the fact that indirect estimates tend to have greater variance than direct estimates, there has been little in the way of formal analysis to support any of these claims. If a collection of five trials is 'observational', why should a single trial be any different? And how can a meta-analysis, which is supposed to be more reliable than a single trial, be observational while a trial is not? Or is a single trial an observational study, too?

Randomisation protects each trial from confounding factors: these are covariates that affect the trial outcome (without necessarily changing the relative treatment effect) and are distributed differently in each arm because of the selection biases that can occur in observational studies. In principle, therefore, each trial delivers an unbiased estimate of the treatment effect for its trial population. Neither a single trial nor a collection of trials is an observational study in the usual sense. The difficulty with trials is not one of confounding, but of effect modification. A random effects meta-analysis model delivers a weighted average of the unbiased trial estimates, and the pooled estimate is therefore also, in a sense, unbiased. If the variation is due to patient population heterogeneity, the problem is that we no longer know what the pooled estimate is an estimate *of*. In the presence of unrecognised effect modifiers, it is not clear what each trial is estimating nor what the pooled estimate means. *The between-study variance in the random effects model actually reflects the uncertain relevance of the trial data to our target population.* This is the rationale for using the predictive distribution rather than the distribution of the mean treatment effect (Section 5.6.2) and why we emphasise in Section 12.3.2 and throughout the book (Chapters 7–9) that between-trial variation should be minimised.

In Chapter 1 we point out that equation $\hat{d}_{BC}^{Ind} = \hat{d}_{AC}^{Dir} - \hat{d}_{AB}^{Dir}$ tells us that indirect estimates inherit their properties exclusively from the direct estimates they are made up from. In the next section we use this simple relationship to sketch out a formal investigation of the issues and provide some preliminary results.

12.3.1   Direct Comparisons

To avoid having to worry about sampling error, our 'thought experiments' will be conducted on the results of infinite-sized trials. Consider indicator Q for the presence of a trial-level effect modifier. When $Q = 0$, the true effect of B relative to A is δ_{AB}; when $Q = 1$, the effect is $\delta_{AB} + \theta$. For simplicity, and without prejudice to the argument, we are going to assume that this effect modifier reflects a trial characteristic that is either 100% present or 100% absent in any trial – for example, $Q = 0$ is primary and $Q = 1$ is a secondary care setting. As usual, if investigators were aware that Q was an effect modifier, separate analysis or covariate adjustment should be used to analyse data from AB trials (Chapter 8).

If Q is an unrecognised covariate, we are immediately in difficulty because it is now unclear what the target parameter for inference is. Let us introduce a new parameter π, the probability that the investigator will happen to choose a $Q = 1$ trial setting. If hundreds of infinite-sized trials were run, we would expect the pooled estimate to be close to $\delta_{AB} + \pi\theta$, as some trials would be conducted in settings where $Q = 1$, while others would have $Q = 0$. Although we probably all feel some discomfort at this definition of the target parameter, it is nevertheless what the meta-analysis delivers. The target parameter is thus dependent on the prevalence of Q in the 'population' of trials. This simply restates what has been known for a long time: there is no population interpretation for a random effects model (Rubin, 1990). Of course, if we knew that 80% of the target population were treated in a primary care setting and we were aware this was an effect modifier, we would adopt $\delta_{AB} + 0.2\theta$ as our target parameter and the proportion of *trials* in which Q was 0 or 1 would become irrelevant (Rubin, 1990).

Under these circumstances no single trial can ever be 'on target': the estimate from every trial is biased. A trial will either estimate δ_{AB}, if $Q = 0$, or $\delta_{AB} + \theta$, if $Q = 1$; it will never estimate the target $\delta_{AB} + \pi\theta$. We can show that the expected error is 0:

$$(1-\pi)\big((\delta_{AB}+\pi\theta)-\delta_{AB}\big)+\pi\big((\delta_{AB}+\pi\theta)-(\delta_{AB}+\theta)\big)=(1-\pi)\pi\theta+\pi\theta(\pi-1)=0$$

However, this represents the error that would be seen 'in the long run' if we conducted hundreds of trials. In practice, when confronted with a single trial, we are unaware of the effect modifier, and we therefore do not know whether the result is an overestimate or an underestimate of the 'long run' expected effect. In these circumstances, the *expected absolute error* is the appropriate statistic that reflects the size of the deviation of the observed estimate from its true long-term value:

$$\begin{aligned}(1-\pi)\big|(\delta_{AB}+\pi\theta)-\delta_{AB}\big|+\pi\big|(\delta_{AB}+\pi\theta)-(\delta_{AB}+\theta)\big| \\ =(1-\pi)\pi\theta+\pi\theta(1-\pi)=2\pi\theta(1-\pi)\end{aligned} \tag{12.1}$$

If $\pi = 0.5$, for example, the expected absolute error is 0.5θ (Table 12.1). This seems at first sight to be a startling result, but in fact it accords with what many commentators have always noted – trials often give the 'wrong' results (Ioannidis, 2005), as a result of random sampling of the effect modifier.

Having looked at the expected absolute error as a measure of bias in a single trial in the presence of an unrecognised effect modifier, we now look at single realisations of meta-analyses consisting of two, three or more trials. Consider a meta-analysis of two trials with $\pi = 0.5$. There is a 25% chance that $Q = 1$, a 25% chance that $Q = 0$ on both trials and 50% that one trial has $Q = 1$ and the other is $Q = 0$. Here the expected absolute error is $\theta/4$ (Table 12.2). As the

Table 12.1 Expected error and expected absolute error in a 'direct comparison' meta-analysis with $N = 1$ RCTs in the presence of an unrecognised effect modifier, present with probability $\pi = 0.5$.

Outcomes	Trial outcomes	Pr(outcome)	Meta-analysis	Error	| Error |
1	δ_{AB}	0.50	δ_{AB}	$-\theta/2$	$\theta/2$
2	$\delta_{AB} + \theta$	0.50	$\delta_{AB} + \theta$	$+\theta/2$	$\theta/2$
Expectation				0	$\theta/2$

Pooled effect target parameter is $\delta_{AB} + \theta/2$.

Table 12.2 Expected error and expected absolute error in a 'direct comparison' meta-analysis with $N = 2$ RCTs in the presence of an unrecognised effect modifier, present with probability $\pi = 0.5$.

Outcome	Trial outcomes	Pr(outcome)	Meta-analysis	Error	| Error |
1	δ_{AB}, δ_{AB}	0.25	δ_{AB}	$-\theta/2$	$\theta/2$
2	$\delta_{AB}, \delta_{AB} + \theta$	0.50	$\delta_{AB} + \theta/2$	0	0
3	$\delta_{AB} + \theta,$ $\delta_{AB} + \theta$	0.25	$\delta_{AB} + \theta$	$\theta/2$	$\theta/2$
Expectation				0	$\theta/4$

Pooled effect target parameter is $\delta_{AB} + \theta/2$.

number of trials increases, the expected absolute error decreases (see Exercise 12.3). This is portrayed in Figure 12.1: with $\pi = 0.5$, the expected absolute error decreases to 0.125θ when the number of trials $M = 10$ and then to 0.08θ at $M = 20$ and 0.065θ at $M = 40$ trials.

Note in passing that the thought experiments extend naturally to continuous trial-level covariates where we would sample Q from a continuous distribution and have a distribution of interaction terms.

We conclude that in the presence of unrecognised effect modifiers, direct comparisons are biased with respect to their 'population' target parameter but that the degree of bias decreases geometrically with the numbers of trials. These calculations provide theoretic support for the widely held intuition that meta-analysis becomes more reliable as the number of trials increases. But they also show that bias – the difference between what would be observed and the true estimate – is directly proportional to the size of the interaction effect θ. This again points to the importance of not only identifying known effect modifiers in advance but also taking deliberate action to limit the *extent* of heterogeneity from unknown sources.

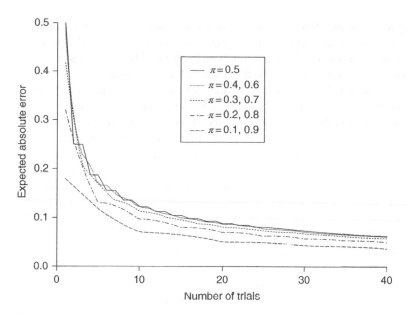

Figure 12.1 Direct comparisons: expected absolute error in meta-analyses, in units of θ, as a function of the number of trials and the population proportion π of trials with a trial-level effect modifier that adds θ to the treatment effect.

12.3.2 Indirect Comparisons

We now extend the same analysis to indirect comparisons: our estimate of the AB effect will be formed from AC and BC trials. Again, to keep it simple but without prejudice to the generality of the argument, we will assume that all trials are of infinite size and that there are always an equal number of AC and BC trials. We begin with a scenario where the effect modifier that acts on the AB effect also impacts the exact same extent on AC comparisons, but it does not impact BC comparisons. A typical scenario would be where A is placebo and B and C are active treatments in the same class. We have $\delta_{AB} = \delta_{AC} - \delta_{BC}$, and the target parameter, of course, is unchanged: $\delta_{AB} + \pi\theta$.

Consider first a single AC and a single BC trial. With $\pi = 0.5$ the AC trial has a 50% chance of being run at the $Q = 0$ setting and 50% $Q = 1$. Therefore the expected pooled estimate is $\delta_{AC} + \theta/2$, and the expected bias is zero. The expected absolute error, however, is again $\theta/2$. Because the BC trial is immune to the effect modifier, the indirect estimate of δ_{AB} inherits its expected absolute error from the AC trial. Clearly, as we go from one each of AC and BC trials to two or more each, the expected absolute error in the indirect estimate will always be exactly the same as the expected absolute error in a direct estimate based on the same number of trials.

In Chapter 1 we pointed out that because $\hat{d}_{BC}^{Ind} = \hat{d}_{AC}^{Dir} - \hat{d}_{AB}^{Dir}$ is an equation, the left-hand side can only be biased if the right-hand side is biased. The previous discussion is just a further illustration: where just one of the direct estimates is biased, the indirect estimate is biased to exactly the same degree.

This suggests that if *both* direct estimates are biased, then the indirect estimate might inherit a double dose of bias. This is precisely what happens if we consider a second indirect comparison scenario. Suppose we now have the same set of treatments (A placebo, B and C active and of the same class), but we now wish to make inferences about δ_{BC} from the AC and AB trials. Starting again with a single AB and a single AC trial, there are four possible outcomes, and the expected absolute error is $\theta/2$, as before. However, as the number of trial pairs increases, the expected absolute error continues to decrease, but the degree of bias is always greater than in direct comparisons (Figure 12.2), because *both* direct estimates can be biased (see Exercise 12.4).

We can summarise these findings as follows. In the presence of unrecognised effect modifiers, both direct and indirect evidence are biased (relative to the target parameter) due to random sampling of the effect modifier. The extent of bias is identical if only one of the sources of direct evidence is subject to the

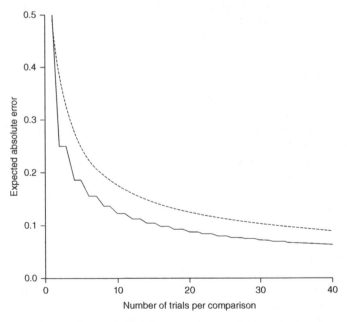

Figure 12.2 Absolute expected error in direct comparisons, or indirect comparisons, where only one of the direct contrasts is subject to an effect modifier present with probability $\pi = 0.5$ (lower curve) and indirect comparisons where both direct contrasts are subject to the same effect modifier (upper curve). In units of the interaction term.

effect modifier. Where both sources of direct evidence are subject to the same effect modifier, the indirect evidence is even more biased, but the notable finding, perhaps, is that in this scenario the direct estimate is unbiased in the presence of an unrecognised effect modifier, because it affects both arms equally. This is, in fact, exactly what we observed in Chapter 8 when exploring covariate effects using meta-regression: with a single observed covariate acting on AB and AC trials, the regression coefficients for the interaction terms in BC trials cancel out.

The scenarios we have explored previously are for 'fixed' effects modifiers, as used in meta-regression (Chapter 8). This form of analysis can be extended to random biases (Chapter 9) and to multiple effect modifiers. Under some circumstances random biases – or to be more precise *their expectations* – may 'cancel out' (Song et al., 2008), making indirect comparisons potentially *less* biased than direct. But in other cases, where the indirect comparison is formed by addition rather than subtraction, the expected biases add as well as their variances, making indirect comparisons substantially worse.

The prediction that direct comparisons between active treatments in the same class will be relatively unbiased is supported empirically by studies that show that active–active comparisons tend to have a lower degree of between-trial heterogeneity (Turner et al., 2012). Conversely, the finding that indirect comparisons are particularly vulnerable to bias when the comparators are in the same class is significant, as this is probably the most common kind of indirect comparison. This situation is portrayed in Figure 12.3, which shows a constant d_{BC} effect, at all levels of the effect modifier, but highly variable results when this is estimated indirectly from AB and AC trials.

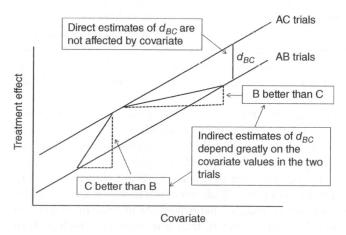

Figure 12.3 Illustration of the difference between direct and indirect estimates formed from two direct comparisons that are *both* affected by an unrecognised effect modifier.

The thought experiments established that, under a wide range of circumstances, meta-analytic estimates based on direct and indirect evidence are equally prone to error in the presence of unrecognised fixed effects modifiers. Under some circumstances, indirect may be more biased, and direct virtually unbiased. Two further important findings are as follows: (i) the degree of bias increases in proportion to the size of the interaction effect associated with the effect modifier and (ii) the degree of bias decreases as the number of studies increases.

12.3.3 Under What Conditions Is Evidence Synthesis Likely to Be Valid?

These last two findings have interesting theoretical and practical implications. On the practical side, investigators may not have much control over how many studies are included in their evidence synthesis, but they are in a position to create conditions under which the risks of invalid inference are curtailed simply by limiting the size of potential interaction effects, which means limiting the degree of clinical heterogeneity at the study inclusion/exclusion stage. Many of the key practical consequences have already been discussed, such as the need to avoid 'lumping', that is, to avoid averaging relative treatment effects over different doses, different co-treatments or over patient groups on first, second or third line therapy. We have in fact argued that, in a decision-making context, it would make no sense to average over these factors in the first place. But even from a strictly evidence synthesis perspective, averaging over different doses is a strange way of pooling evidence, because different doses are, obviously, deliberately *designed* to have different effects. Similarly, patients who have failed on a specific class of treatments are, by definition, already *known* to react to them differently than patients who have not.

Thinking more broadly about effect modifiers and heterogeneity, our thought experiments suggest there would be great benefit in a more systematic exploration to assess vulnerability of inference. This could consist of a systematic review of the literature on potential effect modifiers, particularly IPD analyses of trial data. The distribution of effect modifiers across trials, and especially across contrasts, including severity at the start of trial, should be explicitly tabulated (Jansen and Naci, 2013). Covariate adjustment via meta-regression can then be considered, either as a base-case analysis or as a sensitivity analysis. Careful attention should be paid to indicators of potential heterogeneity. For example, if the absolute effect under a given treatment is similar across trials, this can give some reassurance that a degree of patient homogeneity across trials is in place. Wide variation in study-specific baseline values does not necessarily indicate heterogeneity in relative effects, but it does point to clinical heterogeneity in trial populations, which in turn indicates that estimated effects are more vulnerable to the presence of effect modifiers.

At a theoretical level, the thought experiments tell us that as long as hetero-geneity can be kept at a low level, the validity of both pairwise and network meta-analysis does not really depend on whether the abstract mathematical conditions of exchangeability are met. In fact, if the degree of heterogeneity can be kept low, exchangeability is almost an irrelevance. After all, if the between-trial standard deviation is only 10% of the treatment effect, we might not be too concerned about deviations from exchangeability. Conversely, if the between-trial standard deviation was as high as 50% of the mean treatment effect, which is by no means unusual, we should be quite concerned particu-larly if there are only a small number of trials on influential comparisons, *even if we knew that exchangeability was in place.* As the number of trials increases, we could be less worried about the degree of heterogeneity, but we would then be vulnerable to failures in exchangeability.

We can now distinguish between two reasons why we might observe inconsistency between direct and indirect evidence. The first, due to failure in exchangeability, we might call *intrinsic inconsistency*. In the language of the thought experiments, the probability that $Q = 1$ in AB trials, π_{AB}, is different from the corresponding probability π_{AC} in AC trials. In the second case we have *chance inconsistency*: here the underlying probability that $Q = 1$ is the same in both AB and AC trials, but we obtain a different distribution of effect modifiers through random sampling.

We might summarise this as follows. From a technical point of view, exchange-ability is the only assumption both in network and in pairwise meta-analyses. But the validity, or otherwise, of estimates in both cases depends on the size of interaction effects (heterogeneity) and the number of trials, whether or not the exchangeability assumption is met. Given that it is virtually impossible to ensure that exchangeability is in place, validity can only be ensured by limiting the degree of heterogeneity. Thus, while 'homogeneity' and 'similarity' (Song et al., 2009) are not strictly necessary assumptions, investigators would be well advised to pay them very close attention.

12.4 Empirical Studies of the Consistency Assumption

On the face of it, meta-epidemiological studies testing the heterogeneity assumption using the methods set out in Chapter 7 would seem to be the obvious way of gaining some understanding of the validity of network meta-analysis. In the largest study of this type, looking at 112 triangular networks, significant ($p < 0.05$) inconsistency was reported in 16 (14%, 95% CI 9–22%) networks based on the Bucher method (Song et al., 2011). There are technical difficulties in interpreting this, because under the assumption of consistency there are quite strong constraints on the between-trial variances in loops of evidence (Lu and Ades, 2009), which were not taken into account. This was explored in a more thorough investigation of network inconsistency (Veroniki et al., 2013) looking at 40 networks with four or more treatments containing over 300 evidence loops. The networks were identified from the PubMed system. This paper showed that 9% of triangle loops would be inconsistent using the simple Bucher method adopted by Song et al. (2011), but only 5% when a more stable and consistent estimate of the variance of effects was used. However, using the tests for design inconsistency (Jackson et al., 2014) (see also Chapter 7), they found that eight (20%) of the networks showed evidence of global inconsistency.

Confirming the results of our thought experiments in the previous section, these studies reported a strong negative relation between the risk of inconsistency and the number of studies informing the evidence loop (Song et al., 2011), and a high risk of inconsistency in loops where one of the contrasts is informed by only a single study (Veroniki et al., 2013). As noted in Section 12.3, this does not necessarily indicate a failure in the technical exchangeability assumption: a higher probability of chance imbalances in effect modifiers introduced through random variation is exactly what we would predict from the thought experiments.

The significance of findings from such meta-epidemiological studies is far from clear. If an evidence network is inconsistent, then we know that either the exchangeability assumptions required for valid network meta-analysis were not met (intrinsic inconsistency) or they were met but chance imbalances in effect modifiers were enough to lead to 'inconsistency' being detected.

If the evidence networks are based on pairwise systematic reviews that do not have the same trial inclusion/exclusion criteria, the data generation process might easily deliver a systematic imbalance in effect modifiers, constituting a deviation from exchangeability. This might lead to inconsistency in even the largest networks. On the other hand, if the entire evidence network was assembled under a common inclusion/exclusion protocol, the risk of inconsistency is distinctly lower in large networks, but where small numbers of trials are involved, chance imbalances in effect modifiers are very likely.

Thus, we can only make sense of meta-epidemiological studies if we go back to examine the precise protocols that were followed and look for relationships between the protocols and the findings. The same misgivings must apply to other meta-epidemiological studies of network meta-analysis (Chaimani and Salanti, 2012).

Cochrane reviews are carried out under rigorous protocols that are open to examination. But there appears to be nothing in these protocols that is specifically intended to have the effect of guaranteeing, or helping to guarantee, the exchangeability assumptions that are required for valid evidence synthesis. Nor is there an explicit recognition that clinical heterogeneity must be kept to a minimum, whether to limit the impact of potential deviations from exchangeability or just to ensure the interpretability of the pooled mean. We could, for example, examine the literature identification and selection protocols to see how recognised effect modifiers were to be dealt with. A particular issue is whether the inclusion criteria would result in pooling estimates from different doses, different co-treatments and different formulations or for first, second and third line therapies. As we have seen, 'lumping' over all of these is very common in many Cochrane reviews (see Preface and Chapter 1). It may be no coincidence that inconsistency was found in 16/85 Cochrane reviews compared with 0/27 non-Cochrane reviews ($p = 0.011$) (Song et al., 2011), although this could also be due to the numbers of trials informing each contrast.

In a scientific context where one is asking whether a product or intervention 'in principle' has an effect, lumping may be appropriate given the 'hypothesis testing' rationale (Gotzsche, 2000). In a decision-making context, by contrast, the emphasis is on the coherent estimation of a set of specific relative effects. It would normally be inappropriate to generate effect estimates that were averaged overdose or stage of disease progression, as such estimates have no interpretation for decision makers, and it is debatable as to whether they have any clinical meaning.

It is significant that the meta-epidemiological literature on inconsistency has so far been restricted to systematic reviews and clinical studies and has not looked at the use of network meta-analysis in health technology assessment, for example, in submissions to the NICE technology appraisal process or to similar bodies in other countries. In our experience, evidence networks assembled for decision-making seldom show signs of inconsistency, which we attribute to narrower definitions of treatment and patient population. Empirical studies of inconsistency in this literature would be worthwhile.

12.5 Quality of Evidence Versus Reliability of Recommendation

While previous sections have looked at the validity question from the perspective of someone about to undertake a review and evidence synthesis, we now consider the same questions as they might occur to someone reading a report of the results. Given that a clear statement on the presence or absence of inconsistency will seldom be possible, what can be said about the validity of a network meta-analysis? Once again, a small amount of theory takes us a long way, and before reviewing the proposals that have been made, we begin by reminding ourselves of the formal properties of network estimates.

12.5.1 Theoretical Treatment of Validity of Network Meta-Analysis

In pairwise meta-analysis, whether based on fixed or random effects models, the summary pooled estimates are weighted averages of the treatment effect estimates from the original trials. This same is true in network meta-analysis: each coherent estimate is a linear combination of coefficients (weights) and contrast-specific treatment effects (Lu et al., 2011). For example, a network with AB, AC, BC, AD and BD trials,

$$\hat{\delta}_{AB}^{Coh} = \beta_1\hat{\delta}_{AB} + \beta_2\hat{\delta}_{AC} + \beta_3\hat{\delta}_{BC} + \beta_4\hat{\delta}_{AD} + \beta_5\hat{\delta}_{BD} \tag{12.2}$$

This finding has a number of important implications. The first is that if we are satisfied that the in-going treatment effects are unbiased estimates for the target population, we can be confident that the out-going coherent estimates are also unbiased. The network meta-analysis cannot, for example, 'add bias' that is not already present in the input evidence.

A second implication is that in cases where there are doubts about potential bias in the input study estimates, we should take action at the outset to manage this. Bias adjustment of individual studies is a subjective process, but if a consensus can be reached on the distributions of study-specific biases (Turner et al., 2009), investigators can be confident in a network meta-analysis based on the adjusted trial estimates. Alternatively, the methods of Chapter 9 can be used to adjust out generic biases in the network model, although we should view the results with some caution, as with any form of meta-regression.

A third implication is that, again, in the presence of potential bias, we can use the weight coefficients $(\beta_1, \beta_2, \ldots)$ in equation (12.2) to tell us exactly how much impact a potential bias in, say, the estimate of δ_{BC} could be having on a final estimate $\hat{\delta}_{AB}^{Coh}$. We return to this in Section 12.5.3.

12.5.2 GRADE Assessment of Quality of Evidence from a Network Meta-Analyses

There are two proposals (Puhan et al., 2014; Salanti et al., 2014) on how to develop quality assessments for network meta-analysis. Both start from the quality assessments of pairwise meta-analyses from the Grading of Recommendations Assessment, Development and Evaluation (GRADE) working group (Guyatt et al., 2008).

GRADE assessments of quality cover five *domains*: 'study limitations', approximately risk of internal bias (Guyatt et al., 2011e); 'inconsistency', the term used for heterogeneity (Guyatt et al., 2011c); 'indirectness', which refers to the lack of applicability due to differences in outcome or in target population (Guyatt et al., 2011b); 'imprecision', a high variance in the estimate (Guyatt et al., 2011a); and 'publication bias' (Guyatt et al., 2011d). In each domain, the process assigns an ordered quality score: high, moderate, low and very low. Finally, the five domain-level quality assessments are combined to give an overall quality assessment (Balshem et al., 2011).

For network meta-analysis, the objective of the GRADE assessment is to assign a quality rating score to each and every pairwise effect estimated by the network meta-analysis. The process starts from the five GRADE domain assessments applied to each of the 'direct' estimates. Then a parallel set of GRADE assessments is developed for each of the 'indirect' estimates. Because there are often several non-independent indirect estimates for each comparison, investigators are advised to focus attention on the indirect estimate that is formed from the two direct contrasts that are supported by the most evidence. The quality rating of the indirect evidence is then the lowest of the ratings of the constituent parts, although it may be rated down a level if there is an imbalance in effect modifiers (Puhan et al., 2014).

The final step is to assign a GRADE quality rating to each of the (coherent) network estimates themselves: this is taken to be the highest of the quality rating assigned to the direct and indirect evidence on that contrast, but this can also be downrated one step if the direct and indirect estimates are 'incoherent' (inconsistent in the terminology of this book). Some leeway is allowed on how this inconsistency is assessed: investigators may use node splitting (Chapter 7), or they can pick out the main source of indirect evidence and use the Bucher method. It is also possible for investigators to simply assign the final GRADE rating on the basis of the direct and indirect ratings without a formal examination of inconsistency (Johnston et al., 2014).

Readers will note that in order to assign GRADE quality ratings to a network meta-analysis, it is not necessary to actually carry one out: the entire process can be put together from the original GRADE assessments of the individual pairwise contrasts. Another property is that it does not give an overall quality assessment of the network meta-analysis results, only a set of unrelated assessments of the quality of the estimates for each separate contrast. This takes us back to evaluating a set of unrelated pairwise comparisons (see Chapter 1). There is no specific advice on how a decision maker choosing between S treatments is supposed to use the $S(S-1)/2$ GRADE quality assessments.

It is suggested, however, that clinicians could 'choose a lower ranked treatment with supporting evidence they can trust over a higher ranked treatment with supporting evidence they cannot trust' (Puhan et al., 2014). Although decision-making bodies can always take account of numerous factors 'outside' any formal decision analysis, this approach does not seem to accord with any principles of rational and transparent decision-making.

Looked at in statistical terms, by treating the assessments of the pairwise summaries as independent, the GRADE process fails to recognise that the risk of bias, heterogeneity, relevance and publication bias issues that impact on each contrast in an evidence network are likely to be conceptually and quantitatively similar. Biases caused by known effect modifiers can be removed by covariate adjustment (Chapter 8). Further, as discussed in Chapter 9, random biases associated with risk of bias indicators, or publication biases, can be seen as generic biases operating throughout the network, which can be adjusted for by a suitable model. Treating them as independent generates a weak analysis when compared with explicit modelling of effect modifiers and covariates.

But the main weakness of a quality assessment system is that it is not clear how decision makers should use it. It requires considerable scholarship to understand, collate and document the various internal and external validity biases in order to implement the GRADE assessment of quality of network evidence, or the Cochrane Collaboration risk of bias tool (Higgins and Altman, 2008; Lundh and Gotzsche, 2008; Higgins et al., 2011). However, the resulting 'evidence profiles', quality ratings and accompanying narrative are usually consigned to an appendix and have little or no discernible impact on the final recommendation.

Another method for assigning quality ratings to network evidence has been developed by statisticians associated with the Cochrane Collaboration (Salanti et al., 2014). This starts with the GRADE assessments of the pairwise contrasts at the domain level, as described previously. These are combined into domain-specific assessments of the entire evidence network, using the 'contributions matrix' (König et al., 2013; Krahn et al., 2013). This is the matrix of weight coefficients $(\beta_1, \beta_2, ...)$ from equation (12.2) that determine the influence that each piece of evidence has on each of the estimates (Lu et al., 2011). This is an interesting, although complex, method, and readers are referred to the original papers for details. However, the quality assessments of each of the network estimates that it delivers are, according to its authors, expected to be similar to the assessments given by the GRADE approach. This method also delivers a quality rating of the rankings in the network as a whole. While this is probably a more useful output, it is still unclear how a decision maker would use it.

12.5.3 Reliability of Recommendations Versus Quality of Evidence: The Role of Sensitivity Analysis

The approach we prefer is to deliver to decision makers an evidence synthesis, or a synthesis embedded in a cost-effectiveness analysis or other decision-analytic model, that incorporates all the relevant adjustments and uncertainties, both those due to sampling error and those associated with known effect modifiers, random biases or uncertain relevance.

This is the 'base-case' analysis from which treatment recommendations can be directly derived. As noted in Chapter 5, it delivers a decision that is 'optimal', given the available evidence, but which is not necessarily the 'correct' decision because it is made under uncertainty. Once the base-case analysis has been generated, decision makers may wish to take into consideration factors that are not included in the formal model. This should include an analysis of the robustness of the recommendation to changes in the assumptions. This is the role of sensitivity analyses. In a decision-making context, sensitivity analysis has always been the proper way of raising, and analysing, questions relating the reliability of conclusions. The purpose of a sensitivity analysis is not to assign a quality rating to the analysis nor to overturn the base-case analysis. Its role is to inform decision makers whether plausible changes in their original assumptions, including their assumptions about the evidence, could lead to changes in their final recommendations.

Rather than provide a quality rating to every estimate, the sensitivity analysis can focus attention on what are probably a small number of elements in the evidence ensemble, which are 'driving' the results. The same level of scrutiny and the same degree of scholarship required to assign GRADE ratings to evidence, or to operate the Cochrane risk of bias tool, must now be deployed but are now directed towards providing not a 'quality assessment' but a *quantitative* assessment of the likely biases and their potential impact on the treatment decision. What is required is a structured sensitivity analysis that asks questions of the form: 'if the evidence on contrast XY is biased by an amount β, would that change the treatment decision, and what would the new recommendation be?' A simple method for this kind of sensitivity analysis has been illustrated by Caldwell et al. (2016).

Clearly, the results of such an investigation will depend on the precise quantitative relationships between the two or three treatments with the highest expected score on the objective function used for decision-making, whether efficacy or net economic benefit. The GRADE quality ratings contain no information on the size or direction of treatment effects, let alone the size of differences between treatments, nor on the size of potential biases. Fundamentally, GRADE ratings might tell us how much credence to give each network estimate, but they provide no information on how the treatment recommendation might change if 'better' information was available in different parts of the network. This would seem to be a fruitful area for further research.

12.6 Summary and Further Reading

Network meta-analysis is now routinely used in every area of clinical medicine, in academic papers (Lee, 2014), in clinical guideline development and in re-imbursement decisions in several countries. Nevertheless, doubts about the method continue to be expressed. Some of these concerns relate to whether the assumptions of network meta-analysis are being met, but there has been a degree of confusion about exactly what the assumptions are, leading to alternative formulations and inconsistent use of terminology.

We argue that exchangeability of the trial-specific relative effect parameters is the only technical assumption required and that consistency of the expected treatment effects follows from exchangeability. However, it is virtually impossible to establish that exchangeability holds in any specific instance. The key strategy to ensure validity of inference, whether from direct or indirect data, is to limit the impact of deviations from exchangeability and the impact of chance imbalances in effect modifiers, but holding down clinical heterogeneity. Some suggested 'action points' to achieve this are listed below.

It has been widely stated, or implicitly assumed, that 'direct' evidence is superior to 'indirect' evidence. We have pointed out throughout this book that the properties of indirect estimates can only be inherited from the direct estimates of which they are composed. Our thought experiments in Section 12.3 show that, in general, direct and indirect estimates are equally vulnerable to bias, but that in certain circumstances, where two active treatments of the same class are being indirectly compared via a placebo, indirect comparisons will be more biased and direct comparisons potentially free from bias. These exercises also confirmed that the error that we can expect to observe in syntheses of small numbers of trials is directly proportional to the size of the interactions that drive effect modification. Also, meta-analytic estimates are increasingly vulnerable to bias as the number of trials diminishes.

The thought experiments of Section 12.3 led us to distinguish intrinsic inconsistency, resulting from failure of exchangeability, from chance imbalances in effect modifiers. The distinction throws a particular light on the inconsistency models (see Chapter 7), which some have proposed should be routinely used to allow for the omnipresent risk of inconsistency in networks (Lumley, 2002; Jackson et al., 2014). We can begin by observing that our standard random effects model of Chapters 2 and 4 already instantiate exactly the data generation process that gives rise to chance inconsistency. There is therefore no need to add a second layer of random variation to absorb this. But if this is the case, how should we interpret the extra variation for inconsistency? These additional parameters are also assumed to be exchangeable, but it is difficult to imagine a process that generates real, but random (and so exchangeable), variation in effect modifier distributions across the various designs. For example, one might reasonably expect that the designers of AB, AC and AD trials each consistently draw their effect modifiers from different distributions, whether intentionally or not. But it is much harder to believe that the parameters that define these different distributions are themselves generated randomly and exchangeably.

This takes us back to inconsistency models in which the inconsistency terms are treated as fixed effects rather than exchangeable. These models, however, while useful for detecting global inconsistency, cannot be used for decision-making because the estimated treatment effects and inconsistency terms depend on parameterisation (Higgins et al., 2012; White et al., 2012).

We would never claim that the problems attending network meta-analysis have been overstated. But it does appear that precisely the same problems attend pairwise meta-analysis as well and that this has been under-recognised by many investigators. The advent of network meta-analysis methods has simply brought attention to issues that were previously ignored. In particular, testing for inconsistency in networks has turned out to be a method for detecting either systematic deviations from exchangeability or the presence of chance imbalances in effect modifiers, both of which signal significant heterogeneity in the evidence base.

Of course, we should not have had to wait for network meta-analysis to raise questions about the validity of pairwise meta-analysis. An extensive earlier literature already points out that random effects estimates had no population interpretation (Rubin, 1990) and warns of the dangers of uncritical pooling of clinically and statistically heterogeneous estimates (Greenland, 1994b). Summary estimates from random effects models have even been described as 'meaningless' (Greenland, 1994a; Shapiro, 1994), although we might add that estimates become increasingly meaningful as the degree of between-trial variation is diminished.

In this closing section we bring together strands from the different chapters, and indeed from classic meta-analysis literature, to offer some suggestions about what can be done to help ensure that conclusions from syntheses of direct *or* indirect evidence are secure.

Question formulation, trial inclusion/exclusion and network connectivity

1) Restrict attention to a clinically meaningful target population at a specific stage of their disease.
2) Ensure that all patients included in the trials could be randomised to any of the treatments in the network (Salanti, 2012) (Chapter 1).
3) As a starting point, keep different doses, different co-therapies, as different treatments. Consider class, dose or treatment combination models (Chapter 8) if appropriate.
4) Where trials report multiple outcomes, consider multi-outcome models to ensure robust and coherent decisions based on all the evidence (Chapter 11).

Heterogeneity and bias management

1) Examine the absolute event rates in different trials. If these are heterogeneous, there is a higher risk that unrecognised effect modifiers are present (Song, 1999).
2) Review the general literature, as well as the meta-analysis literature, for potential and known effect modifiers, and consider meta-regression models (Chapter 8).
3) Examine and report on the distribution of effect modifiers across trials and across pairwise contrasts (Jansen and Naci, 2013).
4) Explore the potential effects of quality-related biases including publication bias, and consider bias adjustment models (Chapter 9).
5) Check for inconsistency and report results (Chapter 7).

Reporting

1) Show how many trials inform each contrast through a table or network diagram. Include a description of trials with more than two arms and how they influence the network structure.
2) Report relative treatment effects relative to a reference treatment.
3) Report model fit and methods for model choice.
4) Report heterogeneity.
5) Give a precise reference for the statistical methods used, and supply the computer code and datasets used to allow readers to replicate results.

Those managing or undertaking systematic reviews leading to network meta-analyses can take advantage of a literature on 'checklists', starting of course with the Preferred Reporting Items for Systematic Reviews and Meta-Analyses (PRISMA) requirements (Liberati et al., 2009). Some of these seek to assign a summary numerical assessment (Oxman, 1994). Others, which we believe are more useful, are intended to assist those whose task is to make recommendations based on meta-analysis, or to the technical analysts advising them, or simply to journal editors and reviewers considering papers for publication (Ades et al., 2012, 2013). There is also guidance and a checklist from an ISPOR task force oriented to network meta-analyses (Jansen et al., 2014).

There is a growing literature on simulation studies, but it is beyond the scope of this book to review this. Some of which have concluded that indirect comparisons are biased (Wells et al., 2009; Mills et al., 2011). As noted previously, this is difficult to understand as indirect comparisons cannot be biased unless direct comparisons are also biased. It seems probable that the bias in indirect estimates has indeed been inherited from the well-recognised biases in random effects variance estimators (Böhning et al., 2002) or possibly they are due to the biases resulting from adding a small constant to zero cells. There are also simulation studies claiming to show that the probability(best) outcomes from Bayesian network meta-analysis are biased, because they are sensitive to the number of studies (Kibret et al., 2014). However, it is well known that the posterior distribution of ranks is expected to be sensitive to the number and size of trials in different parts of the network.

There is certainly a scope for well-conducted simulations both on the impact of between-study variance prior distributions on Bayesian meta-analysis (Gelman, 2006) and on the coverage properties of different types of inconsistency detection, which themselves depend critically on how between-study variances are estimated (Veroniki et al., 2013).

The idea of studying the geometrical properties of networks was introduced in 2008 (Salanti et al., 2008a, 2008b), with a number of suggested metrics based on ecological science. These metrics can provide insights into the processes behind the choice of comparators in new trials. Recently, though, it has been suggested that features of network geometry may be related to 'bias' in the evidence network (Salanti, 2012; Hutton et al., 2015; Linde et al., 2016). As noted previously (Section 12.2.1) bias is only introduced in network meta-analysis if the exchangeability assumption is violated or – saying the same thing another way – the missingness of treatments from trials is related to their relative effectiveness. Decisions about which comparators to enter into trials are often made on marketing grounds, but this does not imply that relative effects are biased. Trial designers are free to choose any treatments for inclusion in trials, even on the basis of their expected relative efficacy, without introducing bias.

We conclude with some brief thoughts about future research. In Chapter 1 we showed how a comparison of pairwise estimates and their credible intervals with the corresponding coherent network estimates could help investigators understand the 'drivers' in the analysis. This is also the key role for sensitivity analysis, touched on in Section 12.5, and surely an area that requires further investigation. For general texts on sensitivity analysis, we refer to relevant chapters and tutorial papers in CEA (Briggs et al., 2006, 2012). However, these techniques concern 'forward' MC simulation models in which each parameter is informed by independent sources of evidence. They do not address the complex flow of evidence in Bayesian evidence networks (Madigan et al., 1997).

We noted earlier that the network meta-analysis represents a relative simple evidence network on the scale of the linear predictor in which the influence of each 'input' observation on a coherent network estimate can be characterised as a weighted average of the 'input' contrast estimates (Lu et al., 2011). This idea has been developed further to provide an analysis of inconsistency (Krahn et al., 2013) and a general analysis of information flow in linear networks (König et al., 2013). These algebraic methods might be further adapted to drive sensitivity analyses, or analyses of how an existing evidence network would respond to additional data or which new data to collect to reduce uncertainty in specific parts of the network.

12.7 Exercises

12.1 Starting from the assumption AB and AC effects are each exchangeable, with $\delta_{i,AB} \sim \text{Normal}(d_{AB}, \sigma_{AB}^2)$ and $\delta_{i,AC} \sim \text{Normal}(d_{AC}, \sigma_{AC}^2)$, demonstrate that consistency holds for d_{BC}.

12.2 Show that consistency does not imply exchangeability.

12.3 Using the same approach as in equation (12.1), and, in Tables 12.1 and 12.2, what is the expected absolute error in a direct comparison based on three trials, when $\pi = 0.5$?

12.4 Extend the same methods to study the extent of bias in indirect comparisons where the BC effect from two AB and two AC trials, in the presence of an effect modifier present with probability $\pi = 0.5$, affects treatments B and C equally but not A.

Solutions to Exercises

Chapter 1

1.1 Using the relationships $\hat{d}_{BC}^{Ind} = \hat{d}_{AC}^{Dir} - \hat{d}_{AB}^{Dir}$ and $V_{BC}^{Ind} = V_{AB}^{Dir} + V_{AC}^{Dir}$, we obtain $\hat{d}_{BC}^{Ind} = -1.37$ and $V_{BC}^{Ind} = 0.429$.

1.2 Using the formula $\hat{d}_{BC}^{Pooled} = \dfrac{w_{BC}^{Dir}\hat{d}_{DC}^{Dir} + w_{DC}^{Ind}\hat{d}_{BC}^{Ind}}{w_{BC}^{Dir} + w_{BC}^{Ind}}$, we find $\hat{d}_{BC}^{Pooled} =$

$$\frac{0.47/(0.1)^2 - 1.37/0.429}{(1/0.1)^2 + (1/0.429)} = 0.428$$

1.3 The treatment network is given in Figure 1.1c. Notice that an indirect comparison can be estimated when there is no single common comparator, although the standard error obtained is quite large:

$$\hat{d}_{AD} = \hat{d}_{AB} + \hat{d}_{BC} + \hat{d}_{CD} = -2.8 + 2.7 + 3.0 = 2.9$$

$$s.e_{AD} = \sqrt{V_{AB} + V_{BC} + V_{CD}} = \sqrt{1.42^2 + 1.24^2 + 1.20^2} = 2.23$$

1.4 Analysis 1 comes from a network meta-analysis, and analysis 2 from pairwise comparisons. We do not see a coherence relationship between the posterior means because the odds ratios have skewed posterior distributions. However, we observe coherence in the posterior medians as these will reflect the approximately normal log odds ratios. For example, $2.289 = 1.624 \times 1.411$.

1.5 This can happen when the distribution of the treatment effect is asymmetrical. For example, consider two random variables, X and Y. If X is normally distributed with mean and median 10 and precision 0.25, its

Network Meta-Analysis for Decision-Making, First Edition. Sofia Dias, A. E. Ades, Nicky J. Welton, Jeroen P. Jansen and Alexander J. Sutton.
© 2018 John Wiley & Sons Ltd. Published 2018 by John Wiley & Sons Ltd.
Companion website: www.wiley.com/go/dias/networkmeta-analysis

95% credible limits are 6.1–13.9. Suppose Y is log-normally distributed with mean and precision 1.9 and 1.0, respectively, on the log scale. On the natural scale this means that Y has mean 11, median 6.6 and 95% limits of 0.9–48. But $\Pr(Y > X)$ is 0.35. If these distributions represented efficacy and this was to be the basis of our recommendation, Y would be the optimal choice, but X was more likely to be a better treatment.

Chapter 2

2.1 The code to fit the fixed and random effects models is that given in Section 2.2.3.1. Full code with results, including the required plots, is given in *Ch2_FE_Bi_logit_solution.odc* and *Ch2_RE_Bi_logit_solution.odc* for the fixed and random effects models, respectively.

a) This plot looks very similar to the plot in Figure 2.5, which plots the odds ratios on a log scale. However, this plot is of the log odds ratios on their natural scale, so the line of no effect is set at zero, and the values presented are the natural logs of the values in Figure 2.5. Additional features of Figure 2.5 are mainly for presentation purposes, such as the labels including the treatment names instead of numbers.

b) The main difference between these plots and Figure 2.6 is that the probabilities of being ranked first, second, and so on are represented by bars, instead of lines. However, the vertical and horizontal axes represent the same things, namely, the probabilities of being in each of the ranks, for each treatment. The main reason for using lines instead of bars in Figure 2.6 is for ease of presentation of results of two different models (or different outcomes) on the same graph. However, strictly, the values in Figure 2.6 are not on a continuous scale and should therefore be presented as bars (as in WinBUGS) or discrete points.

2.2 The OR for the fixed effects pairwise meta-analysis (Table 2.2) is 0.79 with 95% CrI (0.63, 1.00), and for the network meta-analysis (Table 2.3), it is 0.73 with 95% CrI (0.60, 0.88). Although the results are similar, the median OR from the network meta-analysis suggests a larger benefit of PTCA compared with Acc t-PA, and the credible interval is narrower and does not include the value of no effect. Results are slightly different because, even though there are 11 studies directly comparing PTCA and Acc t-PA, this comparison is also in multiple loops in the network (Figure 2.1), and therefore there is additional indirect evidence to inform this relative treatment effect. More evidence in a fixed effects model leads to a narrower CrI and to a slightly different median. Should these two results vary a lot, we may suspect that the underlying assumption of homogeneity (and consistency) across trials does not hold. This will be explored further in Chapter 7.

No results are presented in Chapter 1 for comparisons to PTCA, since this treatment was not included in the network. Comparing all the results in Table 2.3 with those given in Chapter 1 (Table 1.3 or 1.4), we note that the majority of results do not change, so the addition of indirect evidence through the new loops formed with PTCA does not influence many of the comparisons. This is because there are very large trials included in the network, which will have the most influence on the posterior estimates. Thus the addition of indirect evidence will have only a minor effect. Comparisons with TNK are unchanged since there is no additional indirect evidence on these comparisons in the extended network (this treatment is connected only as a 'spur' to the network and not involved in any loops).

2.3 New study added to the data in Figure 2.1.

a) An extra circle would be added to the network in Figure 2.1 representing the new treatment, coded 8, and an extra line would be added connecting this new treatment to treatment 3.

b) Connection of the new treatment to the network is only as a 'spur', that is, it does not contribute indirect evidence on any of the other comparisons, so no change is expected to any relative treatment effects, beyond simulation error (MC error). Since only one study is being added, this would also not contribute information on the between-study heterogeneity in a random effects model. Results using the random effects model with the new study added are given in *Ch2_RE_Bi_logit_newstudy.odc*, which is adapted from *Ch2_RE_Bi_logit.odc*, by adding the new data and changing the initial values.

c) Adding this second study would form a loop involving treatments 3, 7 and 8. Thus new indirect evidence would be introduced to the network, which could potentially change all the relative treatment effects. The extent to which relative effects would be affected would depend on the precision (i.e. size) of the new study. If a random effects model was fitted, even though a single study is being added, the between-study heterogeneity could change if the new study had a large impact on the other relative treatment effects. However, note that in any case the relative effects compared with TNK would not change, unless a major change in the heterogeneity substantially affected their precision.

2.4 Note that in this example the outcome is all-cause mortality, but we have both non-pharmacological (surgery) versus pharmacological comparisons and pharmacological versus pharmacological comparisons. Hence two different prior distributions from table 4 of Turner et al. (2015b) may be suitable: the prior distribution suggested for the between-study variance in meta-analyses with non-pharmacological versus pharmacological comparisons is $\sigma^2 \sim \text{lognormal}(-2.92, 1.02^2)$; the prior distribution suggested for the between-study variance in meta-analyses with pharmacological versus pharmacological comparisons is $\sigma^2 \sim \text{lognormal}(-4.18, 1.41^2)$. It is unclear which of these should be chosen. We could argue that the first prior distribution should be chosen since the majority of *studies* included in the network meta-analysis (22 out of 36) compare the surgical intervention (PTCA) with a pharmacological intervention. However, the majority of *comparisons* (7 out of 10) are between pharmacological interventions. We will therefore use both suggested prior distributions. The code to implement the new model is given in *Ch2_RE_Bi_logit_prior.odc*. Note that the code has been changed to allow for a lognormal prior distribution for the between-study variance, with parameters given as data. Thus, the only changes required to use different lognormal distributions are in the list type data. For further details see *Ch2_RE_Bi_logit_prior.odc*.

Using the lognormal $(-2.92, 1.02^2)$ prior distribution causes the posterior distribution of σ to be concentrated around larger values than the distribution in Figure 2.7. The new posterior distribution also has higher median (0.143) and a 95% CrI (0.06, 0.31) that, although narrower, allows for larger values than in Table 2.3. Consequently the relative effect estimates are slightly changed. However the overall conclusion that PTCA is the best treatment is unchanged although the probability that it is indeed the best treatment has reduced slightly to 0.93 from 0.97 (*Ch2_RE_Bi_logit_solution.odc*).

Using the lognormal $(-4.18, 1.41^2)$ prior distribution gives a very similar posterior distribution of σ to that in Figure 2.7, although the new posterior median is slightly higher (0.074) but has a narrower 95% CrI (0.024, 0.21) encompassing lower values than in Table 2.3. Consequently the relative effect estimates are very similar to those in Table 2.3, although their 95% CrI are more precise. The probability that PTCA is the best treatment is 0.98, which is very close to the value obtained when using the uniform prior distribution (*Ch2_RE_Bi_logit_solution.odc*).

In summary, although three very different prior distributions for the between-study heterogeneity were used, the main conclusions are unchanged.

2.5 The code to implement the fixed and random effects models using the risk difference, including results, is given in *Ch2_FE_Bi_RDWarnmodel*.

odc and *Ch2_RE_Bi_RDWarnmodel.odc*, respectively. Note that the initial values have been changed to fit within the bounds of the priors and that for the random effects model, the bounds of the prior distributions of d[k] had to be changed for the model to run. Comparing the risk differences of each treatment relative to treatment 1 with the ones obtained using the logit link code (*Ch2_FE_Bi_logit_solution.odc* and *Ch2_RE_Bi_RDWarnmodel.odc*), we see that they are similar. The between-study heterogeneity on the RD scale is 0.004 with 95% CrI (0.000, 0.017), which although not directly comparable with that obtained using the logit link (as they are on different scales) also suggests that heterogeneity is small. The probabilities that PTCA is the best treatment are also comparable between the logit and RD models as are the overall treatment ranks and conclusions.

The codes implementing the fixed and random effects models using the log relative risk (RR), including results, are given in *Ch2_FE_Bi_logR-RWarnmodel.odc* and *Ch2_RE_Bi_logRRWarnmodel.odc*, respectively. The RR are comparable with those obtained using the logit model, and the overall conclusions are similar. The between study heterogeneity is also small in the random effects log RR model, although again the actual values cannot be compared with those estimated in the logit model since they are on different scales.

Chapter 3

3.1 Febrile neutropenia example:

Summary	Fixed effects model	Random effects model
Posterior mean residual deviances, $\bar{D}_{res}$	73.8	49.6
Posterior mean deviance, $\bar{D}_{model}$	282.3	258.1
Effective number of parameters, p_D	28.0	39.0
Deviance information criteria, DIC	310.3	297.1
Between-study standard deviation, σ: posterior mean (95% credible interval)		0.431 (0.198, 0.724)
Log odds ratio for filgrastim vs no treatment: posterior mean (95% credible interval)	−0.874 (−1.058, −0.690)	−0.948 (−1.287, −0.624)
Log odds ratio for pegfilgrastim vs no treatment: posterior mean (95% credible interval)	−1.509 (−1.811, −1.216)	−1.433 (−1.893, −0.964)
Log odds ratio for lenograstim vs no treatment: posterior mean (95% credible interval)	−0.979 (−1.396, −0.569)	−1.079 (−1.689, −0.498)

a) The random effects model gives a better global fit to the data, as seen by the lower posterior mean residual deviance. The posterior mean residual deviance for the random effects model is close to the observed number of data points 50, suggesting an adequate fit, whereas the fixed effects model gives a posterior mean residual deviance of 73.8 (much higher than 50, indicating lack of fit). The between-study standard deviation is estimated at 0.43, which is reasonably high on a log odds ratio scale, again indicating heterogeneity and a preference for the random effects model. We would prefer a random effects model, although note that the estimated treatment effects aren't very sensitive to choice of fixed or random effects model; however the credible intervals are wider using the random effects model.

b) *The leverage plot was calculated in *Ch3_leverage_febneutro.xlsx* and displayed in the succeeding text for both the fixed and random effects models. The solid curves represent contribution to the *DIC* of 1, 2 and 3, respectively, as they get further from the origin. For the fixed effects model, it can be seen that several of the data points contribute more than 3 to the *DIC*. In particular the point to the far right of the graph corresponds to study 25. For the random effects model, all of the data points contribute less than 3 to *DIC*, with the exception of study 25. The leverage plots are useful to identify possible outliers, such as study 25. Predictive cross-validation for study 25 found a *p*-value of 0.004 (Section 3.4.2), also supporting the view that study 25 is an outlier compared with the other data points, with respect to both the fixed and random effects models.

Febrile neutropenia, fixed effects model

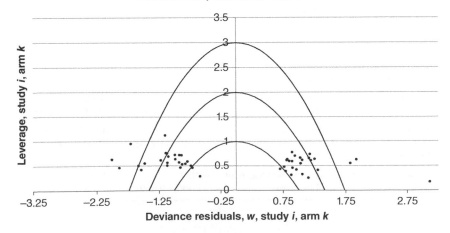

Leverage, study *i*, arm *k*

Deviance residuals, *w*, study *i*, arm *k*

Febrile neutropenia, random effects model

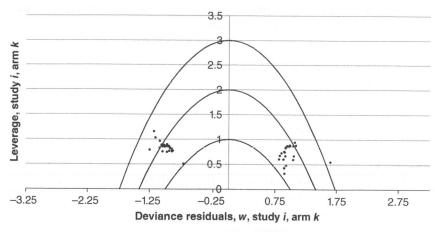

3.2 Move study 45 to the end of the data list. Set ns=49 so that the last study isn't included in the model estimation. Change initial values for mu, so there is one less value. The code is in file *Ch3_thrombo_crossval45.odc.* The cross-validation *p*-value for study 45 is p.cross = 1 – 0.9865 = 0.0135. Taking into account that there are 50 studies that we could have performed cross-validation on, we need to compare the p.cross with 1/(50 + 1) = 0.02, the value of the 50th uniform order statistic. p.cross = 0.0135 is lower than 0.02, suggesting that there is some evidence that study 45 is an outlier.

3.3 Move study 1 to the end of the data list. Set ns = 49 so that the last study isn't included in the model estimation. Change initial values for mu, so there is one less value. The code is in file *Ch3_thrombo_crossval1.odc.*

node	mean	sd	MC error	start	sample
p.cross[2]	0.1386	0.3449	0.0083	20001	20000
p.cross[3]	0.8367	0.3695	0.01576	20001	20000

p.cross [2] = 0.1386 is the probability that the number of events predicted for arm 2 (trt 3) is greater than observed. There is no evidence the comparison of treatments 3 versus 1 in study 1 is an outlier. Because this is the only study comparing treatments 3 versus 1 directly, and the predictions from the remaining studies are made indirectly (via treatment 7, via treatment 9, and so on; see Figure 3.1), this *p*-value gives an alternative test of inconsistency for the comparison 3 versus 1 (see Chapter 7).

p.cross [3] is the probability that the number of events predicted for arm 3 (trt 4) is greater than observed. BUT note that treatment 4 is no longer in the network when study 1 is excluded (see Figure 3.1), so no prediction is possible when study 1 is excluded other than the prior, which is very flat. This *p*-value is therefore meaningless and should be ignored.

3.4 *Add the following lines of code after the line defining r.new[k] to the code in the file *Ch3_thrombo_crossval.odc* (note *a* and *b* are already defined, so that *a*/*b* = arm1 odds ratio):

```
for (i in 1:5){
  r.new.N[i,k]~dbin(p.new[k], N[i])
  OR.N[i,k]<-(r.new.N[i,k]/(N[i]-r.new.N[i,k]))/(a/b)
}
```

Also in the data within the 'list' add in N=c(50,100,150,200,250). The code to produce these results can be found in *Ch3_thrombo_crossval_N.odc*. The results for the predicted odds ratios are shown in the following table and figure. Uncertainty in the prediction decreases with increasing sample size.

Sample size on arm 2 of study 45	Estimated odds ratio and 95% credible interval
$N = 50$	1.30 (0, 5.00)
$N = 100$	1.28 (0, 3.91)
$N = 150$	1.28 (0, 3.91)
$N = 200$	1.28 (0, 3.65)
$N = 250$	1.27 (0, 3.49)

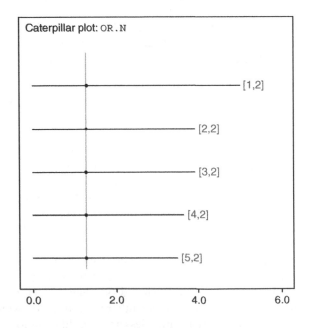

Caterpillar plot: OR.N

Chapter 4

4.1 Exercise modalities to improve weight loss.
  a) The treatment network is a triangle: all three treatments were compared with each other. The code to use is the one with a normal arm-based likelihood and identity link, given in *Ch4_FE_Normal_id.odc* and *Ch4_RE_Normal_id.odc*.
  b) Looking at the values of the weight differences across studies in the data, an upper bound of 10 or 15 should be sufficient.
  c) Solution in file *Ch4_Weight_No_id_RE-3treats.odc*. The upper bound is not large enough and the posterior distribution is constrained.
  d) Solution in file *Ch4_Weight_No_id_RE-3treats.odc*. Increasing the upper bound makes the CrI for the between-study heterogeneity wider. An upper bound of 10 is large enough.
  e) *To obtain an approximate predictive interval for the relative effect of treatment 2 compared with 1, we can calculate its standard deviation as the sum of the standard deviations of the posterior distribution of d[2] and the standard deviation of the between-study heterogeneity distribution sd from the aforementioned code. The standard deviation of the predictive distribution is approximately

$$\sqrt{0.5986^2 + 0.9248^2} = 1.1016$$

  An approximate CrI can be built assuming normality of the relative effect as d[2] $\pm 1.96 \times 1.1016$. Taking the posterior mean of d[2] from the aforementioned code, we obtain an approximate CrI as 1.279 ± 2.1591, that is, from -0.8801 to 3.4381. Note that by assuming approximate normality for the posterior distribution of d[2], we also assume it is symmetric, although we can see in the WinBUGS results that the median is slightly different from the mean.
  f) *Solution in *Ch4_Weight_No_id_RE-pred.odc*. The 95% predictive interval is from -1.649 to 3.597, slightly wider than previously mentioned. The standard deviation of the predictive distribution is 1.273, which is slightly larger than calculated previously.

4.2 Fluoride example on SMD.
  a) Solution in *Ch4_SMD_RE.odc*.
  b) Model fit is poor and there is evidence of between-study heterogeneity, which is moderate on the SMD scale. All interventions, including placebo, appear to be better than no treatment to prevent caries in children. In Chapter 9 we see that model fit is improved when a different scale is used to model the same data.

c) *The code to fit the random effects model with the new prior distribution is given in *Ch4_Ex2_SMD_RE-prior.odc*. Note that the prior distribution suggested is for the log of the between-study *variance* and that the *t*-distribution is parametrised in terms of the inverse of the square of the scale parameter. Results are not sensitive to the prior distribution. This is because in this example there are many studies per comparison and multiple loops of evidence, and therefore the posterior distribution for the between-study heterogeneity is well informed by the data. Thus, the extra information in the prior distribution has no meaningful effect.

4.3 Non-small cell lung cancer: data as log hazard ratios.

a) The models to use are those given in Section 4.4.1. Note that there are multi-arm trials and the covariance can be calculated from the data available using equation (4.12). The data correctly set up for WinBUGS is given with the fixed and random effects code in *Ch4_lnHR_FE.odc* and *Ch4_lnHR_RE.odc*.

b) Network plot: The thickness of lines is proportional to the number of trials making that comparison (given by the numbers on the lines).

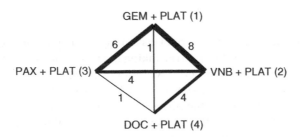

c) Code to fit fixed and random effects models is given in *Ch4_Ex3_FE_lnHR.odc* and *Ch4_Ex3_RE_lnHR.odc*, respectively. Both models fit very well and we prefer the fixed effects model. There is no evidence of difference between any of the treatment strategies on overall survival.

4.4 One product is better on ACR-20 and the other on ACR-50, and on one treatment, more people show a 50% or better improvement than show a

20% or better improvement, which is impossible! In fact, there was not a single coherent network meta-analysis model. Separate analyses have been carried out on the ACR-20, ACR-50 and ACR-70 data. It is therefore impossible to make a coherent decision. To avoid these incoherent results, a single analysis assuming the same treatment effect at each level (such as recommended in Section 4.5.1) would be better, as long as such a model fits the data.

4.5 *Psoriasis example with exchangeable cut-points.
  a) Solution in *Ch4_Psoriasis_RE-diffz.odc*.
  b) *DIC* are similar for the models with exchangeable and fixed z (Table 4.10). Between-study heterogeneity and relative effects are also very similar for both models.

Chapter 5

5.1 Monitor the node T[] (a) in *Ch5_Smoking_CEA.odc* and (b) in *Ch5_Smoking_CEApred.odc*. To obtain the correlation between the parameters in WinBUGS, click on **Correlations** in the **Inference** menu, then type the parameter (i.e. T, d or d.new) in the first box, and press **print**. Results are available in *Ch5_Smoking_CEA.odc* and *Ch5_Smoking_CEApred.odc*.

The predictive relative treatment effects, d.new[], have much higher variances than the mean treatment effects, d[]. Therefore the absolute effects T[] derived from the predictive relative effects are much less correlated. Where the T[] is based on mean effects, the majority of their variance comes from the uncertainty in the baseline model, leading to very high correlation.

5.2 The code with the summary d[] is *Ch5_Smoking_CEAsummary.odc*. This is essentially a 'forward' MC simulation from a distribution informed by data. The effect of ignoring the correlations between the absolute treatment effects is equivalent to carrying out ANOVA for four independent blocks on the MCMC samples of the net benefits from the four interventions rather than a repeated measures ANOVA that takes account of the correlations. With less discrimination between the net benefits on the four strategies, the probabilities that each is cost effective are pushed closer to each other, relative to *Ch5_Smoking_CEA.odc* – see figure with CEACs in the succeeding text.

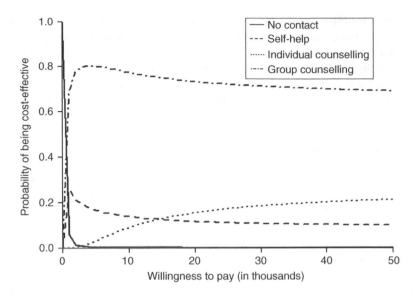

5.3 In *Ch5_Smoking_CEA_summcov.odc*, the correlations between the d[] are entered as data, and the forward MC simulation proceeds from a multivariate normal distribution. The covariances are calculated from the variances of the d[] and the correlations, and the variance–covariance matrix is inverted to obtain a precision matrix.

The posterior correlations, taken from *Ch5_Smoking_CEA.odc*, are as follows:

	d[1]	**d[2]**	**d[3]**
d[1]	1	0.250990	0.338687
d[2]	0.250990	1	0.367861
d[3]	0.338687	0.367861	1

The resulting CEACs are virtually identical to those generated by *Ch5_Smoking_CEA.odc*, suggesting that the multivariate normal approximation to the joint distribution of the d[] was adequate.

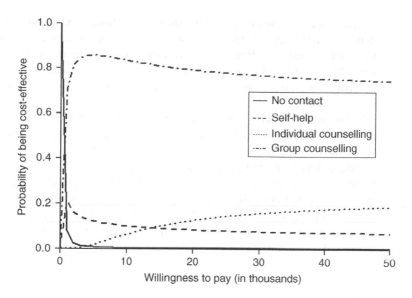

Chapter 6

6.1 EOGBS data.

b) Mantel–Haenszel: OR = 0.11 95% CI (0.03, 0.43)
Inverse variance: OR = 0.12 95% CI (0.03, 0.43)
Compared with Peto: OR = 0.17 95% CI (0.07, 0.39)
So the mean effect size from the Peto method is somewhat smaller
with narrower 95% CI than from the other two methods – which are
in broad agreement.

c) The model and full analysis is available in *Ch6_Ex1c_EOGBS.odc*. The
median value of the posterior distribution for the overall OR is 0.03
with 95% CrI (0.001 to 0.20). This suggests a larger treatment effect
than for any of the frequentist methods.

The estimate of the overall odds ratio is not affected when the prior
distributions are made (even) more diffuse (vague), suggesting that
differences between frequentist methods and the WinBUGS analysis
are not due to the influence of prior distributions.

Given the 'exact' binomial model likelihood specification used in the
WinBUGS code, we have most confidence in this result. This would
suggest that antibiotics are several times more effective for this outcome than suggested by the Peto estimate published in the Cochrane
review.

d) With only four studies included in the analysis, and only one of
these observing the outcome of interest in the antibiotics group, the

data are both sparse and few. This means that there is very little evidence available to estimate the variance between study estimates via extra random effects parameters. This however does not mean that the fixed effects model is theoretically a better model; heterogeneity could well exist, but there is just too little evidence to assess this.

If the binomial likelihood random effects model code, provided in Chapter 2, is used to analyse these data, convergence does appear to be achieved, but the posterior distribution for the between-study standard deviation (sd) parameter is very similar to the Uniform(0,2) prior distribution placed on it (see succeeding text), and thus results are likely to be highly dependent on the specific nature of the prior distribution specified for sd. An alternative would be to use an informative prior distribution for the heterogeneity parameter, as discussed.

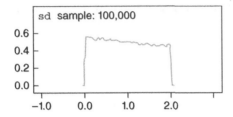

6.2 Fictitious data.
   a) Network excluding study comparisons including arms with zero events:

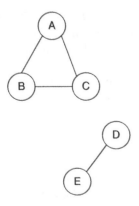

The network is disconnected as no study links treatments D and E to treatments A, B and C.

b) Network including all studies:

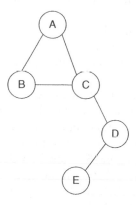

The network is now connected, but the connectedness relies on a C versus D trial with 0 events in the C arm.

c) The model and further output are included in *Ch6_Ex2c_fictitious. odc*. There is poor mixing for mu [7] (the event rate for the reference treatment (A) in the C vs D study with 0 events in the C arm), while all other mu's have good mixing (mu [6] shown in *Ch6_Ex2c_fictitious. odc* as an example).

However the Gelman–Rubin diagnostic for convergence would suggest that convergence of the MCMC sampler would occur if run long enough, but the density for mu [7] looks irregular and unlike all other mu's (mu [8] shown for illustration). A similar story exists for d [4] and d [5] – which relate to treatments D and E (vs A). Some extremely large log odds ratios are contained within the posterior distributions (see *Ch6_Ex2c_fictitious.odc*).

A scatter plot of mu [7] versus d [4] sampled values is revealing. The correlation between them is essentially perfect (–1) (see *Ch6_Ex2c_fictitious.odc*). This would suggest a lack of identifiability between parameters due to the lack of data, and even though the MCMC sampler would appear to have converged, there is instability in the estimation, and parameter estimates are not reliable.

d) There are various ways of 'fixing' the problem observed in (c) that are discussed in Chapter 6. These include the following:

1) Conducting separate analyses for treatments A, B and C and for treatments D and E. Obviously, doing this means relative estimates between treatments in the two networks cannot be estimated in the analysis (the options that follow do allow such comparisons to be made at the expense of making stronger modelling assumptions).

2) Using informative prior distributions for the mu[i]'s (the event rates in the reference treatment in each study) (possibly using external data or expert opinion).

3) Specifying the mu[i]'s to be exchangeable.

4) Adding a continuity correction to study 7.

See *Ch6_Ex2d_fictitious.odc* for a full analysis using approaches (3) and (4).

When using approach (3) there is much improved mixing of MCMC chains for the problematic parameters, and the correlation between d[4] and mu[7] reduces to –0.58, suggesting improved identifiability of the estimated parameters. Considering the results, there is a lot of uncertainty when comparing treatments 4 and 5 (D and E) with 1, 2 or 3 (A, B and C) – that is, making comparisons that 'span' the 'weak' connection between treatments 3 and 4. Intuitively, this is to be expected due to the lack of data on the comparison 3 versus 4 (C vs D).

Next, a continuity correction of 0.5 was added to the events in study 7 (and thus 1 to the total number of patients in each group).

Note that given the number of patients in each arm of study 7 is approximately equal, using a constant of 0.5 is reasonable in this context. Also note, unlike common practice for frequentist software, the continuity correction was applied to study 7 only and not all studies that observe 0 events in one trial arm. That is, it is being applied only where necessary to minimise the influence of such corrections.

The mixing of both d[4] and mu[7] also looked better using this approach, and it produced reasonable parameter estimates. The correlation between d[4] and mu[7] is still high (approx. –0.9), but no longer perfect, suggesting parameters are now identifiable. The uncertainty around odds ratios including treatments 4 or 5 is very large indeed.

This example is a situation where it is important to explicitly describe what analyses approaches have been used and the assumptions they make when reporting the results. Also, given the fragility of the data, a sensitivity analysis exploring the impact of alternative assumptions would seem to be a sensible thing to do (and this could be followed through to the inputs of any decision model the estimates were being used for).

Chapter 7

7.1 Smoking example:

a) The WinBUGS code to fit a random effects inconsistency model to the smoking data as well as all results are given in *Ch7_SmokingIncon.odc*. The posterior mean of the residual deviance and the between-study heterogeneity are very similar for the inconsistency and consistency models. In addition, no data points have any meaningful improvement in their contribution to the residual deviance.

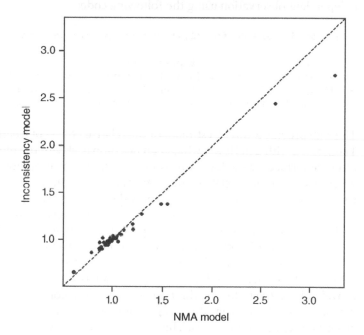

Based on this, we conclude that there is no evidence of inconsistency in this network.

b) The WinBUGS with code added to calculate the indirect estimates according to the Bucher method as well as all results is given in *Ch7_SmokingIncon-Bucher.odc*. *p*-Values for all seven comparisons were calculated using equation (7.3):

Comparison	Prob (from WinBUGS)	*p*-Value
1	0.7285	0.54
2	0.6299	0.74
3	0.6567	0.69
4	0.6139	0.77
5	0.2648	0.53
6	0.4762	0.95
7	0.4479	0.90

These Bayesian *p*-values suggest no evidence of inconsistency.

7.2 Parkinson's example:

a) Arms 2 and 3 of study 3 provide an estimate of the mean difference and its precision, which can be incorporated into WinBUGS as an independent observation using the following code:

```
# calculate mean difference for (2,4) and its precision from data
# in trial 3
prec24 <- 1/var24                    # precision
var24 <- pow(se[3,2],2) + pow(se[3,3],2) # variance in trial 3
mean24 <- y[3,3]-y[3,2]              # mean difference in trial 3
MD24 ~ dnorm(mean24, prec24)        # sample from Normal distribution
```

This will provide a 'direct' estimate of the relative effects of treatment 4 compared with 2 in MD24, which can be compared with the indirect estimate formed by the four-way loop (1, 2, 4, 3). However, note that we are using all three relative effects from a three-arm trial as if they were independent, when in fact we know they are not. The full model and results are given in *Ch7_ParkinsonsFEIncon-Bucherall.odc*. The probability of agreement in this loop is 0.55, which gives a large Bayesian p-value of 0.90, indicating no inconsistency in this loop.

b) *WinBUGS code and results are given in *Ch7_ParkinsonsFEIncon-Bucher2.odc*. We have used the fixed effects adjusted standard errors provided by *netmeta* for the three-arm study and incorporated these data as if they came from three separate two-arm studies with data given as treatment differences. This now provides a direct estimate of d_{24}, which accounts for the correct correlation and consistency between all estimates from the three-arm trial, and can be compared with the indirect evidence in the four-way loop (1, 2, 4, 3). Comparing these results with those in Table 7.6, we see that not only do we now have a direct estimate for treatment 4 compared with 2, but also uncertainty in the other direct estimate influenced by this three-arm trial has slightly increased. Comparing the aforementioned results, we can see that the direct estimate now has more uncertainty, although its median has not changed. In this example, this hardly affects the conclusions as the Bayesian p-value obtained using the adjusted standard errors is still large at 0.94.

7.3 Non-small cell lung cancer example:

a) The WinBUGS code to fit a fixed effects inconsistency model to the non-small cell lung cancer data as well as all results are given in *Ch7_NSCLC_FEIncon.odc*. The posterior mean of the residual deviance is very similar for the inconsistency (10.25) and consistency (9.24) models. In addition, no data points have any meaningful improvement in their contribution to the residual deviance.

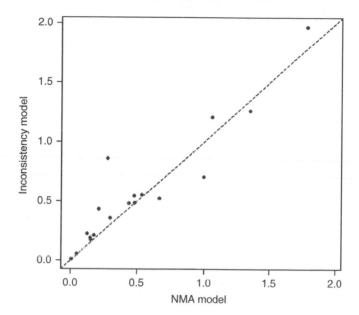

Some data points fit badly (deviance >1.5) in both models, which suggests that the reason for the poor fit is not related to the consistency assumption (as otherwise the fit would improve in the inconsistency model). Based on this, we conclude that there is no evidence of inconsistency in this network.

b) R code for the node-splitting models is given in *Ch7_NSCLC_Ex.R*. The gemtc package states that there are six nodes to split, that is, all pairwise comparisons should be split as they have three independent sources of evidence. Results from the node splits are given as follows:

```
Node-splitting analysis of inconsistency
==========================================
    comparison       p.value       CrI
1   d.1.2            0.8031067
2   -> direct                      0.080 (-0.021, 0.18)
3   -> indirect                    0.10 (-0.051, 0.26)
4   -> network                     0.084 (0.00072, 0.17)
5   d.1.3            0.7755733
6   -> direct                      0.032 (-0.060, 0.12)
```

```
 7   -> indirect                      0.067  (-0.15, 0.29)
 8   -> network                       0.044  (-0.039, 0.13)
 9   d.1.4            0.3582800
10   -> direct                        0.058  (-0.12, 0.24)
11   -> indirect                     -0.053  (-0.21, 0.10)
12   -> network                     -0.0067  (-0.12, 0.11)
13   d.2.3            0.8145867
14   -> direct                       -0.014  (-0.18, 0.16)
15   -> indirect                     -0.041  (-0.19, 0.11)
16   -> network                      -0.040  (-0.15, 0.066)
17   d.2.4            0.3583333
18   -> direct                       -0.12   (-0.25, 0.0031)
19   -> indirect                    -0.0099  (-0.21, 0.19)
20   -> network                      -0.091  (-0.20, 0.016)
21   d.3.4            0.5774400
22   -> direct                        0.019  (-0.34, 0.38)
23   -> indirect                     -0.096  (-0.27, 0.075)
24   -> network                      -0.051  (-0.19, 0.087)
```

This suggests that there is no evidence of inconsistency. The model fit statistics produced by gemtc are as follows:

	Dbar	pD	DIC
d.1.2	8.668866	5.303023	13.97189
d.1.3	7.908329	5.271374	13.17970
d.1.4	8.709978	5.220073	13.93005
d.2.3	8.620864	5.356872	13.97774
d.2.4	8.858382	5.304968	14.16335
d.3.4	8.710022	5.254883	13.96491
consistency	8.548654	4.335465	12.88412

which also suggest no evidence of inconsistency in this network.

Chapter 8

8.1 The code to fit this model is given in *Ch8_BCG_Bi_logit_FE-x1.odc*. This code is similar to that presented in *Ch8_BCG_Bi_logit_FE.odc*, but the covariate terms and corresponding prior distributions have been added, as in *Ch8_BCG_Bi_logit_RE-x1.odc*.

The fixed effects model with the covariate still has a poor fit, with a posterior mean of the residual deviance of 40, $p_D = 14.9$ and $DIC = 55$, which makes it a borderline decision on whether to prefer this model or the random effects model with covariate. Looking at individual data points' contributions to the residual deviance, we can identify two studies (7 and 13) that have a poor fit in the fixed effects model with covariate. We could investigate these further.

8.2 The code to fit this model is given in *Ch8_CZP_Bi_logit_RE-x1.odc*. This code is similar to that presented in *Ch8_CZP_Bi_logit_RE-x1prior.odc*, but it now used two different prior distributions for the between-study heterogeneity.

Using the uniform prior distribution yields an unrealistically wide 95% CrI for the between-study standard deviation. Using the empirically based prior distribution suggested by Turner et al. (2015b) gives more realistic bounds for the between-study standard deviation, although its median is larger and its 95% CrI wider than those obtained using the half-normal prior distribution (Table 8.4). The model with the empirically based prior distribution appears to fit better (probably due to the larger estimated heterogeneity), but there is no meaningful difference in the *DIC* for this model compared with the model using the half-normal prior distribution (Table 8.4).

8.3 The code to fit the three models is given in *Ch8_Pain_No_id_RE-xbasePred.odc*. This code is similar to that presented in *Ch8_CZP_Bi_logit_RE-xbase.odc*, but the likelihood is now normal with an identity link (Chapter 4). Note that the code to calculate the residual deviance has also been changed. The assumptions about common, independent or exchangeable regression coefficients affect the relative effect estimates (which are shown at the mean baseline risk) and model fit. The best fitting model is the one that allows for independent regression coefficients. The results of this model are presented in figure 2 of Achana et al. (2013). In this case, 'baseline risk' is defined as the amount of morphine consumed by the reference group, that is, the patients on placebo. For further details and interpretation, see Achana et al. (2013). The code to fit the model to the original data, correcting the baseline risk, is given in *Ch8_Pain_No_id_RE-xbaseCorrected.odc*. Results are very similar, differing only due to simulation error.

8.4 The code to fit the treatment combination model is given in *Ch8_Mort_FE_Bi_logit-Add.odc*. Note that there are now only two relative effect parameters to estimate (the effect of Volatile and Remote compared with TIVA); hence estimates of relative effects are more precise. Note also that now the relative effect of Remote + TIVA compared with TIVA and Remote + Volatile compared with Volatile is identical since they now both represent the effect of Remote as single therapy, as a consequence of the additive assumptions. The *DIC* for the additive treatment combination model is similar to the unconstrained model, although model fit is slightly worse. Overall the additive model seems reasonably supported by the data.

Chapter 9

9.1 The WinBUGS code for model 1 and model 2 can be found in the files *Ch9_RE_Po_Fluorbias1_sol1.odc* and *Ch9_RE_Po_Fluorbias2_sol1.odc*. Posterior summaries for the model parameters are also available in these files. The posterior mean residual deviance is similar to that for the network meta-analysis model without bias adjustment, and estimated mean bias terms are close to zero with 95% credible intervals that easily contain zero. This suggests that there is little evidence of bias using the indicator inadequate allocation concealment or lack of double blinding.

9.2 The following adjustment is made to the code (see *Ch9_RE_Po_Fluorbias1_sol2.odc*):

```
# bias model prior for variance
kappa ~ dgamma(10,50)
kappa.sq <- pow(kappa,2)
Pkappa <- 1/kappa.sq
# bias model prior for mean
b ~ dnorm(-0.16, 100)
```

Results with and without informative priors for b and κ are given in *Ch9_RE_Po_Fluorbias1_sol2.odc*. It can be seen that although there are some small differences in the estimated values for sd, b and kappa, the relative treatment effect estimates are robust to the use of informative priors. This is partly because there is sufficient evidence in the network meta-analysis to inform these parameters and also because the network meta-analysis evidence is in line with the informative priors from the meta-epidemiological study.

Chapter 10

10.1 The relative treatment effect parameters $d_{1,1k}$ are assumed to be 0 and are therefore removed from the model. As a result the shape is the same for all treatments in the analysis. See *Ch10_exercise1.odc* for the syntax.

10.2   We can assume that the treatment effects regarding the last shape parameter are the same for each intervention relative to a common reference treatment: $d_{1,1k} = d_{1,1}$ for a first-order model or $d_{2,1k} = d_{2,1}$ for a second-order model. Alternatively, we can assume that these treatment effects are exchangeable across interventions, and therefore unstable estimates will shrink towards a common mean: $d_{1,1k} \sim \text{Normal}(D_{1,1}, \sigma^2_{d_{1,1k}})$ for a first-order model or $d_{2,1k} \sim \text{Normal}(D_{2,1}, \sigma^2_{d_{2,1k}})$ for a second-order model. See *Ch10_exercise2_common.odc* and *Ch10_exercise2_shrinkage.odc* as examples.

10.3   Removing non-significant shape-related relative treatment effect parameters from the model effectively implies that the uncertainty regarding changes in the relative treatment effects over time is interpreted as there are no changes in treatment effects over time. This may have a large impact when the point estimate is meaningfully different from zero and the estimated treatment effects need to be extrapolated over time for use in a cost-effectiveness model. Effectively parameter uncertainty is replaced with model structure uncertainty, thereby losing transparency.

Chapter 11

11.1   From equation (11.3) we have $S_A = C_A s_C + (1 - C_A) s_{NC}$, and therefore $S_B = C_B s_C + (1 - C_B) s_{NC}$. Therefore,

$$
\begin{aligned}
d^S_{AB} = S_B - S_A &= C_B s_C + \left(1 - C_B\right) s_{NC} - \left(C_A s_C + \left(1 - C_A\right) s_{NC}\right) \\
&= \left(C_B - C_A\right)\left(s_C - s_{NC}\right) \\
&= d^C_{AB}\left(s_C - s_{NC}\right)
\end{aligned}
$$

showing that treatment effects on survival and coronary patency, if expressed on a risk difference scale, can be modelled without reference to baseline risk on survival or coronary patency.

The synthesis and cross-validation models are given in *Ch11_tpa.odc*. In the synthesis model, the posterior mean of the residual deviance for the six observations was 7.6, a satisfactory fit, although there may be some indication of some lack of fit for the data from the Kennedy trial on survival conditional on patency status (which is assumed independent of treatment). The cross-validation exercise generated a Bayesian p-value of 0.10 that the predicted difference between IVSK and CC in probability of 1-year survival would be less than what was observed in the 20 trials. This represents a two-sided p-value of 0.20. We would therefore conclude that there is no evidence of disagreement between the data sources.

11.2 The synthesis and cross-validation models are given in *Ch11_eogbs.odc*. This file also has code to plot the truncated gamma prior distribution for the between-study variance.

In the full synthesis model, the posterior median relative risks are $1 \to 2$, 0.069; $2 \to 3$, 0.35 and $1 \to 3$, 0.023. Without the synthesis, that is, in the cross-validation model, we have $1 \to 2$, 0.055; $2 \to 3$, 0.064; $1 \to 3$ predicted $= 0.0031$ and $1 \to 3$ observed $= 0.11$. The probability that the observed relative risk is greater than the predicted is 0.942, which equates to a two-sided p-value of 0.12, indicating reasonable consistency between the three evidence sources.

The results show that, as would be expected, the locus of the effect of IVAP is on the $1 \to 2$ transition: it prevents neonatal contamination with GBS. There may be an additional smaller effect in preventing disease in neonates who are contaminated (this is slightly underestimated as our prior distribution is centred on no effect). The net effect $(1 \to 3)$, 0.023 posterior mean relative risk from the synthesis model, indicates the enormous efficacy of IVAP – preventing 97.7% of EOGBS disease. This contrasts with conclusions from recent estimates from Cochrane systematic reviews, which present results for different outcomes separately and use methods that produce estimates biased towards null effects (see Exercise 6.1).

Chapter 12

12.1 If the sets of AB and AC effects are each exchangeable, then $d_{BC} = E[\delta_{i,AC}] - E[\delta_{i,BC}] = d_{AC} - d_{AB}$ and $\sigma_{BC}^2 = Var[d_{BC}] = Var[\delta_{i,AC}] + Var[\delta_{i,BC}] - Cov[\delta_{i,AC}, \delta_{i,AB}] = \sigma_{AB}^2 + \sigma_{AC}^2 - \rho\sigma_{AB}\sigma_{AC}$

12.2 $d_{BC} = d_{AC} - d_{AB}$ does not imply $\sigma_{BC}^2 = \sigma_{AB}^2 + \sigma_{AC}^2 - \rho\sigma_{AB}\sigma_{AC}$, so consistency of the mean effects does not imply exchangeability.

12.3 With three trials and $\pi = 0.5$:

	Trial outcomes	Prob (outcome)	Meta-analysis	Error	|Error|
1	$\delta_{AB}, \delta_{AB}, \delta_{AB}$	0.125	δ_{AB}	$-\theta/2$	$\theta/2$
2	$\delta_{AB}, \delta_{AB}, \delta_{AB} + \theta$	0.375	$\delta_{AB} + \theta/3$	$-\theta/6$	$\theta/6$
3	$\delta_{AB}, \delta_{AB} + \theta, \delta_{AB} + \theta$	0.375	$\delta_{AB} + 2\theta/3$	$+\theta/6$	$\theta/6$
4	$\delta_{AB} + \theta, \delta_{AB} + \theta, \delta_{AB} + \theta$	0.125	$\delta_{AB} + \theta$	$+\theta/2$	$\theta/2$
	Expectation			0	$\theta/4$

12.4 With two trials and $\pi = 0.5$:

	Trial outcomes		Prob outcome	Meta-analysis	Error	| Error |
	AB trials	**AC trials**				
1	δ_{AB}, δ_{AB}	δ_{AC}, δ_{AC}	1/16	δ_{BC}	0	0
2	δ_{AB}, δ_{AB}	$\delta_{AC}, \delta_{AC} + \theta$	1/8	$\delta_{BC} + \theta/2$	$+\theta/2$	$\theta/2$
3	δ_{AB}, δ_{AB}	$\delta_{AC} + \theta, \delta_{AC} + \theta$	1/16	$\delta_{BC} + \theta$	$+\theta$	θ
4	$\delta_{AB}, \delta_{AB} + \theta$	δ_{AC}, δ_{AC}	1/8	$\delta_{BC} - \theta/2$	$-\theta/2$	$\theta/2$
5	$\delta_{AB}, \delta_{AB} + \theta$	$\delta_{AC}, \delta_{AC} + \theta$	1/4	δ_{BC}	0	0
6	$\delta_{AB}, \delta_{AB} + \theta$	$\delta_{AC} + \theta, \delta_{AC} + \theta$	1/8	$\delta_{BC} + \theta/2$	$+\theta/2$	$\theta/2$
7	$\delta_{AB} + \theta, \delta_{AB} + \theta$	δ_{AC}, δ_{AC}	1/16	$\delta_{BC} - \theta$	$-\theta$	θ
8	$\delta_{AB} + \theta, \delta_{AB} + \theta$	$\delta_{AC}, \delta_{AC} + \theta$	1/8	$\delta_{BC} - \theta/2$	$-\theta/2$	$\theta/2$
9	$\delta_{AB} + \theta, \delta_{AB} + \theta$	$\delta_{AC} + \theta, \delta_{AC} + \theta$	1/16	δ_{BC}	0	0
Expectation					0	$3\theta/8$

Appendices

Appendix A   Sample R Code to Obtain Adjusted Standard Errors Using Netmeta (Chapter 7)

Available as electronic file in *AppAnetmetascript.R*.

```
# Full Thrombo example: load data
datanet <- read.csv("DataThromb.csv", header=TRUE, sep=",")
TreatCodes <- read.csv("TreatCodes.csv", header=TRUE, sep=",")
print(TreatCodes)
#
#######################################################
#    Obtain reduced weights for FE model
#######################################################
library(netmeta)
# Gerta Rücker, Guido Schwarzer, Ulrike Krahn and Jochem König (2016).
# netmeta: Network Meta-Analysis using Frequentist Methods. R package
# version 0.9-1. https://CRAN.R-project.org/package=netmeta
p1 <- pairwise(treat=list(t1, t2, t3),
                event=list(r1, r2, r3),
                n=list(n1,n2,n3),
                data=datanet, studlab=study)
print(p1)
# net1 <- netmeta(TE, seTE, treat1,treat2,studlab, data=p1)
print(net1)
# study 1
s1 <- net1$studlab == 1  # choose study 1
net1$seTE[s1]            # Unadjusted standard errors
net1$seTE.adj[s1]        # Adjusted standard errors
# study 6
s2 <- net1$studlab == 6  # choose study 6
net1$seTE[s2]    # Unadjusted standard errors (as given in the data)
net1$seTE.adj[s2]        # Adjusted standard errors
#
```

Network Meta-Analysis for Decision-Making, First Edition. Sofia Dias, A. E. Ades, Nicky J. Welton, Jeroen P. Jansen and Alexander J. Sutton.
© 2018 John Wiley & Sons Ltd. Published 2018 by John Wiley & Sons Ltd.
Companion website: www.wiley.com/go/dias/networkmeta-analysis

```
# DATA FOR WinBUGS: treatment difference and adjusted st. errors
# change sign to be log-OR of treat 2 compared to 1
cbind(t1=c(net1$treat1[s1], net1$treat1[s2]),
      t2=c(net1$treat2[s1], net1$treat2[s2]),
      y2=c(-net1$TE[s1], -net1$TE[s2]),
      se2=c(net1$seTE.adj[s1], net1$seTE.adj[s2]),
      study=c(rep(1,3), rep(6,3)))
#
# END
```

Appendix B   Derivation of Prior Distribution for Heterogeneity Parameter Used in Certolizumab Example: Random Effects Model with Covariate (Chapter 8)

We want to find a half-normal distribution, that is, a Normal(0,m^2) distribution, truncated to take only positive values, which expresses the prior belief that 95% of trials will give odds ratios within a factor of 2 from the estimated median odds ratio. Assuming no treatment effect (without loss of generality), we have

$$\delta \sim \text{Normal}\left(0, \sigma^2\right) \quad \text{and} \quad \sigma \sim \text{Half-Normal}\left(0, m^2\right)$$

and we want to find m that defines a prior distribution for σ such that the 95% CrI for OR = exp(δ) is approximately (0.5, 2).

Implementing this in WinBUGS gives the following code:

```
model{
delta ~ dnorm(0,tau)
tau <- pow(sd,-2)
prec <- pow(m,-2)
sd ~ dnorm(0,prec)I(0,)
OR <- exp(delta)
}
```

Thus, we want to find m such that the 95% CrI for OR is approximately (0.5, 2). We can do this simply by experimentation with different values of m, given as data. When we set m to 0.32: list(m=0.32), we get the following results:

node	mean	sd	MC error	2.5%	median	97.5%	start	sample
OR	1.051	0.4147	0.001756	0.4924	0.9998	1.985	5001	50000
delta	−0.002994	0.3186	0.001378	−0.7084	−1.602E−4	0.6857	5001	50000
sd	0.2554	0.193	8.714E−4	0.01026	0.2155	0.718	5001	50000

which are close enough to the desired values. Thus we chose $\sigma \sim$ Half-Normal$(0, 0.32^2)$ as the prior in the example given in Section 8.4.2 (equation 8.7).

Appendix C   Reconstructing Data from Published Kaplan–Meier Curves (Chapter 10)

C.1   Details for the Guyot IPD Reconstruction Algorithm

Data inputs required are the coordinates extracted from the Kaplan–Meier (KM) curve: times U_j where $U_1 = 0$ and corresponding survival probabilities S_j for $j = 1, \ldots, J$ points on the KM curve. These points should include the times at which numbers at risk are reported below the curve, must capture all steps in the curve, and adjustments to the extracted coordinates to ensure the survival probabilities are decreasing with time. The curve is split into $m = 1, \ldots, nint$ intervals, defined by the points where number at risk are reported below the curves, where $j = lower_m$ and $j = upper_m$ give the extracted coordinates corresponding to the lower and upper ends of interval m, respectively. The reported number at risk at the start of interval m is $nrisk_m$. If reported, the total number of events, *totevents*, is also used by the algorithm.

Here we assume that the number at risk is reported at the start of the study and at least one other time point and that the total number of events is also reported. See Guyot et al. (2012) for adaptations to the algorithm when this is not the case. We assume the number of censored individuals on each interval is not available. We therefore estimate the number of censored individuals on each interval by an iterative process to match the predictions from the inverted KM equations with the reported number at risk at the ends of each interval. We assume that censoring occurs at a constant rate within each of the time intervals, which seems reasonable if the censoring pattern is non-informative (each subject has a censoring time that is statistically independent of their failure time).

The algorithm is made up of the following steps:

Step 1. Form an initial guess for the number censored on interval m. If there were no censored individuals on interval m, then the number at risk at the beginning of interval $m + 1$ would be

$$nrisk_{m+1}^{nocensor} = nrisk_m \times \frac{S_{lower_{m+1}}}{S_{lower_m}}$$

rounded to the nearest integer, where $S_{lower_{m+1}} / S_{lower_m}$ is the probability of being alive at start of interval $m + 1$ given alive at the start of interval m.

An initial guess for the number censored on interval m is the difference between the reported number at risk at the beginning of interval $m + 1$ and the number at risk under no censoring:

$$ nce\hat{n}sor_m = n.risk^{nocensor}_{m+1} - n.risk_{m+1} = \frac{S_{lower_{m+1}}}{S_{lower_m}} \times nrisk_m - n.risk_{m+1} $$

Step 2. Distribute the $c=1,\ldots,nce\hat{n}sor_m$ censor times, $ce\hat{n}t_c$, evenly over interval m:

$$ ce\hat{n}t_c = U_{lower_m} + c \times \left(U_{lower_{m+1}} - U_{lower_m} \right) / \left(nce\hat{n}sor_m + 1 \right) \quad c=1,\ldots,nce\hat{n}sor_m $$

The number of censored observations between extracted KM coordinates j and $j+1$, $c\hat{e}n_j$, is found by counting the number of estimated censor times, $ce\hat{n}t_c$, that lie between time U_j and U_{j+1}.

Step 3. The estimated number of events, $\hat{e}_j$, at each extracted KM coordinate, j, and hence estimated number at risk at the next coordinate, $\hat{n}_{j+1}$, can then be calculated. The KM equations give

$$ \hat{S}_j^{KM} = \begin{cases} 1 & if \ j=1 \\ \hat{S}_{last(j)}^{KM} \times \left(1 - \dfrac{\hat{e}_j}{\hat{n}_j} \right) & \text{Otherwise} \end{cases} $$

where $last(j)$ is the last coordinate where we estimate that an event occurred prior to coordinate j. This is necessary because the KM equations assume that intervals are defined by the event times.
Rearranging gives

$$ \hat{e}_j = \hat{n}_j \times \left(1 - \frac{S_j}{\hat{S}_{last(j)}^{KM}} \right) \quad j = lower_m, \ldots, upper_m $$

rounded to the nearest integer.
The number of patients at risk at each extracted coordinate, j, is then obtained:

$$ \hat{n}_{j+1} = \hat{n}_j - \hat{e}_j - c\hat{e}n_j \quad j = lower_m, \ldots, upper_m $$

where at the start of the interval we set $\hat{n}_{lower_m} = nrisk_m$. This produces an estimated number at risk at the start of the following interval $\hat{n}_{upper_m+1}$.

Step 4. If $\hat{n}_{upper_m+1} \neq nrisk_{m+1}$ then readjust the estimated number of censored observations in interval m to

$$ nce\hat{n}sor_m = nce\hat{n}sor_m + \left(\hat{n}_{upper_m+1} - nrisk_{m+1} \right) $$

Repeat steps 2–3 iteratively until estimated and published number at risk match (i.e. $\hat{n}_{upper_m+1} = nrisk_{m+1}$).

Step 5. If $m+1$ is not the last interval, repeat steps 1–4 for the following interval.

Step 6. In published RCTs, there is generally no number at risk published at the end of the last interval, *nint*. We assume that the censored rate on the last interval equals the average censor rate from the previous intervals, so the estimated censored individuals on the last interval is the average censor rate multiplied by the length of the last interval, constrained to be less than or equal to the number of patients at risk at the start of the last interval. This assumption is written as

$$ncen\hat{s}or_{nint} = \min\left(\frac{U_{upper_{nint}} - U_{lower_{nint}}}{U_{upper_{nint-1}}} \times \sum_{m=1}^{nint-1} ncen\hat{s}or_m; nrisk_{nint}\right)$$

steps 2–3 are then run for the last interval.

Step 7. If the estimated total number of events prior to the beginning of the last interval $\sum_{j=1}^{upper_{nint-1}} \hat{e}_j$ is greater or equal to the reported total number of events, *totevents*, we assume that no more events or censoring occurs

$$\hat{e}_j = 0, c\hat{e}n_j = 0, \hat{n}_j = \hat{n}_{upper_{nint-1}} \quad m = lower_{nint}, \dots, upper_{nint}$$

Step 8. If $\sum_{j=1}^{upper_{nint-1}} \hat{e}_j$ is less than *totevents*, readjust the estimated number of censored observations, $ncen\hat{s}or_{nint}$, by the difference in total number of events

$$ncen\hat{s}or_{nint} = ncen\hat{s}or_{nint} + \left(\sum_{j=1}^{upper_{nint}} \hat{e}_j - totevents\right)$$

Then rerun steps 2–3, 8 for the last interval, *nint*, until the estimated total number of events, $\sum_{j=1}^{upper_{nint}} \hat{e}_j = totevents$, or until $\sum_{j=1}^{upper_{nint}} \hat{e}_j = totevents$, but the total number of censoring in the last interval, $ncen\hat{s}or_{nint}$, becomes equal to zero.

C.2   Details of the Algorithm for Constructing Interval Data

The algorithm described here is based on the method used by Jansen and Cope (2012) to create interval data. The available follow-up time period for each arm of each trial is divided in w sequential short time intervals $[u_1, u_2], (u_2, u_3], \dots, (u_w, u_{w+1}]$ with $u_1 = 0$. For each time interval $m = 1, 2, 3, \dots, w$, the conditional survival probability is calculated based on the scanned survival proportions according to $S(u_{m+1})/S(u_m)$. If there is no survival proportion available for a specific time point, a corresponding estimate for $S(u_m)$ can be obtained by linear interpolation of the first available extracted scanned survival proportions before and after this time point.

Given the reported numbers at risk n_m at the beginning of the mth interval and the event probability for that interval, the actual number of events r_m for each interval will be calculated. As such, it is desirable to have the time intervals defined in such a way that (some) time points u_m are aligned with the time point for which the size of the at-risk population is reported below the published KM curve.

When the population at risk at the beginning of the mth interval is not reported below the KM curve, it can be imputed. First, based on the reported size of the at-risk population at subsequent time points, n_m will be estimated according to

$$n_m^{bc} = \frac{n_{m+1}^{bc}}{S(u_{m+1})/S(u_m)}$$

With this 'backward calculation' approach, we implicitly assume that censoring occurs before the events happen within a time interval. However, this approach is not feasible if there is no information regarding the at-risk population for time intervals, beyond the at-risk population reported at a certain time point. In other words, this approach is only feasibly for intervals up to the latest time point for which an at-risk population is reported. Next, n_m will be estimated according to

$$n_m^{fc} = n_{m-1}^{fc} \frac{S(u_m)}{S(u_{m-1})}$$

The disadvantage of this 'forward calculation' approach is that censoring is ignored and the sample size potentially too large for those intervals. For intervals where both n_m^{bc} and n_m^{fc} were calculated, the actual estimate for the population at risk is calculated as $n_m = \min(n_m^{bc}, n_m^{fc})$ to ensure the sample size at the beginning of each interval is not overestimated. For intervals where n_m^{bc} could not be calculated, $n_m = n_m^{fc}$.

The number of events for each interval is calculated according to

$$r_m = \min\left(n_m\left(\frac{S(u_m) - S(u_{m+1})}{S(u_m)} \right), n_m - n_{m+1} \right)$$

The time intervals do not have to have the same length. In general, the shorter the intervals, the more variation in the hazard function will be picked up, but if the intervals are too short, the number of events will be zero for many subsequent intervals, potentially compromising estimation.

Obviously, the dataset created with the Guyot IPD algorithm can also be used to create interval data, that is, n_m and r_m for the mth interval.

Appendix D   List of RCTs Included in the Illustrative Example in Section 10.10

1 Facon, T., Mary, J. Y., Hulin, C., Benboubker, L., Attal, M., Pegourie, B., Renaud, M., Harousseau, J. L., Guillerm, G., Chaleteix, C., Dib, M., Voillat, L., Maisonneuve, H., Troncy, J., Dorvaux, V., Monconduit, M., Martin, C., Casassus, P., Jaubert, J., Jardel, H., Doyen, C., Kolb, B., Anglaret, B., Grosbois, B., Yakoub-Agha, I., Mathiot, C., Avet-Loiseau, H., & Intergroupe Francophone du Myélome. 2007. Melphalan and prednisone plus thalidomide versus melphalan and prednisone alone and reduced-intensity autologous stem cell transplantation in elderly patients with multiple myeloma (IFM 99-06): a randomised trial. Lancet, 370, 1209–1218.

2 San Miguel, J. F., Schlag, R., Khuageva, N. K., Dimopoulos, M. A., Shpilberg, O., Kropff, M., Spicka, I., Petrucci, M. T., Palumbo, A., Samoilova, O. S., Dmoszynska, A., Abdulkadyrov, K. M., Schots, R., Jiang, B., Mateos, M. V., Anderson, K. C., Esseltine, D. L., Liu, K., Cakana, A., van de Velde, H., Richardson, P. G., & VISTA Trial Investigators. 2008. Bortezomib plus melphalan and prednisone for initial treatment of multiple myeloma. New England Journal of Medicine, 359, 906–917.

3 Hulin, C., Facon, T., Rodon, P., Pegourie, B., Benboubker, L., & Doyen, C. 2009. Efficacy of melphalan and prednisone plus thalidomide in patients older than 75 years with newly diagnosed multiple myeloma: IFM 01/01 trial. Clinical Oncology, 27, 3664–3670.

4 Palumbo, A., Bringhen, S., Liberati, A. M., Caravita, T., Falcone, A., Callea, V., Montanaro, M., Ria, R., Capaldi, A., Zambello, R., Benevolo, G., Derudas, D., Dore, F., Cavallo, F., Gay, F., Falco, P., Ciccone, G., Musto, P., Cavo, M., & Boccadoro M. 2008. Oral melphalan, prednisone, and thalidomide in elderly patients with multiple myeloma: updated results of a randomized controlled trial. Blood, 112, 3107–3114.

5 Wijermans, P., Schaafsma, M., Termorshuizen, F., Ammerlaan, R., Wittebol, S., Sinnige, H., Zweegman, S., van Marwijk Kooy, M., van der Griend, R., Lokhorst, H., Sonneveld, P., & Dutch-Belgium Cooperative Group HOVON. 2010. Phase III study of the value of thalidomide added to melphalan plus prednisone in elderly patients with newly diagnosed multiple myeloma: the HOVON 49 Study. Journal of Clinical Oncology, 28, 3160–3166.

6 Waage, A., Gimsing, P., Fayers, P., Abildgaard, N., Ahlberg, L., & Björkstrand, B. 2010. Melphalan and prednisone plus thalidomide or placebo in elderly patients with multiple myeloma. Blood, 116, 1405–1412.

7 Beksac, M., Haznedar, R., Firatli-Tuglular, T., Ozdogu, H., Aydogdu, I., Konuk, N., Sucak, G., Kaygusuz, I., Karakus, S., Kaya, E., Ali, R., Gulbas, Z., Ozet, G., Goker, H., & Undar, L. 2011. Addition of thalidomide to oral melphalan/ prednisone in patients with multiple myeloma not eligible for transplantation: results of a randomized trial from the Turkish Myeloma Study Group. European Journal of Haematology, 86, 16–22.

8 Owen, R. G., Morgan, G. J., Jackson, H., Davies, F. E., Drayson, M. T., Ross, F. M., Navarro-Coy, N., Gregory, W. M., Szubert, A. J., Rawstron, A. C., Bell, S. E., Heatley, F., & Child, J. A. 2009. MRC Myeloma IX: preliminary results from the non-intensive study. XII International Myeloma Workshop, p. 79. Clinical Lymphoma & Myeloma. DOI:http://dx.doi.org/ 10.1016/S1557-9190(11)70619-9.

9 Mateos, M. V., Oriol, A., Martínez-López, J., Gutiérrez, N., Teruel, A. I., de Paz, R., García-Laraña, J., Bengoechea, E., Martín, A., Mediavilla, J. D., Palomera, L., de Arriba, F., González, Y., Hernández, J. M., Sureda, A., Bello, J. L., Bargay, J., Peñalver, F. J., Ribera, J. M., Martín-Mateos, M. L., García-Sanz, R., Cibeira, M. T., Ramos, M. L., Vidriales, M. B., Paiva, B., Montalbán, M. A., Lahuerta, J. J., Bladé, J., & Miguel, J. F. 2010. Bortezomib, melphalan, and prednisone versus bortezomib, thalidomide, and prednisone as induction therapy followed by maintenance treatment with bortezomib and thalidomide versus bortezomib and prednisone in elderly patients with untreated multiple myeloma: a randomised trial. Lancet Oncology, 11(10), 934–941.

References

Abrams, K. R., Gillies, C. L. & Lambert, P. C. 2005. Meta-analysis of heterogeneously reported trials assessing change from baseline. *Statistics in Medicine*, 24, 3823–3844.

Achana, F. A., Cooper, N. J., Dias, S., Lu, G., Rice, S. J. C., Kendrick, D. & Sutton, A. J. 2013. Extending methods for investigating the relationship between treatment effect and baseline risk from pairwise meta-analysis to network meta-analysis. *Statistics in Medicine*, 35, 752–771.

Achana, F. A., Sutton, A. J., Kendrick, D., Wynn, P., Young, B., Jones, D. R., Hubbard, S. J. & Cooper, N. J. 2015. The effectiveness of different interventions to promote poison prevention behaviours in households with children: a network meta-analysis. *PLoS One*, 10, e0121122.

Ades, A. E. 2003. A chain of evidence with mixed comparisons: models for multi-parameter evidence synthesis and consistency of evidence. *Statistics in Medicine*, 22, 2995–3016.

Ades, A. E., Caldwell, D. M., Reken, S., Welton, N. J., Sutton, A. J. & Dias, S. 2012. NICE DSU Technical Support Document 7. Evidence synthesis of treatment efficacy in decision making: a reviewer's checklist. *Technical Support Document*. Available: http://scharr.dept.shef.ac.uk/nicedsu/technical-support-documents/evidence-synthesis-tsd-series/ (Accessed 2 August 2017).

Ades, A. E., Caldwell, D. M., Reken, S., Welton, N. J., Sutton, A. J. & Dias, S. 2013. Evidence synthesis for decision making 7: a reviewer's checklist. *Medical Decision Making*, 33, 679–691.

Ades, A. E., Lu, G., Dias, S., Mayo-Wilson, E. & Kounali, D. 2015. Simultaneous synthesis of treatment effects and mapping to a common scale: an alternative to standardisation. *Research Synthesis Methods*, 6, 96–107.

Ades, A. E., Lu, G. & Higgins, J. P. T. 2005. The interpretation of random effects meta-analysis in decision models. *Medical Decision Making*, 25, 646–654.

Ades, A. E., Mavranezouli, I., Dias, S., Welton, N. J., Whittington, C. & Kendall, T. 2010. Network meta-analysis with competing risk outcomes. *Value in Health*, 13, 976–983.

Network Meta-Analysis for Decision-Making, First Edition. Sofia Dias, A. E. Ades, Nicky J. Welton, Jeroen P. Jansen and Alexander J. Sutton.
© 2018 John Wiley & Sons Ltd. Published 2018 by John Wiley & Sons Ltd.
Companion website: www.wiley.com/go/dias/networkmeta-analysis

Ades, A. E., Sculpher, M., Sutton, A., Abrams, K., Copper, N., Welton, N. J. & Lu, G. 2006. Bayesian methods for evidence synthesis in cost-effectiveness analysis. *PharmacoEconomics*, 24, 1–19.

Ades, A. E. & Sutton, A. J. 2006. Multiparameter evidence synthesis in epidemiology and medical decision making: current approaches. *Journal of the Royal Statistical Society: Series A (Statistics in Society)*, 169, 5–35.

Akaike, H. 1974. A new look at the statistical model identification. *IEEE Transations on Automatic Control*, 19, 716–723.

Alfirevic, Z., Keeney, E., Dowswell, T., Welton, N. J., Dias, S., Jones, L. V., Navaratnam, K. & Caldwell, D. M. 2015. Labour induction with prostaglandins: a systematic review and network meta-analysis. *BMJ*, 350, h217.

Alfirevic, Z., Keeney, E., Dowswell, T., Welton, N. J., Medley, N., Dias, S., Jones, L. V. & Caldwell, D. M. 2016. Methods to induce labour: a systematic review, network meta-analysis and cost-effectiveness analysis. *BJOG: An International Journal of Obstetrics & Gynaecology*, 123, 1462–1470.

Anon. 2015. *Exchangeable random variables [Online]*. Wikipedia. Available: http://en.wikipedia.org/wiki/Exchangeable_random_variables (Accessed 13 July 2017).

Arends, L. R., Hunink, M. G. M. & Stijnen, T. 2008. Meta-analysis of summary survival curve data. *Statistics in Medicine*, 27, 4381–4396.

Askling, J., Fahrbach, K., Nordstrom, B., Ross, S., Schmid, C. H. & Symmons, D. 2011. Cancer risk with tumor necrosis factor alpha (TNF) inhibitors: meta-analysis of randomized controlled trials of adalimumab, etanercept, and infliximab using patient level data. *Pharmacoepidemiology and Drug Safety*, 20, 119–130.

Bafeta, A., Trinquart, L., Seror, R. & Ravaud, P. 2014. Reporting of results from network meta-analyses: methodological systematic review. *BMJ*, 348, g1741.

Bagnardi, V., Zambon, A., Quatto, P. & Corrao, G. 2004. Flexible meta-regression functions for modeling aggregate dose-response data, with an application to alcohol and mortality. *American Journal of Epidemiology*, 159, 1077–1086.

Baker, S. & Kramer, B. 2002. The transitive fallacy for randomized trials: if A bests B and B bests C in separate trials, is A better than C? *BMC Medical Research Methodology*, 2, 13.

Balshem, H., Helfand, M., Schünemann, H. J., Oxman, A. D., Kunz, R., Brozek, J., Vist, G. E., Falck-Ytter, Y., Meerpohl, J., Norris, S. & Guyatt, G. H. 2011. GRADE guidelines: 3. Rating the quality of evidence. *Journal of Clinical Epidemiology*, 64, 401–406.

Bates, M. J. 1989. The design of browsing and berrypicking techniques for the online search interface. *Online Information Review*, 13, 407–424.

Berger, U., Schäfer, J. & Ulm, K. 2003. Dynamic Cox modelling based on fractional polynomials: time-variations in gastric cancer prognosis. *Statistics in Medicine*, 22, 1163–1180.

Berkey, C. S., Hoaglin, D. C., Mosteller, F. & Colditz, G. A. 1995. A random effects regression model for meta-analysis. *Statistics in Medicine*, 14, 395–411.

Berlin, J. A., Santanna, J., Schmid, C. H., Szczech, L. A. & Feldman, H. I. 2002. Individual patient- versus group-level data meta-regressions for the investigation of treatment effect modifiers: ecological bias rears its ugly head. *Statistics in Medicine*, 21, 371–387.

Bernardo, J. M. & Smith, A. F. M. 1994. *Bayesian Theory*, New York, John Wiley & Sons, Inc.

Bero, L. 2014. Bias related to funding source in statin trials (editorial). *BMJ*, 349, g5949.

Biggerstaff, B. J. & Tweedie, R. L. 1997. Incorporating variability in estimates of heterogeneity in the random effects model in meta-analysis. *Statistics in Medicine*, 16, 753–768.

Böhning, D., Malzahn, U., Dietz, E. & Schlattmann, P. 2002. Some general points in estimating heterogeneity variance with the DerSimonian-Laird estimator. *Biostatistics*, 3, 445–457.

Boland, A., Dundar, Y., Bagust, A., Haycox, A., Hill, R., Mujica Mota, R., Walley, T. & Dickson, R. 2003. Early thrombolysis for the treatment of acute myocardial infarction: a systematic review and economic evaluation. *Health Technology Assessment*, 7, 1–136.

Borenstein, M., Hedges, L. V., Higgins, J. P. T. & Rothstein, H. R. 2009. *Introduction to Meta-Analysis*, Chichester, John Wiley & Sons, Ltd.

Bossard, N., Descotes, F., Bremond, A. G., Bobin, Y., De Saint Hilaire, P., Golfier, F., Awada, A., Mathevet, P. M., Barbier, Y. & Esteve, J. 2003. Keeping data continuous when analyzing the prognostic impact of a tumor marker: an example with cathepsin D in breast cancer. *Breast Cancer Research and Treatment*, 82, 47–59.

Boutitie, F., Gueyffier, F. O., Pocock, S., Fagard, R. & Boissel, J. P. 2002. J-shaped relationship between blood pressure and mortality in hypertensive patients: new insights from a meta-analysis of individual-patient data. *Annals of Internal Medicine*, 136, 438–448.

Bradburn, M. J., Deeks, J. J., Berlin, J. A. & Localio, A. R. 2007. Much ado about nothing: a comparison of the performance of meta-analysis methods with rare events. *Statistics in Medicine*, 26, 53–77.

Braithwaite, R. S., Roberts, M. S., Chang, C. C. H., Goetz, M. B., Gibert, C. L., Rodriguez-Barradas, M. C., Shechter, S., Schaefer, A., Nucifora, K., Koppenhaver, R. & Justice, A. C. 2008. Influence of alternative thresholds for initiating HIV treatment on life expectancy and quality-adjusted life expectancy: a decision model. *Annals of Internal Medicine*, 148, 178–185.

Brand, K. P. & Small, M. J. 1995. Updating uncertainty in an integrated risk assessment: conceptual framework and methods. *Risk Analysis*, 15, 719–731.

Brazier, J. E., Yang, Y., Tsuchiya, A. & Rowen, D. L. 2010. A review of studies mapping (or cross-walking) non-preference based measures of health to generic preference-based measures. *The European Journal of Health Economics: Health Economics in Prevention and Care*, 11, 215–225.

Brennan, A. & Kharroubi, S. A. 2007. Efficient computation of partial expected value of sample information using Bayesian approximation. *Journal of Health Economics*, 26, 122–148.

Briggs, A., Sculpher, M. & Claxton, K. 2006. *Decision Modelling for Health Economic Evaluation*, Oxford, Oxford University Press.

Briggs, A., Weinstein, M., Fenwick, E., Karnon, J., Sculpher, M. & Paltiel, A., on behalf of the ISPOR-SMDM Modeling Good Research Practices Task Force. 2012. Model parameter estimation and uncertainty analysis: a report of the ISPOR-SMDM modeling good research practices task force-6. *Value in Health*, 15, 835–842.

Brooks, S. P. & Gelman, A. 1998. General methods for monitoring convergence of iterative simulations. *Journal of Computational and Graphical Statistics*, 7, 434–455.

Brown, T., Pilkington, G., Bagust, A., Boland, A., Oyee, J., Tudur-Smith, C., Blundell, M., Lai, M., Martin Saborido, C., Greenhalgh, J., Dundar, Y. & Dickson, R. 2013. Clinical effectiveness and cost-effectiveness of first-line chemotherapy for adult patients with locally advanced or metastatic non-small cell lung cancer: a systematic review and economic evaluation. *Health Technology Assessment*, 17, 1–278.

Bucher, H. C., Griffith, L., Guyatt, G. H. & Opravil, M. 1997a. Meta-analysis of prophylactic treatments against *Pneumocystis carinii* pneumonia and toxoplasma encephalitis in HIV-infected patients. *Journal of Acquired Immune Deficiency Syndromes and Human Retrovirology*, 15, 104–114.

Bucher, H. C., Guyatt, G. H., Griffith, L. E. & Walter, S. D. 1997b. The results of direct and indirect treatment comparisons in meta-analysis of randomized controlled trials. *Journal of Clinical Epidemiology*, 50, 683–691.

Bujkiewicz, S., Jones, H. E., Lai, M. C. W., Cooper, N. J., Hawkins, N., Squires, H., Abrams, K. R., Spiegelhalter, D. J. & Sutton, A. J. 2011. Development of a transparent interactive decision interrogator to facilitate the decision making process in health care. *Value in Health*, 14, 768–776.

Bujkiewicz, S., Thompson, J. R., Sutton, A. J., Cooper, N. J., Harrison, M. J., Symmons, D. P. M. & Abrams, K. R. 2013. Multivariate meta-analysis of mixed outcomes: a Bayesian approach. *Statistics in Medicine*, 32, 3926–3943.

Burch, J., Paulden, M., Conti, S., Stock, C., Corbette, M., Welton, N. J., Ades, A. E., Sutton, A., Cooper, N., Elliot, A. J., Nicholson, K., Duffy, S., McKenna, C., Stewart, S., Westwood, M. & Palmer, S. 2010. Antiviral drugs for the treatment of influenza: a systematic review and economic evaluation. *Health Technology Assesment*, 13, 1–290.

Burgess, S., White, I. R., Resche-Rigon, M. & Wood, A. M. 2013. Combining multiple imputation and meta-analysis with individual participant data. *Statistics in Medicine*, 32, 4499–4514.

Button, K. S., Kounali, D., Thomas, L., Wiles, N. J., Peters, T. J., Welton, N. J., Ades, A. E. & Lewis, G. 2015. Minimal clinically important difference on the

Beck Depression Inventory-II according to the patient's perspective. *Psychological Medicine*, 45, 3269–3279.

Caldwell, D. M., Ades, A. E., Dias, S., Watkins, S., Li, T., Taske, N., Naidoo, B. & Welton, N. J. 2016. A threshold analysis assessed the credibility of conclusions from network meta-analysis. *Journal of Clinical Epidemiology*, 80, 68–76.

Caldwell, D. M., Ades, A. E. & Higgins, J. P. T. 2005. Simultaneous comparison of multiple treatments: combining direct and indirect evidence. *BMJ*, 331, 897–900.

Caldwell, D. M., Gibb, D. M. & Ades, A. E. 2007. Validity of indirect comparisons in meta-analysis.[Letter]. *Lancet*, 369, 270–270.

Caldwell, D. M., Welton, N. J. & Ades, A. E. 2010. Mixed treatment comparison analysis provides internally coherent treatment effect estimates based on overviews of reviews and can reveal inconsistency. *Journal of Clinical Epidemiology*, 63, 875–882.

Caldwell, D. M., Welton, N. J., Dias, S. & Ades, A. E. 2012. Selecting the best scale for measuring treatment effect in a network meta-analysis: a case study in childhood nocturnal enuresis. *Research Synthesis Methods*, 3, 126–141.

Carlin, B. & Hong, H. 2014. Bayesian network meta-analysis for safety evaluation. *In:* Jiang, Q. & Amy Xia, H. (eds.) *Quantitative Evaluation of Safety in Drug Development: Design, Analysis and Reporting*, Boca Raton, CRC Press.

Chaimani, A., Higgins, J. P. T., Mavridis, D., Spyridonos, P. & Salanti, G. 2013a. Graphical tools for network meta-analysis in STATA. *PLoS One*, 8, e76654.

Chaimani, A. & Salanti, G. 2012. Using network meta-analysis to evaluate the existence of small-study effects in a network of interventions. *Research Synthesis Methods*, 3, 161–176.

Chaimani, A., Vasiliadis, H. S., Pandis, N., Schmid, C. H., Welton, N. J. & Salanti, G. 2013b. Effects of study precision and risk of bias in networks of interventions: a network meta-epidemiological study. *International Journal of Epidemiology*, 42, 1120–1131.

Chinn, S. 2000. A simple method for converting an odds ratio to effect size for use in meta-analysis. *Statistics in Medicine*, 19, 3127–3131.

Cholesterol Treatment Trialists' Collaboration. 2010. Efficacy and safety of more intensive lowering of LDL cholesterol: a meta-analysis of data from 170,000 participants in 26 randomised trials. *The Lancet*, 376, 1670–1681.

Chootrakool, H. & Shi, J. Q. 2008. Meta-analysis of multi-arm trials using empirical logistic transform. *The Open Medical Informatics Journal*, 2, 112–116.

Chootrakool, H., Shi, J. Q. & Yue, R. 2011. Meta-analysis and sensitivity analysis for multi-arm trials with selection bias. *Statistics in Medicine*, 30, 1183–1198.

Chou, R., Fu, R., Hoyt Huffman, L. & Korthuis, P. T. 2006. Initial highly-active antiretroviral therapy with a protease inhibitor versus non-nucleoside reverse transcriptase inhibitor: discrepancies between direct and indirect meta-analyses. *Lancet*, 368, 1503–1515.

Cipriani, A., Furukawa, T. A., Salanti, G., Geddes, J. R., Higgins, J. P. T., Churchill, R., Watanabe, N., Nakagawa, A., Omori, I. M., McGuire, H., Tansella, M. & Barbui, C. 2009. Comparative efficacy and acceptability of 12 new generation antidepressants: a multiple-treatments meta-analysis. *Lancet*, 373, 746–758.

Cipriani, A., Higgins, J. P. T., Geddes, J. R. & Salanti, G. 2013. Conceptual and technical challenges in network meta-analysis. *Annals of Internal Medicine*, 159, 130–137.

Claxton, K. 1999. Bayesian approaches to the value of information: implications for the regulation of new pharmaceuticals. *Health Economics*, 8, 269–274.

Claxton, K., Eggington, S., Ginnelly, L., Griffin, S., McCabe, C., Philips, Z., Tappenden, P. & Wailoo, A. 2005a. *Pilot Study of Value of Information Analysis to Support Research Recommendations for the National Institute for Clinical Excellence*. Centre for Health Economics, Research Paper, 4, York, University of York, Centre for Health Economics. Available: https://www.york.ac.uk/media/che/documents/papers/researchpapers/rp4_Pilot_study_of_value_of_information_analysis.pdf (Accessed 2 August 2017).

Claxton, K., Lacey, L. F. & Walker, S. G. 2000. Selecting treatments: a decision theoretic approach. *Journal of the Royal Statistical Society: Series A (Statistics in Society)*, 163, 211–226.

Claxton, K. & Posnett, J. 1996. An economic approach to clinical trial design and research priority-setting. *Health Economics*, 5, 513–524.

Claxton, K., Sculpher, M., McCabe, C., Briggs, A., Akehurst, R., Buxton, M., Razier, J. & O'Hagan, A. 2005b. Probabilistic sensitivity analysis for NICE technology assessment: not an optional extra. *Health Economics*, 14, 339–347.

Cohen, J. 1969. *Statistical Power Analysis for the Behavioral Sciences*, New York, Academic Press.

Colbourn, T., Asseburg, C., Bojke, L., Phillips, Z., Claxton, K., Ades, A. E. & Glibert, R. E. 2007a. Prenatal screening and treatment strategies to prevent group B streptococcal and other bacterial infections in early infancy: cost-effectiveness and expected value of information analysis. *Health Technology Assessment*, 11, 1–226.

Colbourn, T. E., Asseburg, C., Bojke, L., Philips, Z., Welton, N. J., Claxton, K., Ades, A. E. & Gilbert, R. E. 2007b. Preventive strategies for group B streptococcal and other bacterial infections in early infancy: cost effectiveness and value of information analyses. *BMJ*, 335, 655–661.

Collett, D. 1994. *Modelling Survival Data in Medical Research*, London, Chapman & Hall.

Collett, D. 2003. *Modelling Survival Data in Medical Research*, Boca Raton, Chapman & Hall/CRC.

Collins, R., Peto, R., MacMahon, S., Herbert, P., Fiebach, N. H., Eberlein, K. A., Godwin, J., Qizilbash, N., Taylor, J. O. & Hennekens, C. H. 1990. Blood pressure, stroke, and coronary heart disease. Part 2, short-term reductions in blood pressure: overview of randomised drug trials in their epidemiological context. *The Lancet*, 335, 827–838.

Cooper, H. & Hedges, L. 1994. *The Handbook of Research Synthesis*, New York, Russell Sage Foundation.

Cooper, N. J., Kendrick, D., Achana, F., Dhiman, P., He, Z., Wynn, P., Le Cozannet, E., Saramago, P. & Sutton, A. J. 2012. Network meta-analysis to evaluate the effectiveness of interventions to increase the uptake of smoke alarms. *Epidemiologic Reviews*, 34, 32–45.

Cooper, N. J., Peters, J., Lai, M. C. W., Juni, P., Wandel, S., Palmer, S., Paulden, M., Conti, S., Welton, N. J., Abrams, K. R., Bujkiewicz, S., Spiegelhalter, D. & Sutton, A. J. 2011. How valuable are multiple treatment comparison methods in evidence-based health-care evaluation? *Value in Health*, 14, 371–380.

Cooper, N. J., Sutton, A. J., Abrams, K. R., Turner, D. & Wailoo, A. 2003. Comprehensive decision analytical modelling in economic evaluation: a Bayesian approach. *Health Economics*, 13, 203–226.

Cooper, N. J., Sutton, A. J., Ades, A. E., Paisley, S. & Jones, D. R. 2007. Use of evidence in economic decision models: practical issues and methodological challenges. *Health Economics*, 16, 1277–1286.

Cooper, N. J., Sutton, A. J., Lu, G. & Khunti, K. 2006. Mixed comparison of stroke prevention treatments in individuals with nonrheumatic atrial fibrillation. *Archives of Internal Medicine*, 166, 1269–1275.

Cooper, N. J., Sutton, A. J., Morris, D., Ades, A. E. & Welton, N. J. 2009. Adressing between-study heterogeneity and inconsistency in mixed treatment comparisons: application to stroke prevention treatments in individuals with non-rheumatic atrial fibrillation. *Statistics in Medicine*, 28, 1861–1881.

Copas, J. B. & Shi, J. Q. 2001. A sensitivity analysis for publication bias in systematic reviews. *Statistical Methods in Medical Research*, 10, 251–265.

Cope, S., Capkun Niggli, G., Gale, R., Lassen, C., Owen, R., Ouwens, M. J. N. M., Bergman, G. & Jansen, J. P. 2012. Efficacy of once-daily indacaterol relative to alternative bronchodilators in COPD: a patient-level mixed treatment comparison. *Value in Health*, 15, 524–533.

Cope, S., Donohue, J. F., Jansen, J. P., Kraemer, M., Capkun-Niggli, G., Baldwin, M., Buckley, F., Ellis, A. & Jones, P. 2013. Comparative efficacy of long-acting bronchodilators for COPD: a network meta-analysis. *Respiratory Research*, 14, 100.

Cope, S. & Jansen, J. P. 2013. Quantitative summaries of treatment effect estimates obtained with network meta-analysis of survival curves to inform decision-making. *BMC Medical Research Methodology*, 13, 1–12.

Corbett, M. S., Rice, S. J. C., Madurasinghe, V., Slack, R., Fayter, D. A., Harden, M., Sutton, A. J., Macpherson, H. & Woolacott, N. F. 2013. Acupuncture and other physical treatments for the relief of pain due to osteoarthritis of the knee: network meta-analysis. *Osteoarthritis and Cartilage*, 21, 1290–1298.

Cox, D. R. & Oakes, D. 1984. *Analysis of Survival Data*, Boca Raton, Chapman & Hall/CRC.

Cranney, A., Guyatt, G., Griffith, L., Wells, G., Tugwell, P., Rosen, C., Osteoporosis Methodology Group & The Osteoporosis Research Advisory Group. 2002. Summary of meta-analyses of therapies for post-menopausal osteoporosis. *Endocrine Reviews*, 23, 570–578.

Critchfield, G. C. & Willard, K. E. 1986. Probabilistic analysis of decision trees using Monte Carlo simulation. *Medical Decision Making*, 6, 85–92.

Dakin, H., Fidler, C. & Harper, C. 2010. Mixed treatment comparison meta-analysis evaluating the relative efficacy of nucleos(t)ides for treatment of nucleos(t)ide-naive patients with chronic hepatitis B. *Value in Health*, 13, 934–945.

Dakin, H. A., Welton, N. J., Ades, A. E., Collins, S., Orme, M. & Kelly, S. 2011. Mixed treatment comparison of repeated measurements of a continuous endpoint: an example using topical treatments for primary open-angle glaucoma and ocular hypertension. *Statistics in Medicine*, 30, 2511–2535.

Daniels, M. J. & Hughes, M. D. 1997. Meta-analysis for the evaluation of potential surrogate markers. *Statistics in Medicine*, 16, 1965–1982.

Dear, K. G. B. 1994. Iterative generalised least squares for meta-analysis of survival data at multiple times. *Biometrics*, 50, 989–1982.

Deeks, J. J. 2002. Issues on the selection of a summary statistic for meta-analysis of clinical trials with binary outcomes. *Statistics in Medicine*, 21, 1575–1600.

Del Giovani, C., Vacchi, L., Mavridis, D., Filippini, G. & Salanti, G. 2013. Network meta-analysis models to account for variability in treatment definitions: application to dose effects. *Statistics in Medicine*, 32, 25–39.

Demiris, N., Lunn, D. & Sharples, L. D. 2015. Survival extrapolation using the poly-Weibull model. *Statistical Methods in Medical Research*, 24, 287–301.

Demiris, N. & Sharples, L. D. 2006. Bayesian evidence synthesis to extrapolate survival estimates in cost-effectiveness studies. *Statistics in Medicine*, 25, 1960–1975.

Dempster, A. P. 1997. The direct use of likelihood for significance testing. *Statistics and Computing*, 7, 247–252.

Dersimonian, R. & Laird, N. 1986. Meta-analysis of clinical trials. *Controlled Clinical Trials*, 7, 177–188.

Dias, S. & Ades, A. E. 2016. Absolute or relative effects? Arm-based synthesis of trial data. *Research Synthesis Methods*, 7, 23–28.

Dias, S., Sutton, A. J., Ades, A. E. & Welton, N. J. 2013a. Evidence synthesis for decision making 2: a generalized linear modeling framework for pairwise and network meta-analysis of randomized controlled trials. *Medical Decision Making*, 33, 607–617.

Dias, S., Sutton, A. J., Welton, N. J. & Ades, A. E. 2011a. NICE DSU Technical Support Document 3. Heterogeneity: subgroups, meta-regression, bias and bias-adjustment. *Technical Support Document*. Available: http://scharr.dept. shef.ac.uk/nicedsu/technical-support-documents/evidence-synthesis-tsd-series/ (Accessed 2 August 2017).

Dias, S., Sutton, A. J., Welton, N. J. & Ades, A. E. 2011b. NICE DSU Technical Support Document 6. Embedding evidence synthesis in probabilistic cost-effectiveness analysis: software choices. *Technical Support Document*. Available: http://scharr.dept.shef.ac.uk/nicedsu/technical-support-documents/evidence-synthesis-tsd-series/ (Accessed 2 August 2017).

Dias, S., Sutton, A. J., Welton, N. J. & Ades, A. E. 2013b. Evidence synthesis for decision making 3: heterogeneity – subgroups, meta-regression, bias and bias-adjustment. *Medical Decision Making*, 33, 618–640.

Dias, S., Welton, N. J. & Ades, A. E. 2010a. Study designs to detect sponsorship and other biases in systematic reviews. *Journal of Clinical Epidemiology*, 63, 587–588.

Dias, S., Welton, N. J., Caldwell, D. M. & Ades, A. E. 2010b. Checking consistency in mixed treatment comparison meta-analysis. *Statistics in Medicine*, 29, 932–944.

Dias, S., Welton, N. J., Marinho, V. C. C., Salanti, G., Higgins, J. P. T. & Ades, A. E. 2010c. Estimation and adjustment of bias in randomised evidence by using mixed treatment comparison meta-analysis. *Journal of the Royal Statistical Society: Series A (Statistics in Society)*, 173, 613–629.

Dias, S., Welton, N. J., Sutton, A. J. & Ades, A. E. 2011c. NICE DSU Technical Support Document 2. A generalised linear modelling framework for pair-wise and network meta-analysis of randomised controlled trials. *Technical Support Document*. Available: http://scharr.dept.shef.ac.uk/ nicedsu/technical-support-documents/evidence-synthesis-tsd-series/ (Accessed 2 August 2017).

Dias, S., Welton, N. J., Sutton, A. J. & Ades, A. E. 2011d. NICE DSU Technical Support Document 5. Evidence synthesis in the baseline natural history model. *Technical Support Document*. Available: http://scharr.dept.shef.ac.uk/nicedsu/ technical-support-documents/evidence-synthesis-tsd-series/ (Accessed 2 August 2017).

Dias, S., Welton, N. J., Sutton, A. J. & Ades, A. E. 2013c. Evidence synthesis for decision making 5: the baseline natural history model. *Medical Decision Making*, 33, 657–670.

Dias, S., Welton, N. J., Sutton, A. J., Caldwell, D. M., Lu, G. & Ades, A. E. 2011e. NICE DSU Technical Support Document 4. Inconsistency in networks of evidence based on randomised controlled trials. *Technical Support Document*. Available: http://scharr.dept.shef.ac.uk/nicedsu/technical-support-documents/ evidence-synthesis-tsd-series/ (Accessed 2 August 2017).

Dias, S., Welton, N. J., Sutton, A. J., Caldwell, D. M., Lu, G. & Ades, A. E. 2013d. Evidence synthesis for decision making 4: inconsistency in networks of evidence based on randomized controlled trials. *Medical Decision Making*, 33, 641–656.

Dominici, F. 2000. Combining contingency tables with missing dimensions. *Biometrics*, 56, 546–553.

Dominici, F., Parmigiani, G., Reckhow, K. H. & Wolpert, R. L. 1997. Combining information from related regressions. *Journal of Agricultural, Biological, and Environmental Statistics*, 2, 313–332.

Dominici, F., Parmigiani, G., Wolpert, R. L. & Hasselblad, V. 1999. Meta-analysis of migraine headache treatments: combining information from heterogenous designs. *Journal of the American Statistical Association*, 94, 16–28.

Donegan, S., Williamson, P., D'Alessandro, U., Garner, P. & Tudor Smith, C. 2013. Combining individual patient data and aggregate data in mixed treatment comparison meta-analysis: individual patient data may be beneficial if only for a subset of trials. *Statistics in Medicine*, 32, 914–930.

Donegan, S., Williamson, P., D'Alessandro, U. & Tudor Smith, C. 2012. Assessing the consistency assumption by exploring treatment by covariate interactions in mixed treatment comparison meta-analysis: individual patient-level covariates versus aggregate trial-level covariates. *Statistics in Medicine*, 31, 3840–3857.

Donegan, S., Williamson, P., Gamble, C. & Tudor-Smith, C. 2011. Indirect comparisons: a review of reporting and methodological quality. *PLoS One*, 5, e11054.

Dorans, N. J., Pommerich, M. & Holland, P. W. (eds.) 2007. *Linking and Aligning Scores and Scales*, New York, Springer.

Doubilet, P., Begg, C. B., Weinstein, M. C., Braun, P. & McNeil, B. J. 1985. Probabilistic sensitivity analysis using Monte Carlo simulation: a practical approach. *Medical Decision Making*, 5, 157–177.

Draper, N. R. & Smith, H. 1998. *Applied Regression Analysis*, New York, John Wiley & Sons, Inc.

Dumouchel, W. 1996. Predictive cross-validation of Bayesian meta-analyses. *In:* Bernardo, J. M., Berger, J. O., Dawid, A. P. & Smith, A. F. M. (eds.) *Bayesian Statistics 5*, Oxford, Oxford University Press.

Eccles, M., Freemantle, N. & Mason, J. 2001. Using systematic reviews in clinical guideline development. *In:* Egger, M., Davey Smith, G. & Altman, D. G. (eds.) *Systematic Reviews in Health Care: Meta-Analysis in Context*, 2nd ed., London, BMJ.

Eddy, D., Hollingworth, W. & Caro, J. 2012. Model transparency and validation: a report of the ISPOR-SMDM modeling good research practices task force-4. *Value in Health*, 15, 843–850.

Eddy, D. M. 1989. The confidence profile method: a Bayesian method for assessing health technologies. *Operations Research*, 37, 210–228.

Eddy, D. M., Hasselblad, V. & Shachter, R. 1992. *Meta-Analysis by the Confidence Profile Method*, London, Academic Press.

Edwards, S. J., Clarke, M. J., Wordsworth, S. & Borrill, J. 2009. Indirect comparisons of treatments based on systematic reviews of randomised controlled trials. *International Journal of Clinical Practice*, 63, 841–854.

Efthimiou, O., Mavridis, D., Riley, R., Cipriani, A. & Salanti, G. 2014. Joint synthesis of multiple correlated outcomes in networks of interventions. *Biostatistics*, 16, 84–97.

Egger, M. & Davey-Smith, G. 1995. Misleading meta-analysis. *BMJ*, 310, 752–754.

Egger, M., Davey-Smith, G. & Altman, D. 2001. *Systematic Reviews in Health Care: Meta-Analysis in Context*, London, BMJ Publishing Group.

Egger, M., Smith, G. D. & Phillips, A. N. 1997. Meta-analysis. Principles and procedures. *BMJ*, 315, 1533–1537.

Elliott, W. J. & Meyer, P. M. 2007. Incident diabetes in clinical trials of antihypertensive drugs: a network meta-analysis. *Lancet*, 369, 201–207.

Engels, E. A., Schmid, C. H., Terrin, N., Olkin, I. & Lau, J. 2000. Heterogeneity and statistical significance in meta-analysis: an empirical study of 125 meta-analyses. *Statistics in Medicine*, 19, 1707–1728.

Fayers, P. M. & Hays, R. D. 2014. Should linking replace regression when mapping from profile-based measures to preference-based measures? *Value in Health*, 17, 261–265.

Felli, J. C. & Hazen, G. B. 1998. Sensitivity analysis and the expected value of perfect information. *Medical Decision Making*, 18, 95–109.

Fiocco, M., Putter, H. & van Houwelingen, J. C. 2009. Meta-analysis of pairs of survival curves under heterogeneity: a Poisson correlated gamma-frailty approach. *Statistics in Medicine*, 28, 3782–3797.

Flacco, M. E., Manzoli, L., Boccia, S., Capasso, L., Aleksovska, K., Rosso, A., Scaioli, G., De Vito, C., Siliquini, R., Villari, P. & Ioannidis, J. P. A. 2015. Head-to-head randomized trials are mostly industry sponsored and almost always favor the industry sponsor. *Journal of Clinical Epidemiology*, 68, 811–820.

Fleiss, J. L. 1994. Measures of effect size for categorical data. *In:* Cooper, H. & Hedges, L. V. (eds.) *The Handbook of Research Synthesis*, New York, Russell Sage Foundation.

Follmann, D., Elliott, P., Suh, I. & Cutler, J. 1992. Variance imputation for overviews of clinical trials with continuous response. *Journal of Clinical Epidemiology*, 45, 769–773.

Franchini, A., Dias, S., Ades, A. E., Jansen, J. & Welton, N. 2012. Accounting for correlation in mixed treatment comparisons with multi-arm trials. *Research Synthesis Methods*, 3, 142–160.

Fresco, C., French, J., Buchan, I. & Keeley, E. C. 2003. Correspondence: primary angioplasty or thrombolysis for acute myocardial infarction? *Lancet*, 361, 1303–1305.

Frison, L. & Pocock, S. J. 1992. Repeated measures in clinical trials: analysis using mean summary statistics and its implications for design. *Statistics in Medicine*, 11, 1685–1704.

Fu, H., Price, K. L., Nilsson, M. E. & Ruberg, S. J. 2013. Identifying potential adverse events dose-response relationships via Bayesian indirect and mixed treatment comparison models. *Journal of Biopharmaceutical Statistics*, 23, 26–42.

Gamble, C. & Hollis, S. 2005. Uncertainty method improved on best–worst case analysis in a binary meta-analysis. *Journal of Clinical Epidemiology*, 58, 579–588.

Gartlehner, G. & Fleg, A. 2010. Pharmaceutical company-sponsored drug trials: the system is broken. *Journal of Clinical Epidemiology*, 63, 128–129.

Gartlehner, G., Morgan, L., Thieda, P. & Fleg, A. 2010. The effect of sponsorship on a systematically evaluated body of evidence of head-to-head trials was modest: secondary analysis of a systematic review. *Journal of Clinical Epidemiology*, 63, 117–125.

Gelman, A. 1996. Inference and monitoring convergence. *In:* Gilks, W. R., Richardson, S. & Spiegelhalter, D. J. (eds.) *Markov Chain Monte Carlo in Practice*, London, Chapman & Hall.

Gelman, A. 2006. Prior distributions for variance parameters in hierarchical models. *Bayesian Analysis*, 1, 515–533.

Gelman, A., Carlin, J. B., Stern, H. S. & Rubin, D. B. 2004. *Bayesian Data Analysis*, Boca Raton, Chapman & Hall/CRC.

Gilks, W. R., Richardson, S. & Spiegelhalter, D. J. 1996. *Markov Chain Monte Carlo in Practice*, London, Chapman & Hall/CRC.

Glenny, A. M., Altman, D. G., Song, F., Sakarovitch, C., Deeks, J. J., D'Amico, R., Bradburn, M. & Eastwood, A. 2005. Indirect comparisons of competing interventions. *Health Technology Assesment*, 9, 1–134.

Gleser, L. J. & Olkin, I. 1994. Stochastically dependent effect sizes. *In:* Cooper, H. & Hedges, L. V. (eds.) *The Handbook of Research Synthesis*, New York, Russell Sage Foundation.

Goeree, R., O'Brien B., Hunt, R., Blackhouse, G., Willan, A. & Watson, J. 1999. Economic evaluation of long term management strategies for erosive oesophagitis. *PharmacoEconomics*, 16, 679–697.

Golder, S., Glanville, J. & Ginnelly, L. 2005. Populating decision-analytic models: the feasibility and efficiency of database searching for individual parameters. *International Journal of Technology Assessment in Health Care*, 21, 305–311.

Goldstein, H., Yang, M., Omar, R. Z., Turner, R. M. & Thompson, S. G. 2000. Meta-analysis using multilevel models with an application to the study of class size effects. *Applied Statistics*, 49, 399–412.

Gotzsche, P. C. 2000. Why we need a broad perspective on meta-analysis. *BMJ*, 321, 585–586.

Goubar, A., Ades, A. E., De Angelis, D., McGarrigle, C. A., Mercer, C. H., Tookey, P. A., Fenton, K. & Gill, O. N. 2008. Estimates of human immunodeficiency virus prevalence and proportion diagnosed based on Bayesian multiparameter synthesis of surveillance data. *Journal of the Royal Statistical Society: Series A (Statistics in Society)*, 171, 541–580.

Govan, L., Ades, A. E., Weir, C. J., Welton, N. J. & Langhorne, P. 2010. Controlling ecological bias in evidence synthesis of trials reporting on collapsed and overlapping covariate categories. *Statistics in Medicine*, 29, 1340–1356.

Greenland, S. 1994a. Can meta-analysis be salvaged? *American Journal of Epidemiology*, 140, 783–787.

Greenland, S. 1994b. Invited commentary: a critical look at some popular meta-analytic methods. *American Journal of Epidemiology*, 140, 290–296.

Greenland, S., Maclure, M., Schlesselman, J., Poole, C. & Morgenstern, H. 1991. Standardized regression coefficients: a further critique and review of some alternatives. *Epidemiology*, 2, 387–392.

Greenland, S., Schlesselman, J. & Criqui, M. 1986. The fallacy of employing standardized coefficients and correlations as measures of effect. *American Journal of Epidemiology*, 123, 203–208.

Grieve, R., Hawkins, N. & Pennington, M. 2013. Extrapolation of survival data in cost-effectiveness analyses: improving the current state of play. *Medical Decision Making*, 33, 740–742.

Grimmett, G. D. R. & Stirzaker, D. R. 1992. *Probability and Random Processes*, Oxford, Oxford University Press.

Guevara, J. P., Berlin, J. A. & Wolf, F. M. 2004. Meta-analytic methods for pooling rates when follow-up duration varies: a case study. *BMC Medical Research Methodology*, 4, 17.

Guyatt, G., Oxman, A., Vist, G., Kunz, R., Falck-Ytter, Y., Alonso-Coello, P., Schünemann, H. & Grade Working Group. 2008. GRADE: an emerging consensus on rating quality of evidence and strength of recommendations. *BMJ*, 336, 924–926.

Guyatt, G. H., Oxman, A. D., Kunz, R., Brozek, J., Alonso-Coello, P., Rind, D., Devereaux, P. J., Montori, V. M., Freyschuss, B., Vist, G., Jaeschke, R., Williams, J. W., Jr., Murad, M. H., Sinclair, D., Falck-Ytter, Y., Meerpohl, J., Whittington, C., Thorlund, K., Andrews, J. & Schünemann, H. J. 2011a. GRADE guidelines: 6. Rating the quality of evidence: imprecision. *Journal of Clinical Epidemiology*, 64, 1283–1293.

Guyatt, G. H., Oxman, A. D., Kunz, R., Woodcock, J., Brozek, J., Helfand, M., Alonso-Coello, P., Falck-Ytter, Y., Jaeschke, R., Vist, G., Akl, E. A., Post, P. N., Norris, S., Meerpohl, J., Shukla, V. K., Nasser, M. & Schünemann, H. J. 2011b. GRADE guidelines: 8. Rating the quality of evidence: indirectness. *Journal of Clinical Epidemiology*, 64, 1303–1310.

Guyatt, G. H., Oxman, A. D., Kunz, R., Woodcock, J., Brozek, J., Helfand, M., Alonso-Coello, P., Glasziou, P., Jaeschke, R., Akl, E. A., Norris, S., Vist, G., Dahm, P., Shukla, V. K., Higgins, J., Falck-Ytter, Y. & Schünemann, H. J. 2011c. GRADE guidelines: 7. Rating the quality of evidence: inconsistency. *Journal of Clinical Epidemiology*, 64, 1294–1302.

Guyatt, G. H., Oxman, A. D., Montori, V., Vist, G., Kunz, R., Brozek, J., Alonso-Coello, P., Djulbegovic, B., Atkins, D., Falck-Ytter, Y., Williams, J. W., Jr., Meerpohl, J., Norris, S. L., Akl, E. A. & Schünemann, H. J. 2011d. GRADE guidelines: 5. Rating the quality of evidence: publication bias. *Journal of Clinical Epidemiology*, 64, 1277–1282.

Guyatt, G. H., Oxman, A. D., Vist, G., Kunz, R., Brozek, J., Alonso-Coello, P., Montori, V., Akl, E. A., Djulbegovic, B., Falck-Ytter, Y., Norris, S. L.,

Williams, J. W., Jr., Atkins, D., Meerpohl, J. & Schünemann, H. J. 2011e. GRADE guidelines: 4. Rating the quality of evidence: study limitations (risk of bias). *Journal of Clinical Epidemiology*, 64, 407–415.

Guyot, P. 2014. *Expected survival time as a summary statistic in economic analysis and evidence synthesis.* PhD thesis, University of Bristol.

Guyot, P., Ades, A. E., Beasley, M., Lueza, B., Pignon, J.-P. & Welton, N. J. 2016. Extrapolation of survival curves from cancer trials using external information. *Medical Decision Making*, 37, 353–366.

Guyot, P., Ades, A. E., Ouwens, M. J. N. M. & Welton, N. J. 2012. Enhanced secondary analysis of survival data: reconstructing the data from published Kaplan-Meier survival curves. *BMC Medical Research Methodology*, 12, 9.

Guyot, P., Welton, N. J., Ouwens, J. N. M. & Ades, A. E. 2011. Survival time outcomes in randomised controlled trials and meta-analyses: the parallel universes of efficacy and cost-effectiveness. *Value in Health*, 14, 640–646.

Haas, D. M., Caldwell, D. M., Kirkpatrick, P., McIntosh, J. J. & Welton, N. J. 2012. Tocolytic therapy for preterm delivery: systematic review and network meta-analysis. *BMJ*, 345, e6226.

Hahn, G. 2001. *BUGS utility for spreadsheets. 1.0.1 ed.* Available: http://faculty.salisbury.edu/~edhahn/bus.htm (Accessed 15 July 2017).

Hardy, R. J. & Thompson, S. G. 1996. A likelihood approach to meta-analysis with random effects. *Statistics in Medicine*, 15, 619–629.

Hasselblad, V. 1998. Meta-analysis of multi-treatment studies. *Medical Decision Making*, 18, 37–43.

Hedges, L. V. 1981. Distribution theory for Glass's estimator of effect size and related estimators. *Journal of Educational Statistics*, 6, 107–128.

Hedges, L. V. & Olkin, I. 1985. *Statistical Methods for Meta-Analysis*, London, Academic Press.

Heiberger, R. M. & Neuwirth, E. 2009. *R Through Excel: A Spreadsheet Interface for Statistics, Data Analysis, and Graphics*, New York, Springer.

Higgins, J. P. T. & Altman, D. G. 2008. Assessing risk of bias in included studies. *In:* Higgins, J. & Green, S. (eds.) *Cochrane Handbook for Systematic Reviews of Interventions Version 5.0.1 [Updated September 2008]*, Oxford, The Cochrane Collaboration.

Higgins, J. P. T., Altman, D. G., Gøtzsche, P. C., Jüni, P., Moher, D., Oxman, A. D., Savović, J., Schulz, K. F., Weeks, L. & Sterne, J. A. C. 2011. The Cochrane Collaboration's tool for assessing risk of bias in randomised trials. *BMJ*, 343, d5928.

Higgins, J. P. T., Deeks, J. J. & Altman, D. G. 2008a. Special topics in statistics. *In:* Higgins, J. & Green, S. (eds.) *Cochrane Handbook for Systematic Reviews of Interventions. Version 5.0.1 [Updated September 2008]*, Oxford, The Cochrane Collaboration.

Higgins, J. P. T., Deeks, J. J., Altman, D. G. & on Behalf of the Cochrane Statistical Methods Group (eds.) 2008b. Special topics in statistics. *In: Cochrane Handbook for Systematic Reviews of Interventions*, Chichester, John Wiley & Sons, Ltd.

Higgins, J. P. T. & Green, S. 2008. *Cochrane Handbook for Systematic Reviews of Interventions Version 5.0.0 [Updated February 2008]*, Chichester, The Cochrane Collaboration, John Wiley & Sons, Ltd.

Higgins, J. P. T., Jackson, D., Barrett, J. K., Lu, G., Ades, A. E. & White, I. R. 2012. Consistency and inconsistency in network meta-analysis: concepts and models for multi-arm studies. *Research Synthesis Methods*, 3, 98–110.

Higgins, J. P. T. & Spiegelhalter, D. J. 2002. Being sceptical about meta-analyses: a Bayesian perspective on magnesium trials in myocardial infarction. *International Journal of Epidemiology*, 31, 96–104.

Higgins, J. P. T. & Thompson, S. G. 2002. Quantifying heterogeneity in a meta-analysis. *Statistics in Medicine*, 21, 1539–1558.

Higgins, J. P. T. & Thompson, S. G. 2004. Controlling the risk of spurious findings from meta-regression. *Statistics in Medicine*, 23, 1663–1682.

Higgins, J. P. T., Thompson, S. G. & Spiegelhalter, D. J. 2009. A re-evaluation of random-effects meta-analysis. *Journal of the Royal Statistical Society: Series A (Statistics in Society)*, 172, 137–159.

Higgins, J. P. T., White, I. R. & Wood, A. 2008b. Imputation methods for missing outcome data in meta-analysis of clinical trials. *Clinical Trials*, 5, 225–239.

Higgins, J. P. T. & Whitehead, A. 1996. Borrowing strength from external trials in a meta-analysis. *Statistics in Medicine*, 15, 2733–2749.

Higgins, J. P. T., Whitehead, A., Turner, R. M., Omar, R. Z. & Thompson, S. G. 2001. Meta-analysis of continuous outcome data from individual patients. *Statistics in Medicine*, 20, 2219–2241.

Hoaglin, D. C., Hawkins, N., Jansen, J., Scott, D. A., Itzler, R., Cappelleri, J. C., Boersma, C., Thompson, D., Larholt, K. M., Diaz, M. & Barrett, A. 2011. Conducting indirect treatment comparison and network meta-analysis studies: report of the ISPOR task force on indirect treatment comparison good research practices – Part 2. *Value in Health*, 14, 429–437.

Holzhauer, B. 2017. Meta-analysis of aggregate data on medical events. *Statistics in Medicine*, 36, 723–737.

Hong, H., Carlin, B. P., Shamliyan, T. A., Wyman, J. F., Ramakrishnan, R., Sainfort, F. & Kane, R. L. 2013. Comparing Bayesian and frequentist approaches for multiple outcome mixed treatment comparisons. *Medical Decision Making*, 5, 702–714.

Hong, H., Chu, H., Zhang, J. & Carlin, B. P. 2016. A Bayesian missing data framework for generalized multiple outcome mixed treatment comparisons. *Research Synthesis Methods*, 7, 6–22.

Hong, H., Fu, H., Price, K. L. & Carlin, B. P. 2015. Incorporation of individual-patient data in network meta-analysis for multiple continuous endpoints, with application to diabetes treatment. *Statistics in Medicine*, 34, 2794–2819.

Hooper, L., Summerbell, C. D., Higgins, J. P. T., Thompson, R. L., Clements, G., Capps, N., Davey Smith, G., Riemersma, R. & Ebrahim, S. 2000. Reduced or modified dietary fat for preventing cardiovascular disease. *Cochrane Database of Systematic Reviews*, 2000(2), CD002137.

Hoyle, M. W. & Henley, W. 2011. Improved curve fits to summary survival data: application to economic evaluation of health technologies. *BMC Medical Research Methodology*, 11, 1–14.

Hunink, M., Glasziou, P., Siegel, J., Weeks, J., Pliskin, J., Elsetin, A. & Weinsein, M. 2001. *Decision Making in Health and Medicine: Integrating Evidence and Values*, Cambridge, Cambridge University Press.

Hutton, B., Salanti, G., Caldwell, D. M., Chaimani, A., Schmid, C. H., Cameron, C., Ioannidis, J. P. A., Straus, S. E., Thorlund, K., Jansen, J. P., Mulrow, C., Catalá-López, F., Gøtzsche, P. C., Dickersin, K., Boutron, I., Altman, D. G. & Moher, D. 2015. The PRISMA extension statement for reporting of systematic reviews incorporating network meta-analyses of health care interventions: checklist and explanations. *Annals of Internal Medicine*, 162, 777–784.

Hutton, B., Salanti, G., Chaimani, A., Caldwell, D. M., Schmid, C., Thorlund, K., Mills, E., Catalá-López, F., Turner, L., Altman, D. G. & Moher, D. 2014. The quality of reporting methods and results in network meta-analyses: an overview of reviews and suggestions for improvement. *PLoS One*, 9, e92508.

Hutton, J. L. 2000. Number needed to treat: properties and problems. *Journal of the Royal Statistical Society: Series A (Statistics in Society)*, 163, 403–419.

ICH Expert Working Group. 2000. *Choice of control group and related issues in clinical trials E10: ICH harmonised tripartite guideline*. Available: http://www.ich.org/fileadmin/Public_Web_Site/ICH_Products/Guidelines/Efficacy/E10/Step4/E10_Guideline.pdf (Accessed 20 July 2000).

Iglehart, J. K. 2009. Prioritizing comparative-effectiveness research: IOM recommendations. *New England Journal of Medicine*, 361, 325–328.

Ioannidis, J. P. A. 2005. Why most published research findings are false. *PLoS Medicine*, 2, e124.

ISIS 4 Collaborative Group. 1995. ISIS-4: a randomised factorial trial assessing early oral captopril, oral mononitrate, and intravenous magnesium sulphate in 58050 patients with suspected acute myocardial infarction. *Lancet*, 345, 669–685.

Jackson, D., Barrett, J. K., Rice, S., White, I. R. & Higgins, J. P. T. 2014. A design-by-treatment interaction model for network meta-analysis with random inconsistency effects. *Statistics in Medicine*, 33, 3639–3654.

Jackson, D., Riley, R. & White, I. R. 2011. Multivariate meta-analysis: potential and promise. *Statistics in Medicine*, 30, 2481–2598.

Jansen, J. P. 2011. Network meta-analysis of survival data with fractional polynomials. *BMC Medical Research Methodology*, 11, 1–14.

Jansen, J. P. 2012. Network meta-analysis of individual and aggregate level data. *Research Synthesis Methods*, 3, 177–190.

Jansen, J. P. & Cope, S. 2012. Meta-regression models to address heterogeneity and inconsistency in network meta-analysis of survival outcomes. *BMC Medical Research Methodology*, 12, 152.

Jansen, J. P., Fleurence, R., Devine, B., Itzler, R., Barrett, A., Hawkins, N., Lee, K., Boersma, C., Annemans, L. & Cappelleri, J. C. 2011. Interpreting indirect treatment comparisons and network meta-analysis for health-care decision making: report of the ISPOR task force on indirect treatment comparisons good research practices – Part 1. *Value in Health*, 14, 417–428.

Jansen, J. P. & Naci, H. 2013. Is network meta-analysis as valid as standard pairwise meta-analysis? It all depends on the distribution of effect modifiers. *BMC Medicine*, 11, 159.

Jansen, J. P. & Trikalinos, T. 2013. PRM237: multivariate network meta-analysis of progression free survival and overall survival. *Value in Health*, 16, A617.

Jansen, J. P., Trikalinos, T., Cappelleri, J. C., Daw, J., Andes, S., Eldessouki, R. & Salanti, G. 2014. Indirect treatment comparison/network meta-analysis study questionnaire to assess relevance and credibility to inform health care decision making: an ISPOR-AMCP-NPC good practice task force report. *Value in Health*, 17, 157–173.

Jansen, J. P., Vieira, M. C. & Cope, S. 2015. Network meta-analysis of longitudinal data using fractional polynomials. *Statistics in Medicine*, 34, 2294–2311.

Johnston, B. C., Kanters, S., Bandayrel, K., Wu, P., Naji, F., Siemieniuk, R. A., Ball, G. D. C., Busse, J. W., Thorlund, K., Guyatt, G., Jansen, J. P. & Mills, E. J. 2014. Comparison of weight loss among named diet programs in overweight and obese adults: a meta-analysis. *JAMA: Journal of the American Medical Association*, 312, 923–933.

Kaltenthaler, E., Tappenden, P., Paisley, S. & Squires, H. 2011. *Identifying and Reviewing Evidence to Inform the Conceptualisation and Population of Cost-Effectiveness Models*, Sheffield, NICE Decision Support Unit.

Keeley, E. C., Boura, J. A. & Grines, C. L. 2003a. Correspondence: primary angioplasty or thrombolysis for acute myocardial infarction? *The Lancet*, 361, 967–968.

Keeley, E. C., Boura, J. A. & Grines, C. L. 2003b. Primary angioplasty versus intravenous thrombolytic therapy for acute myocardial infarction: a quantitative review of 23 randomised trials. *Lancet*, 361, 13–20.

Kew, K. M., Dias, S. & Cates, C. J. 2014. Long-acting inhaled therapy (beta-agonists, anticholinergics and steroids) for COPD: a network meta-analysis. *Cochrane Database of Systematic Reviews*, 26, CD010844.

Kibret, T., Richer, D. & Beyea, J. 2014. Bias in identification of the best treatment in a Bayesian network meta-analysis for binary outcome: a simulation study. *Clinical Epidemiology*, 6, 451–460.

Kleinbaum, D. G. & Klein, M. 2012. *Survival Analysis: A Self-Learning Text*, 3rd ed., New York, Springer.

Kolen, M. J. & Brennan, R. L. 1994. *Test Equating, Scaling and Linking: Methods and Preactices*, New York, Springer.

König, J., Krahn, U. & Binder, H. 2013. Visualizing the flow of evidence in network meta-analysis and characterizing mixed treatment comparisons. *Statistics in Medicine*, 32, 5414–5429.

Krahn, U., Binder, H. & Konig, J. 2013. A graphical tool for locating inconsistency in network meta-analyses. *BMC Medical Research Methodology*, 13, 35.

Kriston, L., Von Wolff, A., Westphal, A., Holzel, L. P. & Harter, M. 2014. Efficacy and acceptability of acute treatments for persistent depressive disorder: a network meta-analysis. *Depression and Anxiety*, 31, 621–630.

Kuss, O. 2015. Statistical methods for meta-analyses including information from studies without any events: add nothing to nothing and succeed nevertheless. *Statistics in Medicine*, 34, 1097–1116.

Lahdelma, R. & Salminen, P. 2001. SMAA-2: stochastic multicriteria acceptibility analysis for group decision making. *Operations Research*, 49, 444–454.

Lambert, P. C., Sutton, A. J., Abrams, K. R. & Jones, D. R. 2002. A comparison of summary patient-level covariates in meta-regression with individual patient data meta-analysis. *Journal of Clinical Epidemiology*, 55, 86–94.

Lambert, P. C., Sutton, A. J., Burton, P., Abrams, K. R. & Jones, D. 2005. How vague is vague? Assessment of the use of vague prior distributions for variance components. *Statistics in Medicine*, 24, 2401–2428.

Latimer, N. 2011. *NICE DSU Technical Support Document 14. Undertaking Survival Analysis for Economic Evaluations Alongside Clinical Trials: Extrapolation with Patient-Level Data*, Sheffield, NICE Decision Support Unit.

Latimer, N. L. 2013. Survival analysis for economic evaluations alongside clinical trials-extrapolation with patient-level data: inconsistencies, limitations and a practical guide. *Medical Decision Making*, 33, 743–754.

Latimer, N. R. & Abrams, K. R. 2014. *NICE DSU Technical Support Document 16. Adjusting Survival Time Estimates in the Presence of Treatment Switching*, Sheffield, NICE Decision Support Unit.

Lauritzen, S. L. 2003. Some modern applications of graphical models. *In:* Green, P. J., Hjort, N. L. & Richardson, S. (eds.) *Highly Structured Stochastics Systems*, Oxford, Oxford University Press.

Lee, A. W. 2014. Review of mixed treatment comparisons in published systematic reviews shows marked increase since 2009. *Journal of Clinical Epidemiology*, 67, 138–143.

Li, J., Zhang, Q., Zhang, M. & Egger, M. 2007. Intravenous magnesium for acute myocardial infarction. *Cochrane Database of Systematic Reviews*, 2(4), CD002755.

Liberati, A., Altman, D. G., Tetzlaff, J., Mulrow, C., Gøtzsche, P. C., Ioannidis, J. P. A., Clarke, M., Devereaux, P. J., Kleijnen, J. & Moher, D. 2009. The PRISMA statement for reporting systematic reviews and meta-analyses of studies that evaluate health care interventions: explanation and elaboration. *PLoS Medicine*, 6, e1000100.

Linde, K., Rücker, G., Schneider, A. & Kriston, L. 2016. Questionable assumptions hampered interpretation of a network meta-analysis of primary care depression treatments. *Journal of Clinical Epidemiology*, 71, 86–96.

Little, R. J. A. & Rubin, D. B. 2002. *Statistical Analysis with Missing Data*, Hoboken, John Wiley & Sons, Inc.

Loke, Y. K., Golder, S. P. & Vandenbroucke, J. P. 2011. Comprehensive evaluations of the adverse effects of drugs: importance of appropriate study selection and data sources. *Therapeutic Advances in Drug Safety*, 2, 59–68.

Longworth, L. & Rowen, D. 2011. *NICE DSU Technical Support Document 10: The use of mapping methods to estimate health state utility values*, Available: http://www.nicedsu.org.uk (Accessed 15 July 2017).

Longworth, L. & Rowen, D. 2013. Mapping to obtain EQ 5D utility values for use in NICE health technology assessments. *Value in Health*, 16, 202–210.

Lu, G. & Ades, A. 2004. Combination of direct and indirect evidence in mixed treatment comparisons. *Statistics in Medicine*, 23, 3105–3124.

Lu, G. & Ades, A. 2006. Assessing evidence consistency in mixed treatment comparisons. *Journal of the American Statistical Association*, 101, 447–459.

Lu, G. & Ades, A. E. 2009. Modelling between-trial variance structure in mixed treatment comparisons. *Biostatistics*, 10, 792–805.

Lu, G., Ades, A. E., Sutton, A. J., Cooper, N. J., Briggs, A. H. & Caldwell, D. M. 2007. Meta-analysis of mixed treatment comparisons at multiple follow-up times. *Statistics in Medicine*, 26, 3681–3699.

Lu, G., Brazier, J. E. & Ades, A. E. 2013. Mapping from disease-specific to generic health-related quality-of-life scales: a common factor model. *Value in Health*, 16, 177–184.

Lu, G., Kounali, D. & Ades, A. E. 2014. Simultaneous multi-outcome synthesis and mapping of treatment effects to a common scale. *Value in Health*, 17, 280–287.

Lu, G., Welton, N. J., Higgins, J. P. T., White, I. R. & Ades, A. E. 2011. Linear inference for mixed treatment comparison meta-analysis: a two-stage approach. *Research Synthesis Methods*, 2, 43–60.

Lumley, T. 2002. Network meta-analysis for indirect treatment comparisons. *Statistics in Medicine*, 21, 2313–2324.

Lundh, A. & Gotzsche, P. C. 2008. Recommendations by Cochrane Review Groups for assessment of the risk of bias in studies. *BMC Medical Research Methodology*, 8, 22.

Lunn, D. 2004. *WinBUGS differential interface (WBDiff)*. Available: http://winbugs-development.mrc-bsu.cam.ac.uk/wbdiff.html/ (Accessed 31 July 2017).

Lunn, D., Jackson, C., Best, N., Thomas, A. & Spiegelhalter, D. 2013. *The BUGS Book*, Boca Raton, CRC Press.

Lunn, D., Spiegelhalter, D., Thomas, A. & Best, N. 2009. The BUGS project: evolution, critique and future directions. *Statistics in Medicine*, 28, 3049–3067.

Lunn, D. J., Thomas, A., Best, N. & Spiegelhalter, D. 2000. WinBUGS – a Bayesian modelling framework: concepts, structure, and extensibility. *Statistics and Computing,* 10, 325–337.

Madan, J., Ades, A. E., Price, M., Maitland, K., Jemutai, J., Revill, P. & Welton, N. J. 2014a. Strategies for efficient computation of the expected value of partial perfect information. *Medical Decision Making,* 34, 327–342.

Madan, J., Chen, Y.-F., Aveyard, P., Wang, D., Yahaya, I., Munafo, M., Bauld, L. & Welton, N. J. 2014b. Synthesis of evidence on heterogeneous interventions with multiple outcomes recorded over multiple follow-up times reported inconsistently: a smoking cessation case study. *Journal of the Royal Statistical Society: Series A (Statistics in Society),* 177, 295–314.

Madan, J., Stevenson, M. D., Ades, A. E., Cooper, K. L., Whyte, S. & Akehurst, R. 2011. Consistency between direct and indirect trial evidence: is direct evidence always more reliable? *Value in Health,* 14, 953–960.

Madigan, D., Mosurski, K. & Almond, R. G. 1997. Graphical explanation in belief networks. *Journal of Computational and Graphical Statistics,* 6, 160–181.

Mantel, N. & Haenszel, W. 1959. Statistical aspects of the analysis of data from retrospective studies of disease. *Journal of the National Cancer Institute,* 22, 719–748.

Marshall, E. C. & Spiegelhalter, D. J. 2003. Approximate cross-validatory predictive checks in disease mapping models. *Statistics in Medicine,* 22, 1649–1660.

Marshall, E. C. & Spiegelhalter, D. J. 2007. Identifying outliers in Bayesian hierarchical models: a simulation-based approach. *Bayesian Analysis,* 2, 409–444.

Mavridis, D. & Salanti, G. 2013. A practical introduction to multivariate meta-analysis. *Statistical Methods in Medical Research,* 22, 133–158.

Mavridis, D., Sutton, A., Cipriani, A. & Salanti, G. 2013. A fully Bayesian application of the Copas selection model for publication bias extended to network meta-analysis. *Statistics in Medicine,* 32, 51–66.

Mavridis, D., Welton, N. J., Sutton, A. & Salanti, G. 2014. A selection model for accounting for publication bias in a full network meta-analysis. *Statistics in Medicine,* 33, 5399–5412.

Mavridis, D., White, I. R., Higgins, J. P. T., Cipriani, A. & Salanti, G. 2015. Allowing for uncertainty due to missing continuous outcome data in pairwise and network meta-analysis. *Statistics in Medicine,* 34, 721–741.

Mawdsley, D., Bennetts, M., Dias, S., Boucher, M. & Welton, N. J. 2016. Model-based network meta-analysis: a framework for evidence synthesis of clinical trial data. *CPT: Pharmacometrics & Systems Pharmacology,* 5, 393–401.

Mayo-Wilson, E., Dias, S., Mavranezouli, I., Kew, K., Clark, D. M., Ades, A. E. & Pilling, S. 2014. Psychological and pharmacological interventions for social anxiety disorder in adults: a systematic review and network meta-analysis. *The Lancet Psychiatry,* 1, 368–376.

McCullagh, P. & Nelder, J. A. 1989. *Generalised Linear Models*, London, Chapman & Hall.

McIntosh, M. W. 1996. The population risk as an explanatory variable in research synthesis of clinical trials. *Statistics in Medicine*, 15, 1713–1728.

Melendez-Torre, G. J., Bonell, C. & Thomas, J. 2015. Emergent approaches to the meta-analysis of multiple heterogeneous complex interventions. *BMC Medical Research Methodology*, 15, 47.

Merrill, R. & Hunter, B. 2010. Conditional survival among cancer patients in the United States. *The Oncologist*, 15, 873–882.

Mills, E. J., Ghement, I., O'Regan, C. & Thorlund, K. 2011. Estimating the power of indirect comparisons: a simulation study. *PLoS One*, 6, e16237.

Mills, E. J., Thorlund, K. & Ioannidis, J. P. A. 2012. Calculating additive treatment effects from multiple randomized trials provides useful estimates of combination therapies. *Journal of Clinical Epidemiology*, 65, 1282–1288.

Misra, S. 2011. *XL2BUGS: excel add-in to convert data for WinBUGS*. Available: http://www.simon.rochester.edu/fac/misra/software.htm (Accessed 15 July 2017).

Moher, D., Hopewell, S., Schulz, K. F., Montori, V., Gøtzsche, P. C., Devereaux, P. J., Elbourne, D., Egger, M. & Altman, D. G. 2010. CONSORT 2010 explanation and elaboration: updated guidelines for reporting parallel group randomised trials. *Journal of Clinical Epidemiology*, 63, e1–e37.

Moreno, S. G., Sutton, A. J., Ades, A. E., Stanley, T. D., Abrams, K. R., Peters, J. L. & Cooper, N. J. 2009a. Assessment of regression-based methods to adjust for publication bias through a comprehensive simulation study. *BMC Medical Research Methodology*, 9, 2.

Moreno, S. G., Sutton, A. J., Turner, E. H., Abrams, K. R., Cooper, N. J., Palmer, T. M. & Ades, A. E. 2009b. Novel methods to deal with publication biases: secondary analysis of antidepressant trials in the FDA trial registry database and related journal publications. *BMJ*, 339, b2981.

MRC Biostatistics Unit. 2015a. *Calling WinBUGS 1.4 from other programs [Online]*. Available: http://www.mrc-bsu.cam.ac.uk/software/bugs/calling-winbugs-1-4-from-other-programs/ (Accessed 13 March 2015).

MRC Biostatistics Unit. 2015b. *DIC: deviance information criterion [Online]*. Available: http://www.mrc-bsu.cam.ac.uk/software/bugs/the-bugs-project-dic/ (Accessed 13 March 2015).

Naci, H., Brugts, J. & Ades, T. 2013a. Comparative tolerability and harms of individual statins. A study-level network meta-analysis of 246 955 participants from 135 randomized, controlled trials. *Circulation: Cardiovascular Quality and Outcomes*, 6, 390–399.

Naci, H., Brugts, J. J., Fleurence, R. & Ades, A. E. 2013b. Dose-comparative effects of different statins on serum lipid levels: a network meta-analysis of 256,827 individuals in 181 randomized controlled trials. *European Journal of Preventive Cardiology*, 20, 658–670.

Naci, H., Dias, S. & Ades, A. E. 2014a. Industry sponsorship bias in research findings: a network meta-analytic exploration of LDL cholesterol reduction in the randomised trials of statins. *BMJ*, 349, g5741.

Naci, H., van Valkenhoef, G., Higgins, J. P. T., Fleurence, R. & Ades, A. E. 2014b. Evidence-based prescribing: combining network meta-analysis with multicriteria decision analysis to choose among multiple drugs. *Circulation: Cardiovascular Quality and Outcomes*, 7, 787–792.

Nam, I.-S., Mengerson, K. & Garthwaite, P. 2003. Multivariate meta-analysis. *Statistics in Medicine*, 22, 2309–2333.

National Clinical Guideline Centre. 2012. *Crohn's Disease: Management in Adults, Children and Young People*, London, The National Clinical Guideline Centre at The Royal College of Physicians.

National Clinical Guideline Centre. 2014. *Lipid Modification: Cardiovascular Risk Assessment and the Modification of Blood Lipids for the Primary and Secondary Prevention of Cardiovascular Disease*, London, National Clinical Guideline Centre.

National Clinical Guideline Centre. 2015. *Type 1 Diabetes in Adults: Diagnosis and Management*, London, National Clinical Guideline Centre.

National Collaborating Centre for Mental Health. 2013. *Social Anxiety Disorder: Recognition, Assessment and Treatment*. National Clinical Guideline Number, 159, London, The British Psychological Society and The Royal College of Psychiatrists.

National Collaborating Centre for Mental Health. 2014. *The Assessment and Management of Bipolar Disorder in Adults, Children and Young People in Primary and Secondary Care, Updated Edition*, National Clinical Guideline Number, 185, London, The British Psychological Society and The Royal College of Psychiatrists.

National Institute for Health and Care Excellence. 2012. *Roflumilast for the Management of Severe Chronic Obstructive Pulmonary Disease*. NICE Technology Appraisal Guidance TA244, London, The National Institute for Health and Care Excellence.

National Institute for Health and Care Excellence. 2013a. *Guide to the Methods of Technology Appraisal 2013*, London, National Institute for Health and Care Excellence.

National Institute for Health and Care Excellence. 2013b. *Hyperphosphataemia in Chronic Kidney Disease*, London, National Institute for Health and Care Excellence.

National Institute for Health and Clinical Excellence. 2008a. *Guide to the Methods of Technology Appraisal*, London, National Institute for Health and Clinical Excellence.

National Institute for Health and Clinical Excellence. 2008b. *Guide to the Methods of Technology Appraisal (Updated June 2008)*, London, National Institute for Health and Care Excellence.

National Institute for Health and Clinical Excellence. 2008c. *Social Value Judgements: Principles for the Development of NICE Guidance*, 2nd ed., London, National Institute for Health and Care Excellence.

National Institute for Health and Clinical Excellence. 2009a. *Advanced Breast Cancer: Diagnosis and Treatment*, NICE Clinical Guidelines, Cardiff, National Collaborating Centre for Cancer.

National Institute for Health and Clinical Excellence. 2009b. *Guide to the Multiple Technology Appraisal Process*, London, National Institute for Health and Clinical Excellence. Available: https://www.nice.org.uk/Media/Default/About/what-we-do/NICE-guidance/NICE-technology-appraisals/Previous%20MTA%20Guide.pdf (Accessed 2 August 2017).

National Institute for Health and Clinical Excellence. 2009c. *Guide to the Single Technology Appraisal (STA) Process*, London, National Institute for Health and Clinical Excellence. Available: https://www.nice.org.uk/Media/Default/About/what-we-do/NICE-guidance/NICE-technology-appraisals/Previous%20STA%20Guide.pdf (Accessed 2 August 2017).

National Institute for Health and Clinical Excellence. 2010. *Certolizumab Pegol for the Treatment of Rheumatoid Arthritis. NICE Technology Appraisal Guidance*, London, National Institute for Health and Clinical Excellence. Available: https://www.nice.org.uk/guidance/TA186 (Accessed 2 August 2017).

National Institute for Health and Clinical Excellence. 2011. *Colorectal Cancer: The Diagnosis and Management of Colorectal Cancer*. Clinical Guideline, 131, Cardiff, National Collaborating Centre for Cancer.

National Institute for Health and Clinical Excellence. 2012. *Single Technology Appraisal (STA). Specification for Manufacturer/Sponsor Submission of Evidence*, London, National Institute for Health and Care Excellence. Available: https://www.coursehero.com/file/18761348/specification-for-manufacturer-sponsor-submission-of-evidence-june-2012.doc (Accessed 7 August 2017).

NICE Collaborating Centre for Mental Health. 2010. *Schizophrenia: Core Interventions in the Treatment and Management of Schizophrenia in Adults in Primary and Secondary Care (Update)*, London, The British Psychological Society and The Royal College of Psychiatrists.

Nixon, R., Bansback, N. & Brennan, A. 2007. Using mixed treatment comparisons and meta-regression to perform indirect comparisons to estimate the efficacy of biologic treatments in rheumatoid arthritis. *Statistics in Medicine*, 26, 1237–1254.

Norton, E., Miller, M., Wang, J., Coyne, K. S. & Kleinman, L. 2012. Rank reversal in indirect comparisons. *Value in Health*, 15, 1137–1140.

Ntzoufras, I. 2009. *Bayesian Modeling Using WinBUGS*, Hoboken, John Wiley & Sons, Inc.

O'Hagan, A. 2003. HSSS model criticism. *In:* Green, P. J., Hjort, N. L. & Richardson, S. (eds.) *Highly Structured Stochastics Systems*, Oxford, Oxford University Press.

Oakley, J. E., Brennan, A., Tappenden, P. & Chilcott, J. 2010. Simulation sample sizes for Monte Carlo partial EVPI calculations. *Journal of Health Economics*, 29, 468–477.

Oba, Y., Sarva, S. T. & Dias, S. 2016. Efficacy and safety of long-acting β-agonist/long-acting muscarinic antagonist combinations in COPD: a network meta-analysis. *Thorax*, 71, 15–25.

Ohlssen, D., Price, K. L., Xia, H. A., Hong, H., Kerman, J., Fu, H., Quartey, G., Heilmann, C. R., Ma, H. & Carlin, B. P. 2014. Guidance on the implementation and reporting of a drug safety Bayesian network meta-analysis. *Pharmaceutical Statistics*, 13, 55–70.

Ouwens, M. J. N. M., Philips, Z. & Jansen, J. P. 2010. Network meta-analysis of parametric survival curves. *Research Synthesis Methods*, 1, 258–271.

Oxman, A. D. 1994. Systematic reviews: checklists for review articles. *BMJ*, 309, 648–651.

Parmar, M. K. B., Torri, V. & Stewart, L. 1998. Extracting summary statistics to perform meta-analyses of the published literature for survival endpoints. *Statistics in Medicine*, 17, 2815–2834.

Parmigiani, G. 2002. *Modeling in Medical Decision Making: A Bayesian Approach*, Chichester, John Wiley & Sons, Ltd.

Parmigiani, G. & Kamlet, M. S. 1993. A cost-utility analysis of alternative strategies in screening for breast cancer. *In:* Gatsonis, C., Hodges, J., Kass, R. & Singpurwalla, N. (eds.) *Case Studies in Bayesian Statistics*, New York, Springer Verlag.

Petrou, S. & Gray, A. 2011. Economic evaluation using decision analytical modelling: design, conduct, analysis, and reporting. *BMJ*, 342, d1766.

Phillippo, D. M., Ades, A. E., Dias, S., Palmer, S., Abrams, K. R. & Welton, N. J. 2016. *NICE DSU Technical Support Document 18. Methods for Population-Adjusted Indirect Comparisons in Submissions to NICE*, Sheffield, ScHARR, University of Sheffield.

Pirolli, P. & Card, S. 1999. Information foraging. *Pyschological Review*, 106, 643–675.

Plummer, M., Best, N., Cowles, K. & Vines, K. 2006. CODA: convergence diagnosis and output analysis for MCMC. *R News*, 6, 7–11.

Poole, D. & Raftery, A. E. 2000. Inference for deterministic simulation models: the Bayesian melding approach. *Journal of the American Statistical Association*, 95, 1244–1255.

Pratt, J. W., Raiffa, H. & Schlaiffer, R. 1995. *Introduction to Statistical Decision Theory*, Cambridge, MA, Massachusetts Institute of Technology.

Prentice, R. L. & Gloeckler, L. A. 1978. Regression analysis of grouped survival data with application to breast cancer data. *Biometrics*, 34, 57–67.

Presanis, A., De Angelis, D., Spiegelhalter, D., Seaman, S., Goubar, A. & Ades, A. 2008. Conflicting evidence in a Bayesian synthesis of surveillance data to estimate HIV prevalence. *Journal of the Royal Statistical Society: Series A (Statistics in Society)*, 171, 915–937.

Presanis, A. M., De Angelis, D., Goubar, A., Gill, O. N. & Ades, A. E. 2012. Bayesian evidence synthesis for a transmission dynamic model for HIV among men who have sex with men. *Biostatistics*, 12, 666–681.

Presanis, A. M., Ohlssen, D., Spiegelhalter, D. J. & De Angelis, D. 2013. Conflict diagnostics in directed acyclic graphs, with applications in Bayesian evidence synthesis. *Statistical Science*, 28, 376–397.

Prevost, T. C., Abrams, K. R. & Jones, D. R. 2000. Hierarchical models in generalised synthesis of evidence: an example based on studies of breast cancer screening. *Statistics in Medicine*, 19, 3359–3376.

Price, M. J. & Briggs, A. H. 2002. Development of an economic model to assess the cost effectiveness of asthma management strategies. *PharmacoEconomics*, 20, 183–194.

Price, M. J., Welton, N. J. & Ades, A. E. 2007. *Parameterisation of treatment effects in Markov models: impact on decision uncertainty and EVPI*. International Society for Clinical Biostatistics, 28th Annual Conference, Alexandroupolis, Greece, 8–11 July 2007.

Price, M. J., Welton, N. J. & Ades, A. E. 2011. Parameterisation of treatment effects for meta-analysis in multi-state Markov models. *Statistics in Medicine*, 30, 140–151.

Puhan, M. A., Schünemann, H. J., Murad, M. H., Li, T., Brignardello-Petersen, R., Singh, J. A., Kessels, A. G. & Guyatt, G. H. 2014. A GRADE working group approach for rating the quality of treatment effect estimates from network meta-analysis. *BMJ*, 349, g5630.

R Development Core Team. 2010. *R: A Language and Environment for Statistical Computing*, Vienna, R Foundation for Statistical Computing.

Rabin, R. & Charro, F. 2001. EQ-D5: a measure of health status from the EuroQol Group. *Annals of Medicine*, 33, 337–343.

Raftery, A. E., Givens, G. H. & Zeh, J. E. 1995. Inference from a deterministic population dynamics model for Bowhead whales (with discussion). *Journal of the American Statistical Association*, 90, 402–430.

Raiffa, H. 1961. *Decision Analysis: Introductory Lectures on Choices Under Uncertainty*, Reading, Addison-Wesley.

Raiffa, H. & Schlaifer, R. 1967. *Applied Statistical Decision Theory*, New York, Wiley Interscience.

Raiffa, H. & Schlaiffer, R. 2000. *Applied Statistical Decision Theory*, New York, Wiley Interscience.

Rhodes, K. M., Turner, R. M. & Higgins, J. P. T. 2015. Predictive distributions were developed for the extent of heterogeneity in meta-analyses of continuous outcome data. *Journal of Clinical Epidemiology*, 68, 52–60.

Riemsma, R., Lhachimi, S. K., Armstrong, N., van Asselt, A. D. I., Allen, A., Manning, N., Harker, J., Tushabe, D. A., Severens, J. L. & Kleijnen, J. 2011. *Roflumilast for the Management of Severe Chronic Obstructive Pulmonary Disease: A Single Technology Appraisal*, York, Kleijnen Systematic Reviews Ltd.

Riley, R. 2009. Multivariate meta-analysis: the effect of ignoring within-study correlation. *Journal of the Royal Statistical Society: Series A (Statistics in Society)*, 172, 789–811.

Riley, R. D., Abrams, K. R., Lambert, P. C., Sutton, A. J. & Thompson, J. R. 2007a. An evaluation of bivariate random-effects meta-analysis for the joint synthesis of two correlated outcomes. *Statistics in Medicine*, 26, 78–97.

Riley, R. D., Lambert, P. C. & Abo-Zaid, G. 2010. Meta-analysis of individual participant data: rationale, conduct, and reporting. *BMJ*, 340, c221.

Riley, R. D., Lambert, P. C., Staessen, J. A., Wang, J., Gueyffier, F. & Boutitie, F. 2008. Meta-analysis of continuous outcomes combining individual patient data and aggregate data. *Statistics in Medicine*, 27, 1870–1893.

Riley, R. D., Simmonds, M. C. & Look, M. P. 2007b. Evidence synthesis combining individual patient data and aggregate data: a systematic review identified current practice and possible methods. *Journal of Clinical Epidemiology*, 60, 431–439.

Riley, R. D. & Steyerberg, E. W. 2010. Meta-analysis of a binary outcome using individual participant data and aggregate data. *Research Synthesis Methods*, 1, 2–17.

Robbins, J., Greenland, S. & Breslow, N. 1986. A general estimator for the variance of the Mantel-Haenszel odds ratio. *American Journal of Epidemiology*, 124, 719–723.

Rothman, K. J., Greenland, S. & Lash, T. L. 2012. *Modern Epidemiology*, Philadelphia, Lippincott, Williams & Wilkins.

Royston, P. & Altman, D. G. 1994. Regression using fractional polynomials of continuous covariates: parsimonious parametric modelling. *Journal of the Royal Statistical Society: Series C (Applied Statistics)*, 43, 429–467.

Royston, P. & Lambert, P. C. 2011. *Flexible Parametric Survival Analysis Using Stata: Beyond the Cox Model*, College Station, Stata Press.

Royston, P. & Parmar, M. K. B. 2002. Flexible parameteric proportional-hazards and proportional-odds models for censored survival data, with application to prognostic modelling and estimation of treatment effects. *Statistics in Medicine*, 21, 2175–2197.

Royston, P. & Parmar, M. K. B. 2013. Restricted mean survival time: an alternative to the hazard ratio for the design and analysis of randomized trials with a time-to-event outcome. *BMC Medical Research Methodology*, 13, 152–152.

Royston, P. & Sauerbrei, W. 2008. *Multivariable Model-Building: A Pragmatic Approach to Regression Anaylsis Based on Fractional Polynomials for Modelling Continuous Variables*, Chichester, John Wiley & Sons, Ltd.

Rubin, D. B. 1988. Using the SIR algorithm to simulate posterior distributions. *In:* Bernardo, J. M., Degroot, M. H., Lindley, D. V. & Smith, A. F. M. (eds.) *Bayesian Statistics 3*, Oxford, Clarendon Press.

Rubin, D. B. 1990. A new perspective. *In:* Wachter, K. W. & Straf, M. L. (eds.) *The Future of Meta-Analysis*, New York, Russell Sage Foundation.

Rücker, G. 2012. Network meta-analysis, electrical networks and graph theory. *Research Synthesis Methods*, 3, 312–324.

Rücker, G. & Schwarzer, G. 2014. Reduce dimension or reduce weights? Comparing two approaches to multi-arm studies in network meta-analysis. *Statistics in Medicine*, 33, 4353–4369.

Rücker, G. & Schwarzer, G. 2016. Automated drawing of network plots in network meta-analysis. *Research Synthesis Methods*, 7, 94–107.

Rücker, G., Schwarzer, G., Krahn, U. & König, J. 2015. *Netmeta: network meta-analysis using frequentist methods*. R package version 0.8-0. Available: https://CRAN.R-project.org/package=netmeta (Accessed 2 August 2017).

Russell, K. & Kiddoo, D. 2006. The Cochrane library and nocturnal enuresis: an umbrella review. *Evidence-Based Child Health: A Cochrane Review Journal*, 1, 5–8.

Sackett, D. L., Straus, S. E. & Richardson, W. S. 2000. *Evidence Based Medicine: How to Practice and Teach EBM*, London, Churchill-Livingstone.

Salanti, G. 2011. *Multiple-treatments meta-analysis of a network of interventions [Online]*. Available: http://www.mtm.uoi.gr/ (Accessed 15 July 2017).

Salanti, G. 2012. Indirect and mixed-treatment comparison, network, or multiple treatments meta-analysis: many names, many benefits, many concerns for the next generation evidence synthesis tool. *Research Synthesis Methods* 3, 80–97.

Salanti, G., Ades, A. E. & Ioannidis, J. P. A. 2011. Graphical methods and numerical summaries for presenting results from multiple-treatment meta-analysis: an overview and tutorial. *Journal of Clinical Epidemiology*, 64, 163–171.

Salanti, G., Del Giovane, C., Chaimani, A., Caldwell, D. M. & Higgins, J. P. T. 2014. Evaluating the quality of evidence from a network meta-analysis. *PLoS One*, 9, e99682.

Salanti, G., Dias, S., Welton, N. J., Ades, A. E., Golfinopoulos, V., Kyrgiou, M., Mauri, D. & Ioannidis, J. P. A. 2010. Evaluating novel agent effects in multiple treatments meta-regression. *Statistics in Medicine*, 29, 2369–2383.

Salanti, G., Higgins, J. P. T., Ades, A. E. & Ioannidis, J. P. A. 2008a. Evaluation of networks of randomised trials. *Statistical Methods in Medical Research*, 17, 279–301.

Salanti, G., Kavvoura, F. K. & Ioannidis, J. P. A. 2008b. Exploring the geometry of treatment networks. *Annals of Internal Medicine*, 148, 544–553.

Salanti, G., Marinho, V. & Higgins, J. P. T. 2009. A case study of multiple-treatments meta-analysis demonstrates that covariates should be considered. *Journal of Clinical Epidemiology*, 62, 857–864.

Samsa, G. P., Reutter, R. A., Parmigiani, G., Ancukiewicz, M., Abrahamse, P., Lipscomb, J. & Matchar, D. 1999. Performing cost-effectiveness analysis by integrating randomised trial data with a comprehensive ecision model: application to treatment of ischemic stroke. *Journal of Clinical Epidemiology*, 52, 259–271.

Saramago, P., Chuang, L.-H. & Soares, M. 2014. Network meta-analysis of (individual patient) time to event data alongside (aggregate) count data. *BMC Medical Research Methodology*, 14, 105.

Saramago, P., Sutton, A. J., Cooper, N. & Manca, A. 2011. RE: Synthesizing effectiveness evidence from aggregate-and individual-patient level data for use in cost-effectiveness modelling. Presented At HESG Winter Conference, C. F. H. E., University of York, York, 5–7 January 2011.

Saramago, P., Sutton, A. J., Cooper, N. J. & Manca, A. 2012. Mixed treatment comparisons using aggregate and individual participant level data. *Statistics in Medicine*, 31, 3516–3536.

Sauerbrei, W., Royston, P. & Look, M. 2007. A new proposal for multivariable modelling of time-varying effects in survival data based on fractional polynomial time-transformation. *Biometrical Journal*, 49, 453–473.

Savovic, J., Harris, R. J., Wood, L., Beynon, R., Altman, D., Als-Nielsen, B., Balk, E. M., Deeks, J., Gluud, L. L., Gluud, C., Ioannidis, J. P. A., Juni, P., Moher, D., Pildal, J., Schulz, K. F. & Sterne, J. A. C. 2010. Development of a combined database for meta-epidemiological research. *Research Synthesis Methods*, 1, 212–225.

Savovic, J., Jones, H., Altman, D., Harris, R., Juni, P., Pildal, J., Als-Nielsen, B., Balk, E. M., Gluud, C., Gluud, L. L., Ioannidis, J. P. A., Schulz, K. F., Beynon, R., Welton, N. J., Wood, L., Moher, D., Deeks, J. J. & Sterne, J. 2012a. Influence of study design characteristics on intervention effect estimates from randomised controlled trials: combined analysis of meta-epidemiological studies. *Health Technology Assessment*, 16, 1–81.

Savovic, J., Jones, H. E., Altman, D. G., Harris, R. J., Juni, P., Pildal, J., Als-Nielsen, B., Balk, E. M., Gluud, C., Gluud, L. L., Ioannidis, J. P. A., Schulz, K. F., Beynon, R., Welton, N. J., Wood, L., Moher, D., Deeks, J. J. & Sterne, J. A. C. 2012b. Influence of reported study design characteristics on intervention effect estimates from randomized, controlled trials. *Annals of Internal Medicine*, 157, 429–438.

Schmid, C. H., Trikalinos, T. A. & Olkin, I. 2014. Bayesian network meta-analysis for unordered categorical outcomes with incomplete data. *Research Synthesis Methods*, 5, 162–185.

Schulz, K. F., Chalmers, I., Hayes, R. J. & Altman, D. G. 1995. Empirical evidence of bias. Dimensions of methodological quality associated with estimates of treatment effects in controlled trials. *JAMA: The Journal of the American Medical Association*, 273, 408–412.

Schwingshackl, L., Dias, S., Strasser, B. & Hoffmann, G. 2013. Impact of different training modalities on anthropometric and metabolic characteristics in overweight/obese subjects: a systematic review and network meta-analysis. *PLoS One*, 8, e82853.

Sculpher, M., Fenwick, E. & Claxton, K. 2000. Assessing quality in decision analytic cost-effectiveness models. A suggested framework and example of application. *Pharmacoeconomics*, 17, 461–477.

Senn, S. 2010. Hans van Houwelingen and the art of summing up. *Biometrical Journal*, 52, 85–94.

Senn, S., Gavini, D. M. & Scheen, A. 2013. Issues in performing a network meta-analysis. *Statistical Methods in Medical Research*, 22, 169–189.

Shachter, R., Eddy, D. M. & Hasselblad, V. 1990. An influence diagram approach to medical technology assessment. *In:* Oliver, R. M. & Smith, J. Q. (eds.) *Influence Diagams, Belief Nets and Decision Analysis*, New York, John Wiley & Sons, Inc.

Shapiro, S. 1994. Meta-analysis/Shmeta-analysis. *American Journal of Epidemiology*, 140, 771–778.

Sidik, K. & Jonkman, J. N. 2007. A comparison of heterogeneity variance estimators in combining results of studies. *Statistics in Medicine*, 26, 1964–1981.

Signorovitch, J. E., Sikirica, V., Erder, M. H., Xie, J. P., Lu, M., Hodgkins, P. S., Betts, K. A. & Wu, E. Q. 2012. Matching-adjusted indirect comparisons: a new tool for timely comparative effectiveness research. *Value in Health*, 15, 940–947.

Simmonds, M. C. & Higgins, J. P. T. 2007. Covariate heterogeneity in meta-analysis: criteria for deciding between meta-regression and individual patient data. *Statistics in Medicine*, 26, 2982–2999.

Simmonds, M. C. & Higgins, J. P. T. 2014. A general framework for the use of logistic regression models in meta-analysis. *Statistical Methods in Medical Research*, 25, 2858–2877.

Simmonds, M. C., Higgins, J. P. T., Stewart, L. A., Tierney, J. F., Clarke, M. J. & Thompson, S. G. 2005. Meta-analysis of individual patient data from randomized trials: a review of methods used in practice. *Clinical Trials*, 2, 209–217.

Singh, J. A., Christensen, R., Wells, G. A., Suarez-Almazor, M. E., Buchbinder, R., Lopez-Olivo, M. A., Ghogomu, E. T. & Tugwell, P. 2009. Biologics for rheumatoid arthritis: an overview of Cochrane reviews. *Cochrane Database of Systematic Reviews*, 2009, CD007848.

Smaill, F. M. 1996. Intrapartum antibiotics for Group B streptococcal colonisation. *Cochrane Database of Systematic Reviews*, 1996, CD000115. Available: http://onlinelibrary.wiley.com/doi/10.1002/14651858.CD000115/full (Accessed 2 August 2017).

Smith, B. J. 2005. *The BOA package. 1.1.5 ed.* Available: http://www.public-health. uiowa.edu/boa (Accessed 15 July 2017).

Smith, T. C., Spiegelhalter, D. J. & Thomas, A. 1995. Bayesian approaches to random-effects meta-analysis: a comparative study. *Statistics in Medicine*, 14, 2685–2699.

Soares, M. O., Bojke, L., Dumville, J., Iglesias, C., Cullum, N. & Claxton, K. 2011. Methods to elicit experts' beliefs over uncertain quantities: application to a cost effectiveness transition model of negative pressure wound therapy for severe pressure ulceration. *Statistics in Medicine*, 30, 2363–2380.

Soares, M. O., Dumville, J., Ades, A. E. & Welton, N. J. 2014. Treatment comparisons for decision making: facing the problems of sparse and few data. *Journal of the Royal Statistical Society: Series A (Statistics in Society)*, 177, 259–279.

Soares, M. O., Welton, N. J., Harrison, D. A., Peura, P., Shankar Hari, M., Harvey, S. E., Madan, J., Ades, A. E., Palmer, S. J. & Rowan, K. M. 2012. An evaluation of the feasibility, cost and value of information of a multicentre randomised controlled trial of intravenous immunoglobulin for sepsis (severe sepsis and septic shock): incorporating a systematic review, meta-analysis and value of information analysis. *Health Technology Assessment*, 16, 1–186.

Song, F. 1999. Exploring heterogenity in meta-analysis: is the L'Abbe plot useful. *Journal of Clinical Epidemiology*, 52, 725–730.

Song, F., Altman, D., Glenny, A.-M. & Deeks, J. 2003. Validity of indirect comparison for estimating efficacy of competing interventions: evidence from published meta-analyses. *BMJ*, 326, 472–476.

Song, F., Harvey, I. & Lilford, R. 2008. Adjusted indirect comparison may be less biased than direct comparison for evaluating new pharmaceutical interventions. *Journal of Clinical Epidemiology*, 61, 455–463.

Song, F., Loke, Y.-K., Walsh, T., Glenny, A.-M., Eastwood, A. J. & Altman, D. G. 2009. Methodological problems in the use of indirect comparisons for evaluating healthcare interventions: survey of published systematic reviews. *BMJ*, 338, b1147.

Song, F., Xiong, T., Parekh-Bhurke, S., Loke, Y. K., Sutton, A. J., Eastwood, A. J., Holland, R., Chen, Y.-F., Glenny, A.-M., Deeks, J. J. & Altman, D. G. 2011. Inconsistency between direct and indirect comparisons of competing interventions: meta-epidemiological study. *BMJ*, 343, d4909.

Spiegelhalter, D., Thomas, A., Best, N. & Lunn, D. 2007. *WinBUGS user manual version 1.4 January 2003. upgraded to version 1.4.3.* Available: https://www.mrc-bsu.cam.ac.uk/software/bugs/the-bugs-project-winbugs/ (Accessed 15 July 2017).

Spiegelhalter, D. J. 2003. Incorporating Bayesian ideas into health-care evaluation. *Statistical Science*, 19, 156–174.

Spiegelhalter, D. J., Abrams, K. R. & Myles, J. 2004. *Bayesian Approaches to Clinical Trials and Health-Care Evaluation*, New York, John Wiley & Sons, Inc.

Spiegelhalter, D. J., Best, N. G., Carlin, B. P. & van der Linde, A. 2002. Bayesian measures of model complexity and fit. *Journal of the Royal Statistical Society: Series B (Statistical Methodology)*, 64, 583–616.

Spiegelhalter, D. J., Best, N. G., Carlin, B. P. & Van Der Linde, A. 2014. The deviance information criterion: 12 years on. *Journal of the Royal Statistical Society: Series B (Statistical Methodology)*, 76, 485–493.

Spineli, L. M., Higgins, J. P. T., Cipriani, A., Leucht, S. & Salanti, G. 2013. Evaluating the impact of imputations for missing participant outcome data in a network meta-analysis. *Clinical Trials*, 10, 378–388.

Stern, H. S. & Cressie, N. 2000. Posterior predictive model checks for disease mapping models. *Statistics in Medicine*, 19, 2377–2397.

Sterne, J. A. C., Bradburn, M. J. & Egger, M. 2001. Meta-analysis in Stata. *In:* Egger, M., Davey Smith, G. & Altman, D. G. (eds.) *Systematic Reviews in Health Care: Meta-Analysis in Context*, London, BMJ Books.

Stettler, C., Allemann, S., Wandel, S., Kastrati, A., Morice, M. C., Schomig, A., Pfisterer, M. E., Stone, G. W., Leon, M. B., De Lezo, J. S., Goy, J. J., Park, S. J., Sabate, M., Suttorp, M. J., Kelbaek, H., Spaulding, C., Menichelli, M., Vermeersch, P., Dirksen, M. T., Cervinka, P., De Carlo, M. D., Erglis, A., Chechi, T., Ortolani, P., Schalij, M. J., Diem, P., Meier, B., Windecker, S. & Juni, P. 2008. Drug eluting and bare metal stents in people with and without diabetes: collaborative network meta-analysis. *BMJ*, 337, a1331.

Stettler, C., Wandel, S., Allemann, S., Kastrati, A., Morice, M. C., Schomig, A., Pfisterer, M. E., Stone, G. W., Leon, M. B., De Lezo, J. S., Goy, J. J., Park, S. J., Sabate, M., Suttorp, M. J., Kelbaek, H., Spaulding, C., Menichelli, M., Vermeersch, P., Dirksen, M. T., Cervinka, P., Petronio, A. S., Nordmann, A. J., Diem, P., Meier, B., Zwahlen, M., Reichenbach, S., Trelle, S., Windecker, S. & Juni, P. 2007. Outcomes associated with drug-eluting and bare-metal stents: a collaborative network meta-analysis. *Lancet*, 370, 937–948.

Stewart, L. A. & Clarke, M. J. 1995. Practical methodology of meta-analyses (overviews) using updated individual patient data. *Statistics in Medicine*, 14, 2057–2079.

Stinnett, A. & Mullahy, J. 1998. Net health benefits: a new framework for the analysis of uncertainty in cost-effectiveness analyses. *Medical Decision Making*, 18, S68–S80.

Stone, N., Robinson, G. J., Lichtenstein, A. H., Bairey Merz, C. N., Blum, C. B., Eckel, R. H., Goldberg, A. C., Gordon, D., Levy, D., Lloyd-Jones, D. M., McBride, P., Sanford Schwartz, J., Shero, S. T., Smith, S. C., Watson, K. & Wilson, P. W. F. 2013. 2013 ACC/AHA guideline on the treatment of blood cholesterol to reduce atherosclerotic cardiovascular risk in adults. *Circulation*, 129, S1–S45.

Strong, M., Oakley, J., Brennan, A. & Watson Breeze, P. 2015. Estimating the expected value of sample information using the probabilistic sensitivity analysis sample. A fast non-parametric regression-based method. *Medical Decision Making*, 35, 570–583.

Sturtz, S., Ligges, U. & Gelman, A. 2005. R2WinBUGS: a package for running WinBUGS from R. *Journal of Statistical Software*, 12, 1–16.

Sutton, A. J. 2002. *Meta-analysis methods for combining information from different sources evaluating health interventions.* PhD, University of Leicester.

Sutton, A. J. & Abrams, K. R. 2001. Bayesian methods in meta-analysis and evidence synthesis. *Statistical Methods in Medical Research*, 10, 277–303.

Sutton, A. J., Abrams, K. R., Jones, D. R., Sheldon, T. A. & Song, F. 2000. *Methods for Meta-Analysis in Medical Research*, London, John Wiley & Sons, Ltd.

Sutton, A. J., Cooper, N. J., Abrams, K. R., Lambert, P. C. & Jones, D. R. 2005. A Bayesian approach to evaluating net clinical benefit allowed for parameter uncertainty. *Journal of Clinical Epidemiology*, 58, 26–40.

Sutton, A. J. & Higgins, J. P. T. 2008. Recent developments in meta-analysis. *Statistics in Medicine*, 27, 625–650.

Sutton, A. J., Kendrick, D. & Coupland, C. A. C. 2008. Meta-analysis of individual- and aggregate-level data. *Statistics in Medicine*, 27, 651–669.

Sweeting, M. J., De Angelis, D., Hickman, D. & Ades, A. E. 2008. Estimating HCV prevalence in England and Wales by synthesising evidence from multiple data sources: assessing data conflict and model fit. *Biostatistics*, 9, 715–734.

Sweeting, M. J., Sutton, A. J. & Lambert, P. C. 2004. What to add to nothing? Use and avoidance of continuity corrections in meta-analysis of sparse data. *Statistics in Medicine*, 23, 1351–1375.

Sweeting, M. J., Sutton, A. J. & Lambert, P. C. 2006. Correction. *Statistics in Medicine*, 25, 2700–2700.

Tan, S. H., Cooper, N. J., Bujkiewicz, S., Welton, N. J., Caldwell, D. M. & Sutton, A. J. 2014. Novel presentational approaches were developed for reporting network meta-analysis. *Journal of Clinical Epidemiology*, 67, 672–680.

Taylor, R. S. & Elston, J. 2009. The use of surrogate outcomes in model-based cost-effectiveness analyses: a survey of UK health technology assessment reports. *Health Technology Assesment*, 13(8), iii, ix–xi, 1–50.

Tervonen, T. & Lahdelma, R. 2007. Implementaing stochastic multi-criteria acceptibility analysis. *European Journal of Operations Research*, 178, 500–513.

Tervonen, T., Naci, H., Van Valkenhoef, G., Ades, A. E., Angelis, A., Hillege, H. L. & Postmus, D. 2015. Applying multiple criteria decision analysis to comparative benefit-risk assessment: choosing among statins in primary prevention. *Medical Decision Making*, 35, 859–871.

Thompson, J., Palmer, T. & Moreno, S. 2006. Bayesian analysis in Stata with WinBUGS. *The Stata Journal*, 6, 530–549.

Thompson, K. M. & Evans, J. S. 1997. The value of improved national exposure information for perchloroethylene (Perc): a case study for dry cleaners. *Risk Analysis*, 17, 253–271.

Thompson, S., Ekelund, U., Jebb, S., Lindros, A. K., Mander, A., Sharp, S., Turner, R. & Wilks, D. 2011. A proposed method of bias adjustment for meta-analyses of published observational studies. *Internation Journal of Epidemiology*, 40, 765–777.

Thompson, S. G., Smith, T. C. & Sharp, S. J. 1997. Investigating underlying risk as a source of heterogeneity in meta-analysis. *Statistics in Medicine*, 16, 2741–2758.

Thorlund, K., Mills, E. J., Wu, P., Ramos, E., Chatterjee, A., Druyts, E. & Goadsby, P. J. 2014. Comparative efficacy of triptans for the abortive treatment of migraine: a multiple treatment comparison meta-analysis. *Cephalalgia*, 34, 258–267.

Tierney, J. F., Stewart, L. A., Ghersi, D., Burdett, S. & Sydes, M. R. 2007. Practical methods for incorporating summary time-to-event data into meta-analysis. *Trials*, 8, 1–16.

Trikalinos, T. & Olkin, I. 2008. A method for the meta-analysis of mutually exclusive binary outcomes. *Statistics in Medicine*, 27, 4279–4300.

Trikalinos, T. & Olkin, I. 2012. Meta-analysis of effect sizes reported at multiple time points: a multivariate approach. *Clinical Trials*, 9, 610–620.

Trinquart, L., Chatellier, G. & Ravaud, P. 2012. Adjustment for reporting bias in network meta-analysis of antidepressant trials. *BMC Medical Research Methodology*, 12, 150.

Tu, Y.-K. 2014. Use of generalized linear mixed models for network meta-analysis. *Medical Decision Making*, 34, 911–918.

Tudor Smith, C., Marson, A. G., Chadwick, D. W. & Williamson, P. R. 2007. Multiple treatment comparisons in epilepsy monotherapy trials. *Trials*, 8, 34.

Tudor Smith, C., Williamson, P. R. & Marson, A. G. 2005. Investigating heterogeneity in an individual patient data meta-analysis of time to event outcomes. *Statistics in Medicine*, 24, 1307–1319.

Turner, N. L., Dias, S., Ades, A. E. & Welton, N. J. 2015a. A Bayesian framework to account for uncertainty due to missing binary outcome data in pairwise meta-analysis. *Statistics in Medicine*, 34, 2062–2080.

Turner, R. M., Davey, J., Clarke, M. J., Thompson, S. G. & Higgins, J. P. T. 2012. Predicting the extent of heterogeneity in meta-analysis, using empirical data from the Cochrane Database of Systematic Reviews. *International Journal of Epidemiology*, 41, 818–827.

Turner, R. M., Jackson, D., Wei, Y., Thompson, S. G. & Higgins, J. P. T. 2015b. Predictive distributions for between-study heterogeneity and simple methods for their application in Bayesian meta-analysis. *Statistics in Medicine*, 34, 984–998.

Turner, R. M., Omar, R. Z., Yang, M., Goldstein, H. & Thompson, S. G. 2000. A multilevel model framework for meta-analysis of clinical trials with binary outcomes. *Statistics in Medicine*, 19, 3417–3432.

Turner, R. M., Spiegelhalter, D. J., Smith, G. C. S. & Thompson, S. G. 2009. Bias modelling in evidence synthesis. *Journal of the Royal Statistical Society: Series A (Statistics in Society)*, 172, 21–47.

UKPDS. *UK Prospective Diabetes Study, [Online]*. Oxford, University of Oxford. Available: http://www.dtu.ox.ac.uk:8000/UKPDS/ (Accessed 2 August 2017).

Van Houwelingen, H. C., Arends, L. R. & Stijnen, T. 2002. Advanced methods in meta-analysis: multi-variate approach and meta-regression. *Statistics in Medicine*, 21, 589–624.

Van Houwelingen, H. C., Zwinderman, K. H. & Stijnen, T. 1993. A bivariate approach to meta-analysis. *Statistics in Medicine*, 12, 2273–2284.

Van Valkenhoef, G. & Ades, A. E. 2013. Evidence synthesis assumes additivity on the scale of measurement: response to "rank reversal in indirect comparisons". *Value in Health*, 16, 449–451.

Van Valkenhoef, G., Dias, S., Ades, A. E. & Welton, N. J. 2016. Automated generation of node-splitting models for assessment of inconsistency in network meta-analysis. *Research Synthesis Methods*, 7, 80–93.

Van Valkenhoef, G. & Kuiper, J. 2016. *GeMTC: network meta-analysis using bayesian methods. R package. 0.8.1 ed.: CRAN*. Available: https://CRAN.R-project.org/package=gemtc (Accessed 17 July 2017).

Van Valkenhoef, G., Lu, G., De Brock, B., Hillege, H., Ades, A. E. & Welton, N. J. 2012a. Automating network meta-analysis. *Research Synthesis Methods*, 3, 285–299.

Van Valkenhoef, G., Tervonen, T., De Brock, B. & Hillege, H. 2012b. Algorithmic parameterization of mixed treatment comparisons. *Statistics and Computing*, 22, 1099–1111.

Van Valkenhoef, G., Tervonen, T., Zhao, J., De Brock, B., Hillege, H. & Postmus, D. 2012c. Multi-criteria benefit-risk assessment using network meta-analysis. *Journal of Clinical Epidemiology*, 65, 394–403.

Vanni, T., Karnon, J., Madan, J., White, R. G., Edmunds, W. J., Foss, A. M. & Legood, R. 2011. Calibrating models in economic evaluation: a seven-step approach. *Pharmacoeconomics*, 29, 35–49.

Veroniki, A. A., Mavridis, D., Higgins, J. P. & Salanti, G. 2014. Characteristics of a loop of evidence that affect detection and estimation of inconsistency: a simulation study. *BMC Medical Research Methodology*, 14, 1–12.

Veroniki, A. A., Soobiah, C., Tricco, A. C., Elliott, M. J. & Straus, S. E. 2015. Methods and characteristics of published network meta-analyses using individual patient data: protocol for a scoping review. *BMJ Open*, 5, e007103.

Veroniki, A. A., Straus, S. E., Soobiah, C., Elliott, M. J. & Tricco, A. C. 2016. A scoping review of indirect comparison methods and applications using individual patient data. *BMC Medical Research Methodology*, 16, 1–14.

Veroniki, A. A., Vasiliadis, H. S., Higgins, J. P. T. & Salanti, G. 2013. Evaluation of inconsistency in networks of interventions. *Internation Journal of Epidemiology*, 42, 332–345.

Victor, N. 1995. Indications and contra-indications for meta-analysis. *Journal of Clinical Epidemiology*, 48, 5–8.

Vieira, M. C., Cope, S. & Jansen, J. P. 2013. PRM13: network meta-analysis of longitudinal data. *Value in Health*, 16, A15.

Warn, D. E., Thompson, S. G. & Spiegelhalter, D. J. 2002. Bayesian random effects meta-analysis of trials with binary outcomes: method for absolute risk difference and relative risk scales. *Statistics in Medicine*, 21, 1601–1623.

Warren, F. C., Abrams, K. R. & Sutton, A. J. 2014. Hierarchical network meta-analysis models to address sparsity of events and differing treatment classifications with regard to adverse outcomes. *Statistics in Medicine*, 33, 2449–2466.

Wei, Y. & Higgins, J. P. T. 2013a. Estimating within-study covariances in multivariate meta-analysis with multiple outcomes. *Statistics in Medicine*, 32, 1191–1205.

Wei, Y. & Higgins, J. P. T. 2013b. Bayesian multivariate meta-analysis with multiple outcomes. *Statistics in Medicine*, 32, 2911–2934.

Weinstein, M. C., O'Brien, B., Hornberger, J., Jackson, J., Johannesson, M., McCabe, C., Luce, B. R. & ISPOR Task Force on Good Research Practices-Modeling Studies. 2003. Principles of good practice for decision analytic modeling in health care evaluation: report of the ISPOR Task Force on Good Research Practices-Modeling Studies. *Value in Health*, 6, 9–17.

Wells, G. A., Sultan, S. A., Chen, L., Khan, M. & Coyle, D. 2009. *Indirect Evidence: Indirect Treatment Comparisons in Meta-Analysis*, Ottawa, Canadian Agency for Drugs and Technologies in Health.

Welton, N. J. & Ades, A. E. 2005. Estimation of Markov chain transition probabilities and rates from fully and partially observed data: uncertainty propagation, evidence synthesis and model calibration. *Medical Decision Making*, 25, 633–645.

Welton, N. J. & Ades, A. E. 2012. Research decisions in the face of heterogeneity: what can a new study tell us? *Health Economics*, 21, 1196–1200.

Welton, N. J., Ades, A. E., Caldwell, D. M. & Peters, T. J. 2008a. Research prioritisation based on expected value of partial perfect information: a case study on interventions to increase uptake of breast cancer screening. *Journal of the Royal Statistical Society: Series A (Statistics in Society)*, 171, 807–841.

Welton, N. J., Ades, A. E., Carlin, J. B., Altman, D. G. & Sterne, J. A. C. 2009a. Models for potentially biased evidence in meta-analysis using empirically based priors. *Journal of the Royal Statistical Society: Series A (Statistics in Society)*, 172, 119–136.

Welton, N. J., Caldwell, D. M., Adamopoulos, E. & Vedhara, K. 2009b. Mixed treatment comparison meta-analysis of complex interventions: psychological interventions in coronary heart disease. *American Journal of Epidemiology*, 169, 1158–1165.

Welton, N. J., Cooper, N. J., Ades, A. E., Lu, G. & Sutton, A. J. 2008b. Mixed treatment comparison with multiple outcomes reported inconsistently across trials: evaluation of antivirals for treatment of influenza A and B. *Statistics in Medicine*, 27, 5620–5639.

Welton, N. J., Madan, J. J., Caldwell, D. M., Peters, T. J. & Ades, A. E. 2014. Expected value of sample information for cluster randomised trials with binary outcomes. *Medical Decision Making*, 34, 352–365.

Welton, N. J., Soares, M. O., Palmer, S., Ades, A. E., Harrison, D., Shankar-Hari, M. & Rowan, K. M. 2015. Accounting for heterogeneity in relative treatment effects for use in cost-effectiveness models and value-of-information analyses. *Medical Decision Making*, 35, 608–621.

Welton, N. J., Sutton, A. J., Cooper, N. J., Abrams, K. R. & Ades, A. E. 2012. *Evidence Synthesis for Decision Making in Healthcare*, New York, John Wiley & Sons, Inc.

Welton, N. J., White, I., Lu, G., Higgins, J. P. T., Ades, A. E. & Hilden, J. 2007. Correction: interpretation of random effects meta-analysis in decision models. *Medical Decision Making*, 27, 212–214.

Welton, N. J., Willis, S. R. & Ades, A. E. 2010. Synthesis of survival and disease progression outcomes for health technology assessment of cancer therapies. *Research Synthesis Methods*, 1, 239–257.

White, I. R. 2015. Network meta-analysis. *Stata Journal*, 15, 951–985.

White, I. R., Barrett, J., Jackson, D. & Higgins, J. 2012. Consistency and inconsistency in network meta-analysis: model estimation using multivariate meta-regression. *Research Synthesis Methods*, 3, 111–125.

White, I. R., Higgins, J. & Wood, A. M. 2008a. Allowing for uncertainty due to missing data in meta-analysis. Part 1: Two-stage methods. *Statistics in Medicine*, 27, 711–727.

White, I. R., Wood, A., Welton, N. J., Ades, A. E. & Higgins, J. P. T. 2008b. Allowing for uncertainty due to missing data in meta-analysis. Part 2: Hierarchical models. *Statistics in Medicine*, 27, 728–745.

Whitehead, A. 2002. *Meta-Analysis of Controlled Clinical Trials*, Chichester, John Wiley & Sons, Ltd.

Whitehead, A., Bailey, A. J. & Elbourne, D. 1999. Combining summaries of binary outcomes with those of continuous outcomes in meta-analysis. *Journal of Biopharmaceutical Statistics*, 9, 1–16.

Whitehead, A., Omar, R. Z., Higgins, J. P. T., Savaluny, E., Turner, R. M. & Thompson, S. G. 2001. Meta-analysis of ordinal outcomes using individual patient data. *Statistics in Medicine*, 20, 2243–2260.

Whitehead, A. & Whitehead, J. 1991. A general parameteric approach to the meta-analysis of randomised clinical trials. *Statistics in Medicine*, 10, 1665–1677.

Whyte, W., Walsh, C. & Chilcott, J. 2011. Bayesian calibration of a natural history model with application to a population model for colorectal cancer. *Medical Decision Making*, 31, 625–641.

Wood, L., Egger, M., Gluud, L. L., Schulz, K., Juni, P., Altman, D., Gluud, C., Martin, R. M., Wood, A. J. G. & Sterne, J. A. C. 2008. Empirical evidence of bias in treatment effect estimates in controlled trials with different interventions and outcomes: meta-epidemiological study. *BMJ*, 336, 601–605.

Woods, B. S., Hawkins, N. & Scott, D. A. 2010. Network meta-analysis on the log-hazard scale, combining counts and hazard ratio statistics accounting for multi-arm trials: a tutorial. *BMC Medical Research Methodology*, 10, 54.

Woolacott, N., Hawkins, N. S., Mason, A., Kainth, A., Khadjesari, Z., Vergel, Y. B., Misso, K., Light, K., Chalmers, R., Sculpher, M. & Riemsma, R. 2006. Etanercept and efalizumab for the treatment of psoriasis: a systematic review. *Health Technology Assessment*, 10, 1–233.

Yan, J. & Prates, M. 2013. *Package rbugs: Fusing R and OpenBugs. 0.5-9 ed.* Available: http://cran.r-project.org/web/packages/rbugs/index.html (Accessed 17 July 2017).

Yu, X., Baade, P. & O'Connell, D. 2012. Conditional survival of cancer patients: an Australian perspective. *BMC Cancer*, 12, 8.

Yuan, Y. & Little, R. J. A. 2009. Meta-analysis of studies with missing data. *Biometrics*, 65, 487–496.

Yusuf, S., Koon, T. & Woods, K. 1993. An effective, safe, simple, and inexpensive intervention. *Circulation*, 87, 2043–2046.

Yusuf, S., Peto, R., Lewis, J., Collins, R. & Sleight, P. 1985. Beta blockade before and after myocardial infarction: an overview of randomised trials. *Progress in Cardiovascular Diseases*, 27, 335–371.

Zangrillo, A., Musu, M., Greco, T., Di Prima, A. L., Matteazzi, A., Testa, V., Nardelli, P., Febres, D., Monaco, F., Calabrò, M. G., Ma, J., Finco, G. & Landoni, G. 2015. Additive effect on survival of anaesthetic cardiac protection and remote ischemic preconditioning in cardiac surgery: a bayesian network meta-analysis of randomized trials. *PLoS One*, 10, e0134264.

Zelen, M. 1971. The analysis of several 2 X 2 tables. *Biometrika*, 58, 129–137.

Zhang, J., Carlin, B. P., Neaton, J. D., Soon, G. G., Nie, L., Kane, R., Virnig, B. A. & Chu, H. 2014. Network meta-analysis of randomized clinical trials: reporting the proper summaries. *Clinical Trials*, 11, 246–262.

Zhang, Z. & Wang, L. 2006. *Use BAUW to convert data.* Available: http://www.psychstat.org/us/article.php/52.htm (Retrieved 26 January 2011).

Zorzela, L., Golder, S., Liu, Y., Pilkington, K., Hartling, L., Joffe, A., Loke, Y. & Vohra, S. 2014. Quality of reporting in systematic reviews of adverse events: systematic review. *BMJ*, 348, f7668.

Index

Network Meta-Analysis for Decision-Making, First Edition. Sofia Dias, A. E. Ades, Nicky J. Welton, Jeroen P. Jansen and Alexander J. Sutton.
© 2018 John Wiley & Sons Ltd. Published 2018 by John Wiley & Sons Ltd.
Companion website: www.wiley.com/go/dias/networkmeta-analysis

Printed and bound by CPI Group (UK) Ltd, Croydon, CR0 4YY

10/10/2024

14572031-0001